BEYOND EARTH:

THE FUTURE OF HUMANS

IN SPACE

Bob Krone, Ph.D., Editor

Foreword by Edgar Mitchell, Sc.D., Captain, USN (Ret)
Apollo 14 Lunar Module Pilot

Langdon Morris and Kenneth Cox, Ph.D.,
Associate Editors

An Aerospace Technology Working Group
Sponsored Book

An Apogee Space Book

Library of Congress Cataloging in Publication Data
ISBN # 1-894959-41-8
ISBN13 # 978-1-894959-41-4
Bob Krone, Ph.D., Editor, 1930-

Beyond Earth: The Future of Humans in Space

Includes index.

1. Space Exploration. 2. Space Habitation. 3. Human Migration to Space. 4. Space Governance. 5. Space Law. 6. Ethics for Humans in Space. 7. Space Technology. 8. International Cooperation for Space. 9. Leadership for Space. 10. BioTech Revolutions for Space. 11. Strategic Alternatives for Space. 12. Views of Global Leadership on Space. 13. Theory for Humans in Space. 14. Intelligence in Space. 15. Childrens' Visions of Space. 15. First Lunar City. 16. Spaceports. 17. Space Settlements. 18. Space Operations. 19. Education for Space. 20. Psychology for Space. 21. Space and Earth's Biosphere. 22. Making Space Popular. 23. Music and Arts in Space. 24. Space and Commerce. 25. Risks and Safety for Space. 26. Research for the Future of Space. 27. Meaning of Space. 28. Oases in Space. 29. Space and Evolution.

Printed in Canada
CGPublishing, Inc, Box 62034, Burlington, Ontario, Canada
www.apogeebooks.com

This book is dedicated to the future.

CONTENTS

* PART III. SCIENCE, TECHNOLOGY, ENGINEERING, AND MANAGEMENT FOR SPACE 161

* PART IV. STRATEGY & SYNTHESIS 253

FOREWORD

By Edgar Mitchell, Sc.D., Captain, US Navy (Ret)
Apollo 14 Astronaut

Earth's adventures into space are not yet fifty years old as this account is being written. Flat Earth thinkers and doubters not withstanding, we have been to our nearest neighbor in space, the Moon, and initiated explorations there. I assert this from first hand knowledge, as one who has made that wonderful journey. It is understandable, however, since more than half of Earth's current population were not yet born at the time of the Apollo flights to the moon, that a certain mystique, a mythology, and even doubt, has recently arisen about those first efforts at spacefaring voyages from Earth's surface. To some they seem incredible.

If we think back five centuries to the period when the first circumnavigations of the globe occurred, we ask, were fifteenth century shipbuilders and seamen holding lengthy discussions and conferences, writing learned papers and books, debating about the risks and opportunities of global exploration, market expansion, trade, and tourism? And the problems attendant there to? I doubt it. A few visionary explorers and entrepreneurs, even dating back to the Phoenicians, and acting mostly alone, certainly caught the idea of exploration via the Earth's oceans and they led the way. But then it did not take the resources of an entire nation, and years of new science and technological development to make the leap. A sturdy ship, a willing crew, and enough provisions for a few months were sufficient for an explorer to get started toward distant lands on Earth.

And even at the beginning of space exploration in October 1957, our knowledge of the cosmos was severely limited. No humans had risen beyond twenty thousand meters, and understanding the larger cosmos was still in its infancy. The visionary astronomer, Sir Fred Hoyle (1915 - 2001), observed at the beginning of the space age that when he first saw a picture of Earth from deep space, his view of life would not be the same again. And he was correct. The pictures of Earth from deep space have touched us all deeply, and stirred profound questions about our existence, our origins, and of course our future.

Though the decision to explore our lunar neighbor was motivated fundamentally by political considerations, it nevertheless constituted a rapid expansion of human experience and embodied the timeless quest for greater knowledge, in this case as available only through the genius of twentieth century science and technology. It also continued the human destiny, evident since our earliest days, of venturing further into the unknown, and establishing our presence there. Never before, however, have the challenges been so great, the risk so unrelenting, the necessity so imperative, nor the reward so compelling - for now that we know about the limited life cycle of every solar system, in the long term our venture into space has to do with the ultimate survival of our species.

Yes, we will go deeper into our solar system, and beyond in due course, of that I am sure. The latter will take new science and new understanding, but the former is mostly a matter of technology, political will, and resources. While we lost the momentum for human deep space exploration following the initial lunar missions, perhaps it is just as well. For one can make a strong argument that the exploration of our solar system, and beyond, should be a cooperative effort of our entire planetary civilization, not just one nation. Clearly, however, we are not yet ready to undertake such adventures as a united world, and we can but hope that just maybe we will learn to settle our differences short of war, before we export that brand of insanity outside of our atmosphere.

Available to us still, however, are the detailed records of our initial accomplishments in space, and the living talents of many of the men and women who have brought us through these initial fifty years. The personal knowledge, experiences and skill of these individuals are invaluable for understanding what should be done next in order to continue the quest toward future extraterrestrial goals.

It is toward that end that many of the pioneers in space exploration have come together to present this volume, to examine key questions and issues that we confront as our Earth-originated civilization continues its quest to explore and understand the cosmos through robots and by human space flight. The legacy that we, the participants in the greatest space adventure of the twentieth century, can leave to those that follow, is our experience, our best thinking and advice to those that pick up where we left off, in order to continue the adventure. I believe that this volume informs and inspires their efforts.

PREFACE

This book is about science, not science fiction.

It is about space science as a means to achieve the lofty goals of new knowledge, economic benefit, and peace. And it is about space as a destination and an adventure that can, by means of the striving for it and through the outcomes that may be achieved in reaching it, bring multifold benefits to all of humanity.

It's been almost fifty years since the dawn of the space age, as marked by the launch of Sputnik in 1957. Now, in 2006, we can realistically anticipate that humans will one day live and work in space, and as we contemplate that day we know well that we, too, may have a part to play in the planning and preparation required to accomplish this remarkable vision. This book is the joint effort of more than forty co-authors, who contributed their experiences, insights, and ideas for the benefit of people all around the world who are interested in a peaceful future for humans in space.

Many of the co-authors spent their entire careers as participants in the global space community, and together they have a combined total of roughly one thousand person-years of experience in space science. They worked on every major space program this country has ever mounted, and most of the minor ones as well. Readers will find brief descriptions of the authors following each chapter.

These are accomplished individuals with expertise based not only on theory, but on hard-won experience. They are extraordinary individuals who come from Australia, Asia, the Middle East, Europe, and North America. They have felt the elation of successful launches, successful missions, and successful splashdowns; and they have suffered the devastation of failure, have felt the anguish of watching colleagues die in pursuit of knowledge and adventure beyond Earth. They have labored in the vast NASA bureaucracy, and in small start-up companies, and today they continue to work on space; they design, manage, engineer, teach, write, advocate, negotiate ... within the halls of NASA, in skunkworks searching for the next breakthrough, and in universities educating future leaders of the ongoing space movement.

No matter where in space science they work, they have another thing in common, for they certainly agree that it is the human destiny as well as our responsibility to go to space, and it is, of course, precisely because we agree about the importance of space that we have worked together to write this book. We are also a diverse and opinionated group, quite capable of deep disagreement, and quite willing to engage in heartfelt discussion on many issues, as you will see in these pages.

Thus, among the 36 chapters of this book you will find a wide variety of conceptual, philosophical, and technical viewpoints. Some chapters are highly personal, exploring the psychological dimensions of space, while others deal with engineering topics, assessing the technical merits of various alternative approaches and issues. Still others concern governance, management, biology, physiology, law, strategy, vision, beliefs, attitudes, and challenges that we face as we consider our future in space.

So how, you may well ask, did these people come together? The answer is that all the authors are participants in an extraordinary, and perhaps unique organization called ATWG, the Aerospace Technology Working Group. ATWG was established in 1990 as forum for collaboration among NASA, its contractors, and academia to support effective communication and the development of new ideas and technologies. It's bi-annual meetings have been held around the country, always under the leadership of its founding director, Dr. Ken Cox. Ken tells the full story in Chapter One, so suffice it to say for our purposes in this Preface that after ten years under the NASA banner, ATWG moved outside the agency to become an independent organization. Its bi-annual meetings have continued without interruption, involving leading space experts in compelling presentations and rich dialog about every conceivable aspect of space science.

And in this rich milieu of concepts and discoveries, it gradually became obvious that in light of the ongoing turmoil at NASA, which you will also read about in these pages, there was a compelling need for a new vision of the future of humans in space to be written, one that integrated the visionary, managerial, philosophical, and technical perspectives.

And so we have taken it upon ourselves to write this book, and we offer it as a testament to our shared commitment that this is absolutely something which humanity must do, and which humanity certainly can do, for a myriad of reasons and via a variety of means which are the very subjects of the chapters that follow.

Humans will go permanently to space not only for adventure or for opportunity, but primarily for the betterment of all humankind; research conducted in recent decades shows conclusively that we can anticipate tremendous benefits when we set about to face the many challenges, exploit the vast resources, and come to understand the profound capabilities that are uniquely to be found in space.

We began this adventure in authorship with three assumptions that have not been changed by the year of intellectual exchanges that has gone into the preparation of this book.

First, the urge for flight is part of our human nature. Perhaps it is in our genes, but from wherever it originates, it is undoubtedly our need to explore and our unquenchable curiosity about the universe that drives us to space.

Second, even if these urges were ignored, the continual improvement of the quality of life for the human race on Earth, and perhaps even our ultimate survival, may hinge on the success of human exploration and habitation of space.

And third, we are all aware that this is a critical time for the space movement, and for all of us. Human society around the world is in turmoil, and the prospects for our future are frightening. But we remain optimistic that we will overcome these challenges, and we see clearly that our generation can use the opportunity presented by our outward expansion into the solar system to design a rewarding and exciting future for human collaboration, and to capitalize on the lessons learned from the venture into space to redirect human history on Earth toward peace and cooperation.

And as we do so, even the significant steps that we've taken since the dawn of the space age in 1957, including getting into orbit, going to the Moon, and building space stations, will in retrospect seem to be tiny ones compared to what lies ahead. Migrating into space will test us beyond anything we have previously accomplished, as we are destined to face challenges both fantastically breathtaking and supremely dangerous.

※※※

Beyond Earth: The Future of Humans in Space describes a vision that is accessible to anyone interested in humankind's great space adventure, the human settlement of the Solar System. Scholars, scientists, engineers, managers, astronauts, artists, authors, and university professors engage with the important questions that shape our unique circumstance: Why does space matter to us? What can we use it for? How can we get there efficiently? What will ordinary life be like in space? What will our homes be like on the Moon? On Mars? In orbit? Will we play? Will we love? It also documents, in Appendix A, a considerable list of research questions and hypotheses about the still unknown or partially known aspects of the future of humans in space.

Space offers many unique opportunities for us to solve many of the problems we face on Earth. A hundred years of effort enabled us to solve the problems of flight in our atmosphere, and for fifty years that urge, combined with our curiosity about the universe, has propelled us into space. Now, as the 21st Century begins, we know with certainty that human exploration of, and migration to, Earth's moon and our Solar System must occur.

In the end, there are two fundamental reasons why now is the time for the serious planning for an off-Earth, space faring civilization. Both are rooted in the compelling need for survival:

Benefits for Earth from space are significant - improvements for Earth's biosphere, energy appetite, health sciences, industries, and limited resources. These will have revolutionary societal, political, and physical impacts.

Threats to humanity emanate from countless natural causes, Earth-based and space-based, or from human-produced phenomena such as population growth and pollution; weapons for mass-killings; hurricanes and climate change; disease pandemics; gamma-ray bursts; asteroid-Earth collisions; Earthquakes and volcanoes. Many of these can be avoided, reduced in impact, or completely eliminated by the exploitation of the new capabilities that our journeys into space will require us to generate.

This is the conviction of those who contributed to this book, a conviction that comes from the basic conclusion that space challenges, resources, and environments are essential for the improvement of the quality of life for humans on the Earth; and perhaps even for the ultimate survival of our human race.

Bob Krone, Langdon Morris, and Kenneth Cox, Editors
1 April 2006

Please note: The views, ideas, and opinions expressed in these chapters are those of the individual authors only, and are not necessarily official positions of the organizations for which they work.

ACKNOWLEDGMENTS
by the Editors

Bob Krone:

Editing this work has been the most rewarding intellectual experience of my three careers. From the book's origin in October 2004 to the completion of the manuscript in 2006, the project rapidly gained momentum as the members of the Aerospace Technology Working Group (ATWG) engaged with one another to examine the key issues and prepare their various chapters

The impressive spectrum of diverse professional expertise that constitutes ATWG provided the perfect environment for the networking of intellectual energy, and Ken Cox provided invaluable leadership to keep this project on a continuous build.

Langdon Morris oversaw editing of all the chapters, engaging in multiple iterations with each author or author team, resulting in a remarkable clarity of thought and expression. Thomas Diegelman, co-author of Chapter 25, also volunteered his NASA expertise to document the iteration of thirty-six chapters and the multiple images for the book.

And perhaps the single most remarkable fact about a book of this complexity was that every one of the forty-two authors, including the Foreword author, Astronaut Dr. Edgar Mitchell, contributed their time and talents out of love for the subject and commitment to its underlying principles, and did so without compensation.

We realized nearly from the beginning that we were participating in a publication effort with far reaching positive implications for the global space community, and we even dare to hope that it has implications for all of humanity. It was a privilege and an education to be sharing ideas with this exceptional group of global scientists, engineers, managers, writers, professors, religious, government, and business leaders who contributed to Chapter 3, *"Views of Global Leadership."* It was an inspiring experience from original book idea in October 2004 to the final publication.

The Apogee Books Space Press was a pleasure to work with. Richard and Rob Godwin accepted the book on a simple premise, and throughout provided professional counsel and rapid responses to every question. Ric Connors was very helpful and innovative in the advance design and execution of multiple marketing efforts.

On the personal side, I should like to thank my wife, Sue Krone, who was an unwavering supporter and also added valuable inputs throughout the process.

Ken Cox:

I would like to honor the experience of creative collaboration that occurred between Bob Krone, Langdon Morris and myself in the editing of this book involving Earth, space and all life in the universe. We all shared a passion for frontiers, as well as a deep appreciation of history learnings of the past. Bob Krone's energy and wisdom in starting this book had a flavor of intuitive genius blended with a lifetime of strong academic experience. Langdon Morris' background as an author of multiple books provided unusual energy for synergistic interactions with the individual chapter authors. Both Bob's and Langdon's contributions were exceptional.

I would like to thank some of my mentor's within NASA as I traveled the excitement of being involved in Apollo, shuttle and station. Max Faget and John Hanaway provided me significant opportunities to work both Apollo and Shuttle. Dick Truly in his role as NASA Adminstrator encouraged me to initiate the Aerospace Technology Working Group in the early nineties. And finally, Henry Pohl absolutely was a gentleman and scholar to work for and be associated with during the shuttle/station development.

I especially would like to acknowledge and thank my wife, Kay and my son and daughter, Chris and Karen, for their long support and patience throughout my career. I can still recall the many times I called home from scenic "Downey."

Langdon Morris:

It was Bob Krone's brilliant insight to propose this book, his leadership that pulled it together, and his thoughtful editing that built its structure; it is Ken Cox's remarkable leadership of ATWG that created a forum in which so many exceptionally gifted space scientists have engaged with one another for so many years on so many fascinating topics. It has been my privilege to work with both of them in the development of this book, and with the many authors who willingly engaged in thoughtful dialog as we looked for the best way to express their insights and ideas.

I would like to acknowledge and appreciate my family, who tolerated my many long hours at the computer during the course of the many evenings of this project. Indeed, let us acknowledge all the volunteer authors for their marvelous contributions to this work, and all their family members - husbands, wives, sons, daughters, fathers, and mothers - who supported them during the many days and evenings that they dedicated to the preparation of this book.

Finally, to the readers of this book, let me say that I hope that you have as much fun reading it as we did writing it! And I hope as well that it inspires a commitment in you, a commitment to actively support the journey to space as a journey of peace and for peace, an adventure whose great potentials and benefits are manifested to support the increased well being of all of humanity. This has been our goal.

PART I.

BEGINNINGS

The stars, the planets, space and the vastness of the universe have captured the imagination of nearly everyone who has ever gazed into the night sky. In Part I we consider three dimensions of this impact.

Dr. Kenneth Cox describes the genesis and history of the Aerospace Technology Working Group, the organization that has sponsored this book; Langdon Morris explores the meanings of space as we begin the Third Millennium, and Martin Schwab compiles a marvelous collection of thoughts about space from world leaders in government, business, academia, and faith.

Chapter 1

The Aerospace Technology Working Group

By Kenneth J. Cox

We search for ways to connect our dreams to the reality of the present and the promise of the future.

The Aerospace Technology Working Group, ATWG (or as it was called at that time, SATWG), was conceived at NASA in 1989 and formed in 1990. It is an affiliation of world class thinkers, space scientists, engineers, academics, business consultants, environmentalists and authors whose mission is to support global collaboration and action for the development of space science, space commerce, and permanent space habitats. Its members have many hundreds of years of combined experience in space science, and widespread personal networks that permit synergy and opportunity to emerge across national boundaries.

From a philosophical perspective, ATWG members consider Earth, Space and Humanity as a single integrated system. The organization provides input to NASA officials and other space interests on such issues as energy and power sources available from both Earth and space, integral investment strategies for the future, technical development strategies in space science, and system and mission design. It also works to develop innovative approaches to bureaucratic and governance processes that would otherwise undermine progress on major initiatives.

ATWG helps leaders formulate and execute breakthrough projects that require unprecedented collaborative teaming, projects that must be accomplished through the involvement of other individuals, agencies, companies, and nations. We view these projects as learning laboratories for discovering and developing the depth of leadership, teamwork, and management necessary for breakthrough outcomes.

Since the beginning we have been dedicated to the principles of collaboration, teamwork, learning, and the positive uses of space for the benefit of all humanity, and we continue in that spirit today. ATWG has held 33 biannual meetings without fail since it was established, and numbers 34 and 35 will occur this year, 2006. (1)

This chapter briefly tells the story of ATWG, emphasizing an overview of our vision, values, and aspirations.

I was appointed to head ATWG as a 25-year NASA veteran, and although I have subsequently retired from NASA I continue to lead this organization of extraordinary individuals, many of whom have conspired to create this book.

But the roots of ATWG began to take root much earlier. As a young NASA engineer who had already worked five years in the Apollo program, I was part of the team that helped bring Apollo 13 back to Earth safely after one of its oxygen tanks exploded in the command module en route to the moon in April 1970. In those moments of tension and uncertainty I learned what men and women can do when they are filled with courage, humanity, and purpose.

This was a pivotal experience for me, and for many others at NASA, and it remains with me always. In fact, it led me twenty years later to the formation of the ATWG, the Aerospace Technology Working Group, to promote the future of humans in space. It was collaboration, intensity, action, and the integration of science and spirit that brought the Apollo 13 astronauts back safely, just as these same qualities were the foundation of all of NASA's successful Apollo missions, and indeed they contribute to the best that human civilization has ever accomplished across the millennia.

These qualities are also the genesis of this book, *Beyond Earth: The Future of Humans in Space*, and the qualities that the many authors of this book wish to evoke in our future space endeavors, the quest for the permanent human settlement of space.

✳✳✳

Accomplishments and Challenges

Three key phenomena of the last 60 years that significantly influenced the subsequent course of human society are the science and technology methodologies and artifacts that were achieved during and since World War II, digital computing and the information age, which is also one of those war outcomes, and the Apollo era with humans traveling to and returning from Outer Space that built on both prior trends.

The challenge today that is faced by every space faring nation, and indeed by all of humanity, is to complete the inspired possibility of Apollo by establishing permanent human habitats in Space. This will assuredly happen, but knowing that we are still left with huge relevant questions that need to be asked involving Why? How? and When?

The challenges are immense. In Chapter 5 of this book Yehezkel Dror states them most eloquently.

"The repercussions of moving into space are largely inconceivable, efforts to predict them on the basis of a very different past being extremely doubtful."

"Bureaucracies, as correctly understood, are essential for large scale human action, whether by business or governance. The correct question, therefore, is not how to reduce the roles of bureaucracy, but how to change core features of bureaucracy so as to make them fit the requirements of a new epoch including human movement into space."

"No single country, including the USA, has the material, scientific and technological, human and political resources for moving humanity into space."

"......unless undertaken by strong global governance, human movement into space is sure to involve conflictive action, including militarization, with dismal consequences."

"De facto global governance dominated by the USA in cooperation with other willing major states may do the job while being more feasible in the foreseeable future. But, however structured, without strong global governance, human settlement of space has no chance."

ATWG has been my personal journey from Apollo to the challenges and immense opportunities we face today. On this journey, I have come to see that our future in space, as well as our future as a species here on Earth, requires that we balance our interest in science and technology with our commitment and compassion for humanity. We cannot explore outer space without exploring inner space at the same time.

ATWG's History

ATWG began this way: In the summer of 1989 NASA Administrator Richard Truly decided to establish an Advanced Development Planning Group to encourage collaboration between the NASA Operation field centers and the NASA Research field centers in the area of technology readiness and operational development. Six different working groups were considered - operations, avionics, entry, propulsion, power, and structures. All six related directly to the mission operation and design of flight vehicles. Also identified as important were the areas of pre-launch payload operations, launch and tracking operations, and landing/turnaround operations.

The topic of Avionics includes analog and digital computer hardware, operational and application driven software, actuators, sensors, communication and tracking systems, power, and multiple connecting devices.

An early collaborative effort was the Space Transportation Avionics Technology Symposium held in

Williamsburg, Virginia November 7-9, 1989. The results are documented in NASA Conference Publication #3081 in 1990. (2) The success at this Symposium led to the formation of the Strategic Avionics Technology Working Group (SATWG) in January 1990.

The initial themes for SATWG were focused on science, technology and engineering, particularly on cost reduction, flight capability enhancement, reliability and safety, and operations efficiency for the Space Shuttle. Some specific areas included eliminating hydraulics, automated rendezvous and docking, operationally efficient payloads, and vehicle health management.

The first official SATWG meeting was hosted by General Dynamics in San Diego in November 1990, and the second was hosted by Boeing in Seattle in February 1991. During this time period, sixteen small SATWG seed studies were developed in avionics requirements, architecture, process improvement, technology transition, vehicle health management, technology and guidance, navigation, and control technology. The goal of these studies was to provide "technology bridging" between the technology developers, the system integrators, and the operational users, who had previously been working in separate silos with insufficient contact between them.

In late 1991, a SATWG Industry Group was formed to enhance the collaborative efforts, and the positive result was that new paradigms were developed that emphasized "doing business" rather than "doing government." This included quality function techniques and Integrated Product Teams.

In a collaborative process initiative, a list of ten top challenges was developed. These challenge items included operability, affordability, dependability, suitability, management cultural change, multi-program initiatives, concurrent engineering techniques, incentives, innovations, partnering and networking. We made progress slowly, and while the industry participants sincerely attempted to utilize the results of these initiatives, their ability to effectively collaborate within competitive boundaries was not well developed. On the public sector side, bureaucratic inefficiencies in both the federal government and academia hampered our ability to put solutions into effect.

In 1993 the focus turned to enterprise integration, team USA, and global space commerce. Ten major trends were identified, including the end of the cold war, global concept development, a constrained fiscal environment, public and private sector relationship shifts, planetary sustainability issues, expansion of information systems, phasing "cold warriors" into "trade warriors", space commercial market development, international space cooperation, and international space competition. (3)

We worked on these topics, and many others, in the framework of the ATWG Forum meetings and between meetings in small working teams, often informal and self-organized.

During this period, we had a direct experience of bureaucratic meddling, as pressure from NASA headquarters impacted us directly when someone in the hierarchy insisted that we forgo the "strategic" designation in our name or our funding would be decreased. We had more interest in doing the work than retaining the label, so we readily complied, and thus SATWG (Strategic Avionics Technology Working Group) became ATWG (Aerospace Technology Working Group).

As the years of dedicated work by many extraordinary men and women unfolded, I began to develop an uneasy feeling. I observed that the amount of action leading to results was minimal even though the quality of the thinking was very good, and significant time was being invested. I saw that bureaucratic and governance processes often undermined real progress, and often the distinction between lip service and genuine commitment was lost.

Thus, our list of issues and topics continues to be valuable today, but now we need to move beyond just ideas on how to improve or add to the overall top ten challenges, and put the emphasis on taking real action on the affordable and sustainable development of space.

By the late 1990s I also came to see that the future of space justifiably belonged to people all over the world and not only the strongest nations. I realized that if we keep heading in the direction we are going, we will likely end up with "Star Wars" rather than a "United Federation of Planets," with space as a venue of conflict rather than one of peaceful development for the benefit of all. And I came to see that global cooperation for the future of space could serve humanity, but only if a 100 percent individual and collective commitment to action was present could we bring our grand visions into reality.

Looking Ahead

I am now convinced that that Earth, Outer Space and all Life must be considered as integrated elements that comprise the actual system in the Universe. The underlying questions of our situation today concern our ability to learn, our wisdom and awareness, and the actionable processes needed to shape the future of human civilization. We may not succeed, but a growing group of us believe it is irresponsible not to try.

Thus, I want this book to help inspire people all over the world to provide the necessary leadership and commitment to attain shared visions for space, even as this must occur within today's diverse and turbulent environments. Our challenge is to adapt to increasing chaos and complexity in order to develop mature relationships, and to identify and understand the significant patterns that lead to positive change.

The important attributes and characteristics to cultivate have primarily to do with our own inner capabilities. These include: 1) integrity, congruence, and discipline, 2) adapting, unifying, sharing and risk taking, and 3) listening, looking, intuiting and reflecting. We see, thus, that the necessary change occurs within ourselves first, and that we build upon our inner awareness to manifest positive change through the thoughtful application of science and technology.

It's been a long way around to reach this understanding, but as you will see in the chapters that follow, there is an astonishing consistency in the recognition that our inner qualities are as important to the foundation upon which a space program of unprecedented complexity must rely as our technical and managerial capabilities. Remember, many of these authors are accomplished rocket scientists!

✻✻✻

The members of ATWG are committed to helping transform the space programs of every space faring nation into an Earth/space movement that integrates permanent settlements with personal adventure, science, commerce, ecology, and holistic well being for everyone and everything on the Earth.

The prospect of humans living permanently in the barren environment of Outer Space carries with it many uncertainties. However, there is a meaningful possibility that humans may design our own evolution, as John Stewart suggests in Chapter 22, and that new collaborative endeavors, together with new unfolding evolutionary insights and new science and technology breakthroughs will excite the heroic qualities in all humans. If this happens, it will be because the "inner warriors" in all of us remain committed to attaining the best possible outcomes for all the world's peoples.

The intellectual energy and insight created by our networking to develop this *Beyond Earth: The Future of Humans in Space* book has provided a catalyst for us to broaden the traditional scope of ATWG by adding to its projects:

"The Space & Earth Development Alliance (SEDA)"

This unfolding Alliance concept was first formulated in early spring of 2006 and will focus on the global themes and concepts captured in *Beyond Earth*. It will be an international umbrella activity to foster the improvement of all life on earth through capabilities generated by the future of humans in space.

Let me close this first chapter by expressing my deep appreciation to the many hundreds of individuals who have committed their time, energy, and insights to AWTG activities over these many years, and to the individuals who have participated in the extraordinary project to produce this remarkable book. This book expresses our shared commitment to achieving the wonderful possibilities that a focused space program may achieve for the benefit of all of humanity, and you can be sure that as ATWG goes forward, this intent is as clearly held in our minds as it is the goal of our actions. We hope you will join with us in this remarkable endeavor!

✻✻✻

You can learn more about ATWG at www.atwg.org. Many ATWG Forum meetings are open to the public.

(1) A complete list of the ATWG meetings, themes, and locations is presented in Appendix B.

(2) Space Transportation Avionics Technology Symposium, Williamsburg, Virginia, NASA Conference Publication 3081, NASA 1990.

(3) For more detail on ATWG in the 1990s, see Robbie E. Davis-Floyd, *"Commercializing Outer Space: The SATWG Stories."* Late Editions VII, Para-Sites: A Casebook against Cynical Reason, George Marcus ed. University of Chicago Press, 1999.

About the Author

Dr. Kenneth J. Cox earned his bachelor's degree in 1953 and his master's degree in 1956 in electrical engineering from the University of Texas/Austin, and his PhD at Rice University in 1966.

In 1963 he joined NASA to develop the flight control system for the Little Joe II Booster Vehicle. Later, Dr. Cox became the Technical Manager for the Apollo Digital Control Systems, which included the Lunar Module, the Command Module and the Command/Service Module, the first spacecraft to fly with a digital flight control system. From 1971 to 1987 he served as the Space Shuttle Technical Manager for the Integrated Guidance, Navigation and Control System, focused on developing new ways of utilizing a common set of digital computers to support all phases of flight. From 1977 to1979 he supported the Apollo-Soyez joint mission between USSR and the USA in the Integrated Flight Control Systems. From 1987 to 1995 Dr. Cox was Chief of the Avionics Systems Division, and directed early technology and digital system concept developments for the International Space Station. He later was named the Chief of the Navigation, Control and Aeronautics Division.

From 1995 to 2003 Dr. Cox was Chief Technologist for the NASA Johnson Space Center, with the responsibility to promote government, industry and academia networking and collaborating for outer space related activities. He helped initiate an Avionics Technology Integration Group in the early '90's, the group that evolved into the ATWG. He also served on the Atlas Centaur Accident Investigation Board in 1997.

He retired from NASA in 2003 after forty years of service to focus on ATWG and the peaceful development of space.

He speaks frequently at industry conferences. He conducted an AIAA Distinguished Lecture Series for six years, and other major conference speaking engagements include the World Future Society, the Creative Problem Solving Institute, Society for Automotive Engineering, the Science and Consciousness Conference, the Space Frontier Foundation, and many others. He is recipient of the AIAA Mechanics and Control of Flight Award, AIAA Digital Avionics Award, and the NASA Medal for Exceptional Engineering.

Chapter 2

The Meaning of Space

By Langdon Morris

If you go far away from the lights of the city on a clear and moonless night and you gaze into the sky, you will behold the awe-inspiring vastness of space; millions of stars and galaxies will twinkle at you across incomprehensible distances, beckoning you to come, to explore, to discover! What is this universe, you may ask yourself. And what is our place in it? How was it made? Who else lives here?

These are timeless questions that even our most distant ancestors asked, for the very nature of space calls to us in ways that we cannot escape. It confronts us with some of the most basic and fundamental questions about our existence, our past, and our destiny; and the depths of space, its vastness and incomprehensibility, embody for us a set of meanings that are unlike any other.

For example, space, the concept of outer space and its sheer enormity, makes us wonder how rare life might be in the universe; it makes us think about the possibility of voyages across the unlimited black emptiness from our world to others. It manifests the apparent outcome of billions of years of history emanating from a single explosion of incomprehensible power, and all of this combines to make space one of the most provocative and compelling concepts among all the enduring icons that burn within human consciousness.

Among the approximately 6.5 billion people who are alive today, only a tiny handful are likely to ever venture into space during the course of their lifetimes. Fifty, one hundred, perhaps five hundred, or maybe even 5000 will have the opportunity to escape Earth's gravity and then gaze back upon our small planet.

What will this experience mean to them? For these few, the idea of "outer space" will surely have a unique and transcendent meaning, for it might well be the unique and defining experience of a lifetime, an experience without parallel, an epiphany.

For the rest of us, however, the meaning of space doesn't have to be any less profound even though it will necessarily lack the vital dimension of experience. This is so because space carries meanings that are unique and unmatched by any other aspect of the human experience, either on Earth or off of it we can all partake of these meanings regardless of whether our status is astronaut or armchair astronaut.

Still the very words I've used to convey the nature of space - "vast," "inconceivable," "incomprehensible," "emptiness" - these words and any others that we might choose are still inadequate to express the scale and scope of space, and we find in the end that our language is simply insufficient for the task. The meaning of space is beyond the capacity of our expression, as it is also in many very real ways far beyond our comprehension.

And yet we must try to understand what it does mean, and we must try to express that meaning, for space is the context and container of life on Earth, of our life. It is our local neighborhood in this tiny part of the galaxy we call home.

Space is also, perhaps our destiny, for although a mass exodus to space may yet be centuries in the future, we who are living today cannot escape the idea that a human diaspora will eventually bring human civilization to numerous bodies throughout the solar system. We must also consider the possibility that an active human presence in space is not only our challenge, but also our responsibility. And finally, it is certainly true that whatever we see in space, or imagine that we see, is merely a reflection of ourselves, and the incomprehensible scale of space makes it simply a mirror of the incomprehensible depth within ourselves.

These are four meanings of space - the local neighborhood; human destiny; responsibility; and the psychology of space - that we will explore in this chapter.

Welcome to the Neighborhood

Five decades ago space travel was a dream, but then with billions of dollars and the capabilities of thousands of brilliant and dedicated scientists and engineers, humans launched the space age. And so it is in our own lifetimes that the meaning of space has changed from the domain of science fiction, to that of science. Space is no longer just the realm of imagination or possibility, but of actuality, and in this process our neighborhood has grown.

While forty years is a mere instant on the Earth's vast scale of billions of years, in the context of human history and human civilization forty decades is a significant span of time, and during this period as we have witnessed the transformation of space from a fantasy to a reality, we have also seen a transformation in the reality of human life back on Earth. We see ourselves differently now, for we have looked back from the moon to see the tiny blue ball, and this perspective has nurtured a much greater appreciation for our extraordinary planet. We know and do things differently now, for space science has transformed meteorology (with weather satellites), agriculture (with mapping and analysis from space), transportation (with GPS), communication (with radio, television, and cell satellites), medicine (with telemetry) as well as astronomy, physics, all of the materials sciences, rocketry, and also, sadly, warfare.

Spacecraft have charted all the major bodies in our solar system, and human footprints still dot the face of the Moon (along with the debris we've left there, and on other neighbors as well). Within a few years Voyager will cross the conceptual boundary between the solar system and what lies beyond, and continue on its journey into the unknown.

Forty years along, the space movement has also affected our attitudes and expectations, for it has given us some perspective on how small the Earth is, and how unique life might be. A little more knowledge about our neighborhood has indeed been transformational for the human race.

The Human Destiny in Space

It is our destiny, certainly, to imagine our future in space. The most successful movie series of all time, Star Wars, has made a billionaire of its creator by combining the timeless human themes of deceit, intrigue, love, anger, fear, and greed, with an unbounded imagination about hyper-space travel, weapons of unimaginable power, regular trade between planets, galactic empires and wars, and even knights with mystical powers. For the aficionado of Star Wars or any other type of science fiction, space means an infinite and uncharted domain unbound by everyday rules where you can invent anything you want.

For the aficionado of science however, space is also full of possibilities, for it means an infinite and largely uncharted domain full of wonderful puzzles and mysteries. Every branch of science stands to learn from experiments conducted in science, and we are just beginning to explore what can be done.

But what will be done? What is our destiny in space?

Humans will surely return to the moon, to live and to work, and thereby establish human civilization permanently beyond Earth. We will colonize Mars, and perhaps asteroids and other planets as well, and extend our presence far from our original home. Space-based commerce will become a commonplace, and space tourism a significant industry. Well, we might do all that. And we might not.

We each have our own opinion about what's going to happen, but of course only time will truly tell the tale. Until then, we can entertain ourselves with speculation, and by engaging in the actual work of turning our visions into reality, and in so doing shaping the future to more closely match our desires and expectations.

Nevertheless, global interest in the space movement indicates that it is neither a fad nor a boondoggle, but a serious scientific, commercial, and perhaps also a nationalistic undertaking. For example, in addition to the Russian space program, still active as a remnant of the Cold War, the Chinese space program has stated its commitment to land a man on the moon, India also has an active space program, as do Japan and the nations of Europe. What do all these nations seek in space?

For one thing, they all want to be involved in whatever space commerce turns out to be, if it turns out to be important. There is also a military dimension, for space could represents the highest, most advantageous ground; hence, the renewed competition to reach the moon could be the beginning of a new age of empire, a space-arms-race.

So is it our destiny to use the minerals and energy in space to enhance life on Earth, and so extend the human presence elsewhere in the solar system? Is it our destiny to colonize the solar system and extend our nationalist impulses beyond the atmosphere? Is it our destiny to war in space? Or is it only our destiny to imagine these things?

Redefining Our Responsibility

Dreams that our grandparents and great grandparents imagined, today's space scientists and astronauts have accomplished. In so doing they have redefined reality.

They have engaged in a profound learning process, for every voyage has been a venture into many dimensions of the unknown, an unparalleled opportunity to learn about science, about technology, about the Earth, and about ourselves. For in space we can learn things about all of these dimensions that we cannot learn on Earth, which gives us the opportunity to gain knowledge that could improve life on Earth. In this respect, then, establishing a stable and permanent presence in space in order to perform permanent science is not just a luxury, but perhaps also a responsibility.

Stated somewhat differently, now that we can voyage in space, we must voyage in space. Our unbounded curiosity engages with our unbounded capacity to create problems for ourselves, and off we go to space to seek solutions that are unattainable on Earth.

The Very Distant Mirror

Space is an infinite and suggestive canvas onto which we project our hopes, our fears, our desires, and our imaginations. In this respect, space means what we want it to mean. And as so few of us will actually ever venture into space, our fantasies will remain safely protected from the cold hand of reality's limitations.

In the domain of our imaginations, then, space is an archetype, an icon, or symbol that refers to what's "out there;" but what we imagine is "out there" is, for most of us, a mirror of what's "in here." So what's "out there" is probably quite different for each of us.

For some, space is an irresistible challenge, the "place" to which we will one day voyage en mass, and in so doing will define the destiny of humanity. For these people, space is a calling, a dream, a possibility.

For others, space is a frightening unknown, a nightmare of airless, frozen blackness.

For some it is a mystery, perhaps the ultimate mystery, beckoning without pause, posing questions and riddles that we cannot possibly answer, but which we are compelled to wonder about nevertheless.

Clearly, then, space has so many different meanings because as a mirror, what we see - or imagine that we see - in our idea of space is largely a reflection of that which is inside of ourselves. Dreams and nightmares are projected from within us onto the nearly blank canvas of space. In this respect, space is merely what we imagine it to be; what we want it to be; what we hope it to be; or what we fear it to be.

This reminds us that space may not actually be the "final frontier" that the writers of Star Trek imagined it to be. Oh, it is a frontier to be sure, a vast and challenging one. But as also a mirror, space shows us that the true frontiers are not only the ones we find without, but also those we find within.

Space, as a concept, merely reflects that which is already inside of us, hidden and largely unconscious, evoked perhaps in dreams, or in symbols, reactions, or images that reside deep below the threshold of self-awareness. Space, then, is the abstraction onto which we project our unconscious fears and desires; we see outside what actually lies hidden within.

Summary

Space may be our destiny, and our descendants may one day look back upon our era as the earliest dawning of a human diaspora that eventually spreads our presence far into the solar system, and beyond. Or perhaps our journeys into space will remain the province of an elite few, chosen for their bravery, their scientific skill, their fortitude, their wealth, or their thirst for adventure.

Perhaps space will be militarized, the home to fighting battalions of the space corps; an outcome that certainly would be a tragedy.

Perhaps we will learn to exploit space to solve some of our most pressing problems here on Earth; could it provide unlimited energy? Untold riches? New raw materials?

But in whatever form we voyage into space, whether it is by means of ever more clever robots, or by a small cadre of specialist-adventurers, or if indeed space becomes a realistic destination for the common tourist, miner, or bartender, when we go to space one thing always remains the same, and that is us.

We take ourselves to space in whatever form we go, and in so doing we may confront the sublime mysteries of the universe as we strive to touch the face of God. So what we will surely find in space, no matter who it is that goes, and by what means, and for whatever purpose, is ourselves. Our dreams. Our fears. Our desires. For the truth is that space is not the only final frontier; we ourselves are the final frontier, and all the mysteries without are mirrored by all the mysteries within. However we go to space, what we will ultimately confront is the sublime profoundness of the human spirit, as well as the pettiness, fears, desires, and greed of the human psyche.

Hence, the greatest challenges of human life are not only the mysteries of the universe that tempt us from without, but the mysteries of ourselves that lie within the human consciousness, human knowledge, and human compassion.

Going to space will teach us as much about ourselves as it does about the universe of which we are a part; and thus our adventures in space may be the most important thing that humanity has ever accomplished. For the greatest dangers that human society faces now are those that humans create, and if the voyage into space can help us to understand ourselves better then it may be the most significant and worthwhile undertaking in the history of civilization.

To fulfill this destiny is to fully grasp the profound meaning of space, and the opportunity that our unique moment in history offers. Will we be astute enough to grasp it? This is the essence of our challenge.

So what is the meaning of space? Contemplating on space puts us into contact with our own mortality and raises the possibility of immortality; space reminds us of our constrained experience of time and space, and makes us wonder about the unconstrained expanse of infinity. Space is the unlimited mirror of our own limited selves. Space is the profound counterpart to our sometimes banal, Earth-bound, day-to-day existence. It reminds us to never, ever forget the search for enduring truths that transform the dully mundane into transcendent inspiration and hope. This is the meaning of space, and as we venture to explore it, it is we who are transformed.

About the Author

Langdon Morris is a consultant and author. He is a partner of InnovationLabs LLC (www.innovationlabs.com) a consulting firm that develops and applies advanced methods in the fields of innovation, strategy, and organization. He is an accomplished facilitator, and has worked in this field for more than 20 years with large companies and small start-ups throughout the world. He is Senior Practice Scholar at the University of Pennsylvania Ackoff Center where he is leading a research project on "Breakthrough Business Models," and Senior Fellow of the Economic Opportunities Program of the Aspen Institute. He has taught Business Strategy at the Ecole Nationale des Ponts et Chaussées, Paris, with courses in "Corporate Transformation" and "Complex Systems Management," and the Universidad de Belgrano, Buenos Aires, where he taught "Creating the Future."

He is a member of ATWG and the Scientific Committee of Business Digest, Paris and a frequent contributor. He is a former Contributing Editor of Knowledge Management magazine, to which he also contributed the cover story of the February 1999 issue. He is author or co-author of numerous white papers and four acclaimed books, including Managing *the Evolving Corporation; The Knowledge Channel; Fourth Generation R&D: Managing Knowledge, Technology, and Innovation (with William L. Miller);* and *The War for America.*

[Note: Fourth Generation R&D includes a detailed interview with Ken Cox in which he tells the fascinating story of his involvement in the successful return of the damaged Apollo 13 spacecraft following the explosion en route to the moon.]

Langdon has appeared on dozens of radio and TV outlets nationwide, and speaks frequently at corporate and industry meetings around the world.

Chapter 3

Views of Global Leadership:
Government, Business, Academia, and Faith

By Martin Schwab

As commerce, education, and the rapid transition of thought and matter, by telegraph and steam have changed everything, I rather believe that the great Maker is preparing the world to become one nation, speaking one language, a consummation which will render armies and navies no longer necessary.

President Ulysses S. Grant, 1873. [1]

Introduction

The theme of this chapter is *integration* of different views for human activity *beyond Earth* from leaders around our globe, across professions, and through time. The following compilation of extended quotes can be used as an inspirational guide by other leaders in government, business, academia, and faith. It is these leaders who have the potential by working together and with general populations to chart a new course for human relations that fully utilizes outer space without limiting any of our inherent technical or human capabilities. The path of venturing together in outer space has been at our doorstep for many decades. The time has come to enable our human pioneers from all nations to go beyond the Earth-moon system on our behalf.

Leaders of Government

Ministry of Foreign Affairs of the People's Republic of China - June 30, 2004

One of the chief objectives of this chapter is to promote better understanding of views between the West and China. As this book is written primarily by authors who reside and work in the West, it is appropriate to start this chapter with an official view of the leadership of China for the future of humans in space:

Outer space belongs to mankind. It is not only beneficial to enhancing international security and stability but also in the common interests of all states to ensure the peaceful uses of outer space and to prevent the weaponization of an arms race in outer space. China has always stood for the peaceful uses of outer space and against introducing weapons into outer space. Under current circumstances, especially with the development of outer space technologies and of international security situation, the risk of weaponization of outer space does exist. This does not accord with the interests of any state. Therefore the international community should take effective protective measures, including negotiation of relevant international legal instruments, to ensure the uses of outer space for peaceful purposes only. In June 2002, China, together with Russia, Vietnam, Indonesia, Belarus, Zimbabwe and Syria submitted to the Conference of [sic] Disarmament in Geneva a working paper entitled "Possible Elements for a Future International Legal Agreement on the Prevention of the Deployment of Weapons in Outer Space, and the Threat or Use of Force Against Outer Objects," which has gained positive responses from many states. China is ready, together with the international community, to make unremitting efforts to maintain the peace and tranquility of outer space. [2]

The critical point that China and perhaps the rest of the international community, including many in the West do not typically address is the duty of U.S. leaders to maintain assets in space to defend its homeland from a ballistic missile attack, not necessarily from China. In essence, the counterargument to the above view of China is that unless there is true peace and tranquility on Earth, it is irresponsible for U.S. leaders to limit continued development of potential capabilities in space or anywhere else. Of course, the argument of the right to prepare for self-defense applies to any sovereign state or polity, even to "rogue states." It is commonly understood among security strategists that one reason why war has perpetuated itself throughout human history is because the shield enables the sword which yields anxiety, even if the sword is not used. The rest of the views in this chapter hint at how peace on Earth, through a common human endeavor in space might be achieved, as the need for this endeavor is perhaps more inherent in human nature than violence. In this way, lasting peace and tranquility in space (which requires much more than the absence of hostile postures) might be achieved as well.

President Dr. A.P.J. Abdul Kalam (India)

At the 90th Session of the Indian Science Congress in January 2003, President Dr. A.P.J. Abdul Kalam of India (a position distinct from the Prime Minister of India) addressed many human needs, including the creation of a knowledge society in the 21st century, clean energy for future generations, space solar power to enable water desalination, international cooperation against dangerous asteroids, commercialization of space access and socio-economic development, and cited poverty as a dynamic of terrorism and violence. President Kalam went on to say:

Above all, we must recognize the necessity for the world's space community to avoid terrestrial geo-political conflict to be drawn into outer space, thus threatening the space assets belonging to all mankind. This leads on to the need for an International Space Force made up of all nations willing to participate and contribute to protect world space assets in a manner, which will enable peaceful use of space on a global cooperative basis without the looming threat of conflict on Earth. I am sure India would contribute its best to the creation and sustenance of such an International Space Force. [3]

We will return to this far reaching idea for an International Space Force by President Kalam after we examine a few words spoken by the current leaders of the two main actors of the Cold War, President George W. Bush and President Vladimir Putin. When reading these selections, the reader is invited to keep an eye toward the "art of the possible" concerning relations between the U.S. and China, two countries that could very well develop a new cold war during the 21st century. The concept of the "art of the possible" was used during a 2002 conversation between former Assistant U.S. Secretary of State Richard Armitage and former NASA administrator Sean O'Keefe to describe the direction of engagement and cooperation that the U.S. should take in regard to Sino-American relations in outer space. [4]

President George W. Bush (USA)

While the two elections of President George W. Bush have revealed deep political divides among U.S. citizens and within the transatlantic alliance, it must also be noted that President Bush is the first president since John F. Kennedy to have set NASA and Congress in a clear direction, to our "Moon, Mars and Beyond." [5] President Kennedy's declaration to "put a man on the Moon in this decade," was a space vision articulated within the Cold War reality of competing ideologies presented to the world by the U.S. and the Soviet Union. President George W. Bush's space exploration vision also co-exists within geopolitical realities on Earth. With a little bit of imagination, Moon, Mars and Beyond can be conducted with cognizance toward the more immediate objective of "winning hearts and minds" across our globe, a way to bring the current global War on Terror to a rapid close.

President George W. Bush Speaking with the crew of Space Shuttle Discovery on August 2, 2005, the first successful human spaceflight by the U.S. since the Space Shuttle Columbia tragedy on February 1, 2003:

THE PRESIDENT: Good morning. Thank you for taking my phone call. I just wanted to tell you all how proud the American people are of our astronauts. I want to thank you for being risk-takers for the sake of exploration. I want to welcome our Japanese and Australian and Russian friends. And I wish you Godspeed in your mission. I know you've got very important work to do ahead of you. We look forward to seeing the successful completion of this mission. And, obviously, as you prepare to come back, a lot of Americans will be praying for a safe return. So it's great talking to you. Thanks for being such great examples of courage for a lot of our fellow citizens.

COMMANDER COLLINS: Thank you very much, Mr. President. We want to tell you that we really enjoy what we're doing, we really believe in our mission, and we believe in space exploration and getting people off the planet and seeing what's out there. So the steps that we're taking right now are really worth it, and we want everybody to know that. And thank you very much for taking the time out of your busy schedule to talk to us.

THE PRESIDENT: Well, listen, I want to thank you, Commander, and thank your fellow astronauts there. I agree with you — I think what you're doing is really important. And you've got a strong supporter for your mission here in the White House. I will tell you Laura went down and watched the launch in Florida, with my little brother, Jeb, and came back all excited about the energy that — there on the East Coast of Florida. But we're with you, and wish you all the very best. Thanks for taking my phone call. Now get back to work.

COMMANDER COLLINS: Thank you very much, sir. We did fly over Texas today and had a good look at it. It was beautiful. Have a good day.

THE PRESIDENT: Thank you (laughter). [6]

From the President's Vision for Space Exploration, January 14, 2004:

*We need to see and examine and touch for ourselves. And only human beings are capable of adapting to the inevitable uncertainties posed by space travel...We do not know where this journey will end, yet we know this: **human beings are headed into the cosmos** [bold added]. And along this journey we'll make many technological breakthroughs. We don't know yet what those breakthroughs will be, but we can be certain they'll come, and that our efforts will be repaid many times over. We may discover resources on the moon or Mars that will boggle the imagination, that will test our limits to dream....**We'll invite other nations** [bold added] to share the challenges and opportunities of this new era of discovery. The vision I outline today is a journey, not a race, and I call on other nations to join us on this journey, in a **spirit of cooperation and friendship** [bold added]...May God bless.* [7]

From the National Security Strategy of the United States of America, 2002:

...no nation can build a safer, better world alone...The magnitude of our shared responsibilities makes our disagreements look so small...We have our best chance since the rise of the nation-state in the 17th century to build a world where the great powers compete in peace instead of prepare for war... [8]

President Vladimir Putin (Russia)

In April 2004, Russian President Vladimir Putin reaffirmed his support for the demilitarization of space, but added that Russia [like China] must be ready to counter moves by "others." He did not mention the U.S. specifically.

*For many years, space has been part of military-political rivalry...Now we must do everything to **demilitarize** [bold added] space and turn it into the arena of peaceful cooperation... We will be striving to prevent space from being an arena of military-political confrontation, but we all understand very well that this situation still exists now and will continue to exist for quite a long time...* [9]

What is most interesting is that Putin, at least in this *Associated Press* article and allowing for translation error, calls for the complete demilitarization of space, not merely for countries to pledge to not weaponize space. Despite pleading for peaceful cooperation, in order to maintain Russia's nuclear deterrent capability, in February 2004 Russia tested a new weapon designed to elude missile defense systems such as the one the U.S. is developing, which could ultimately have components in space. [10]

Thomas Hobbes (17th century philosopher)

Before we venture into the darkness of Hobbesian philosophy let us return for a moment to President Kalam's quote from the beginning of this chapter calling for an International Space Force to guard against weaponization of space. Assuming that the quality of space surveillance, inspection, and verification increases with more partners and more capabilities over time, a critical question remains: *Should* a supranational organization be allowed to override the sovereignty of a single nation-state or coalition of nation-states in the interests of global security and harmony? How can war, poverty and global insecurity be alleviated in our world without creating a global hegemon that impinges on the liberty of the human spirit?

While Thomas Hobbes was not a political leader, his philosophy helped to shape our understanding of the very relationship between people and government and continues to explain the balance of power relationship among nation-states today. In his famous 1651 work, Hobbes posits *Leviathan*, a super-powerful government force that dominates all. Leviathan, as explained in more detail by Hobbes below is in a formal sense non-existent in our world today, with the possible exception of the U.S. military. Some strategists conclude that the U.S. military tries to act as Leviathan and is thus the main problem in our world system. Other strategists believe that the U.S. military needs to preserve and extend its Leviathan status in space and on Earth, for the good of peace and prosperity in our world system.

These are complicated philosophical and strategic questions. However, our world community needs to fast reach a conclusion on this age-old intellectual divide. Here's why. Recent war games conducted by the U.S. military have revealed an escalatory nature of conflict when weapons are used to destroy a nation-state's technical means, or "spy satellites." [11] These satellites are also used to verify peaceful deterrent postures of other nation-states' ballistic missiles in an increasingly multi-polar world. Under many different scenarios, human civilization is in jeopardy with the advent of this type of confusing conflict in the space-Earth arena, whether the U.S. is involved in a given incident or not.

In *Leviathan*, Hobbes argues:

*For the laws of nature (as justice, equity, modesty, mercy and in sum, **doing to others as we would be done to**) of themselves, without the terror of some power, to cause them to be observed, are contrary to our natural passions, that carry us to partiality, pride, revenge and the like...in the first place, I put for a general inclination of all mankind, a perpetual and restless desire of power after power, that ceaseth only in death...During the time men live without a common power to keep them all in awe, they are in that condition which is called war; and such a war, as is of every man, against every man...Where there is no common power, there is no law, where no law, no injustice. Force and fraud, are in war the cardinal virtues.* [12]

Hobbes may or may not be right about force and fraud being cardinal virtues in war, but he is undeniably correct about the positive utility that fear can provide to social order. However, we may not need Hobbes' monolithic Leviathan of government, backed by a global military power, to promote order in our world. Peace need not be imposed on us through the power of hierarchy in space or on Earth. The prestige of participation in *a common global endeavor* in space could very well substitute for the power to keep "all in awe" through brutish coercive force exercised by governments.

Even if Earth-bound governments were able to construct sophisticated, detailed, and transparent agreements, codes, and treaties, and even an international rapid reaction force against space weaponization, do human beings really want their governments to do any of this? While treaties may have slowed the process of human self-destruction using different classes of weapons, history has shown repeatedly that treaties themselves, as constituted among nation-states, are inherently irrelevant *over time*. A hypothesis for contemplation: In the age of space exploration, treaties devised to balance power on Earth are unnatural interferences to the unique agencies that outer space presents for lasting cooperation among human groups, including nation-states.

President John F. Kennedy (USA)

President John F. Kennedy was the first global leader who dared humankind to reach *beyond Earth*. Soon thereafter, the Kennedy administration and hundreds of millions of others stared deep into the abyss of the Hobbesian doctrine of mutually assured destruction during the Cuban Missile Crisis. President Kennedy's spoken record may be a more relevant guide to the 21st century political environment of space exploration than the philosophy of Thomas Hobbes.

From address at Rice University on the Nation's Space Effort, Houston, Texas, September 12, 1962:

....Many years ago the great British explorer George Mallory, who was to die on Mount Everest, was asked why did he want to climb it. He said, "Because it is there." Well, space is there, and we're going to climb it, and the moon and the planets are there, and new hopes for knowledge and peace are there. And, therefore, as we set sail we ask God's blessing on the most hazardous and dangerous and greatest adventure on which man has ever embarked. [13]

Cuban Missile Crisis, October 16-28, 1962

With the significant exception of the phrase "new hopes for knowledge and peace are there," President Kennedy's comments above, spoken before the Cuban Missile Crisis, reflect the U.S. tradition of "manifest destiny," perhaps in an effort to keep the U.S. Congress excited about human space exploration. By contrast, President Kennedy's address at the United Nations following the Cuban Missile Crisis includes the concept of "human destiny."

From address before the 18th General Assembly of the United Nations, New York, September 20, 1963:

*Space offers no problems of sovereignty; by resolution of this Assembly, the members of the United Nations have foresworn any claim to territorial rights in outer space or on celestial bodies, and declared that international law and the United Nations Charter will apply. Why, therefore, should man's first flight to the moon be a matter of **national competition** [bold added]? Why should the United States and the Soviet Union, in preparing for such expeditions, become involved in immense duplications of research, construction, and expenditure? Surely we should explore whether the scientists and astronauts of our two countries - indeed **of all the world** [bold added]- cannot work together in the conquest of space, sending someday in this decade to the moon not the representatives of a single nation, but the representatives of all of our countries...A global system of satellites could provide communication and weather information for all corners of the Earth.... The United Nations, building on its successes and learning from its failures, must be developed into a genuine **world security system** [bold added]. But peace does not rest in charters and covenants alone. It lies in the hearts and minds of all people. And if it is cast out there, then no act, no pact, no treaty, no organization can hope to preserve it without the support and the wholehearted commitment of all people. So let us not rest all our hopes on parchment and on paper; let us strive to build peace, a desire for peace, a willingness to work for peace, in the hearts and minds of all our people. I believe that we can. I believe the problems of human destiny are not beyond the reach of human beings...* [14]

President Kennedy's notion of a "world security system" in which all countries take part (as distinct from Hobbes' Leviathan) is comparable in structure and tone with President Kalam's current idea, described earlier for an "International Space Force" that multiple countries could maintain.

President Richard M. Nixon (USA)

The idea that peace on Earth can be furthered by a common human endeavor in space is a vision shared by leaders across the ideological spectrum. While President Nixon's legacy in regard to an expanded mission for NASA is not favorable, it was during the Nixon administration that the "handshake in space" between U.S. and Soviet space pioneers - the 1975 Apollo-Soyuz mission - was planned. During ongoing armed conflict in Vietnam and with internal political division erupting into violence in the U.S., Nixon was well aware of the historic importance of the Moon landings.

Interplanetary Conversation [sic] from the White House to Apollo 11 Crew on the Moon, July 20, 1969 [15]

THE PRESIDENT: ... For one priceless moment in the history of man, all of the people on this Earth are truly one, one in their pride in what you have done and one in our prayers that you will return safely to Earth.

ASTRONAUT ARMSTRONG: ...Thank You, Mr. President. It is a great honor and privilege for us to be here representing not only the United States, but men of peaceable nations, men with an interest and curiosity, and men with a vision for the future. It is an honor for us to be able to participate here today. [16]

On October 15, 2003, the Chinese people experienced this same universal pride with their historic first human spaceflight, and accepted accolades from around our world. Today, the questions of lasting social reform within China and political harmony between China and the U.S. remain unanswered. How might the promise of outer space be exploited by both sides to affect a predictable and preferable outcome of this critical bilateral relationship in our world system? The next section offers one suggestion.

The United Nations

From the Third United Nations Conference on the Exploration and Peaceful Uses of Outer Space (UNISPACE III), The Space Millennium: Vienna Declaration on Space and Human Development, July 30, 1999:

*... action should be taken: ... To improve the international coordination of activities related to near-Earth Objects, harmonizing the worldwide efforts directed at identification, follow-up observations and orbit prediction, while **at the same time** [bold added] giving consideration to developing a common strategy that would include future activities [countermeasures] related to near-Earth Objects ...* [17]

The obvious and primary responsibility for the leaders of the U.S., China, and all nations to undertake together is the rapid construction of countermeasures against the never-ending hazard to Earth from near Earth objects (NEOs) on Earth crossing orbits. It only takes one unseen large or very fast moving and dense asteroid or comet - one missed object by our sporadically funded professional and amateur astronomers - to wreck the prospects of continued human evolution, on Earth *and into the Cosmos*.

Much has been said and presented using PowerPoint presentations on this particular global threat. Perhaps the most engaging event on this topic to date was the 2004 Planetary Defense Conference: Protecting Earth from Asteroids, held in Orange County, California from February 23-26, 2004, sponsored by the American Institute of Aeronautics and Astronautics (AIAA) and The Aerospace Corporation. Though not technically "global leaders," senior scientists and policy analysts from the U.S., Europe and Russia presented ideas for countermeasures to NEOs. Most of the presentations were recorded on digital video and are available online at:
http://www.aero.org/support/planetarydefense/index.html as of January 20, 2006. [18]

The State of California (USA)

California is itself is one of the largest and most globally integrated economies in our world system, and deserves the title of "global leader." It is also one of the closest U.S. states in geographical terms to China.

California has produced more bipartisan leadership on the issue of planetary defense than any other U.S. state. On March 24, 1993, the late U.S. Representative, George Brown (Democrat, California) held Congressional hearings on the NEO threat before the House Committee on Science, Space, and Technology. At those hearings, he said:

If some day in the future we discover well in advance that an asteroid that is big enough to cause a mass extinction is going to hit the Earth, and then we alter the course of that asteroid so that it does not hit us, it will be one of the most important accomplishments in all of human history. [19]

In 2002, U.S. Representative Dana Rohrabacher (Republican, California) commented:

So this is a real threat, but it is not something we have to fear. It is something we have to look at and try to find a way to identify threats. It is called Home Planet Defense. We need to pay some attention to it; and then if an asteroid does threaten us, we will be able to identify it far in advance and deter it from its path so it would not hurt the people of the world. [20]

Leaders of Business

Elon Musk, CEO, SpaceX (USA)

While Elon Musk's comments below articulate a national rather than a global perspective, his ideas are representative of space entrepreneurs from around our world.

From testimony before the Joint Hearing on Commercial Human Spaceflight, U.S. Senate Science, Commerce and Technology Subcommittee on Space; U.S. House of Representatives Subcommittee on Space and Aeronautics, July 24, 2003:

It is despairing to consider that the cost and reliability of access to space have barely changed since the Apollo era over three decades ago. Yet in virtually every other field of technology, we have made great strides in reducing cost and increasing capability, often in ways we did not dream existed....To address this problem, we must create a fertile environment for new space access companies that brings to bear the same free market forces that have made our country the greatest economic power in the world. We are at a crucial turning point today. The recent entrepreneurial activity in space (my company perhaps included) shows promise, but is still embryonic and fragile. It is very important that our government in all its forms proactively adopt a nurturing and supportive approach to new launch vehicle developments, whether orbital or sub-orbital, manned or unmanned....If you doubt that we can possibly see such progress in space access, please reflect for a moment that the Internet, originally a DARPA [U.S. Defense Advanced Research Projects Agency] funded project, showed negligible growth for over two decades until private enterprise entered the picture. At that point, growth accelerated by more than a factor of ten....

The market for satellite delivery, while significant, has limitations in size and application. I suspect the far larger market in the long term is serving people that wish to travel to space for enjoyment. For many people, as shown by a number of marketing studies, this is the fulfillment of a lifelong dream and they are willing to spend a substantial portion of their savings to see that dream realized... If we believe humanity should one day expand to the stars, then people must have some way to see for themselves what space is all about. They must share in its wonders and experience firsthand its meaning. And, in so doing, open the doorway to space for all. [21]

Jean-Marie Luton, Chairman and Jean-Yves Le Gall, Chief Executive Officer, Arianespace (France)

There is no better representative of the views of international space business than Arianespace of France. Its Ariane 5 is generally regarded as the most reliable launcher on Earth. The following statements, written by the above leaders of Arianespace are taken from their 2004 annual report, and are couched in somewhat different language than the other selections in this chapter. Still, they reveal for us a sense of responsibility to a specific global region (Europe), and a sense of global outreach and teamwork, a template perhaps for the *future of humans in space*:

...our tradition of success...has led international operators to choose our launch services for around two-thirds of the commercial satellites now in orbit...Successful launches in 2004 provided further confirmation of the maturity of the standard Ariane 5G launcher. The Ariane 5G performed three missions, boosting Rosetta on a ten-year voyage toward the comet Churyumov-Gerasimenko; Anik F2, the world's largest communications satellite; and the Helios 2A military satellite, along with six auxiliary payloads. The heavy-lift Ariane 5 ECA made a successful return to flight on February 12, 2005, establishing its credibility and confirming Ariane 5's critical role at the core of the strategic challenges facing Europe...giving Europe invaluable new assets. Arianespace's launch services offer is three-pronged, comprising the Ariane 5 heavy lifter, the Soyuz medium launcher and the Vega small launcher. This family of launchers will be operational at the Guiana Space Center in French Guiana starting in 2008...The launch backup agreement set up with Boeing Launch Services and Mitsubishi Heavy Industries through the Launch Services Alliance is an additional commercial advantage for Arianespace...The launch of DirecTV 7SD by Sea Launch [an international subsidiary of Boeing] in May 2004 provided clear proof that the Alliance is now bringing concrete benefits to our customers...

We take a long-term vision of our business, based on innovation of course, but above all on pragmatism. We analyzed our situation in depth in 2004, taking actions to guarantee that our company will continue to play a major international role in the coming years. In collaboration with partners from around the world, Arianespace is already building solid, broad-based foundations for the future: consolidating Ariane 5's reliability, serving the International Space Station, participating in future space exploration programs, developing the Launch Services Alliance, designing the next generation of launchers and much more. [22]

John B. Higginbotham, Founder and Managing Director, SpaceVest Capital and Director Emeritus, Space Foundation (USA)

John B. Higginbotham leads SpaceVest Capital, a major U.S. venture capital firm "investing in advanced technologies to provide compelling solutions for rapidly growing new markets." He asks whether nation-states or even citizens within nation-states want rocket boosters to become so cheap and accessible that intermediate-range ballistic missile technology can be purchased on the open market for around $15,000. The summer before China made its historic human spaceflight, Higginbotham recognized that China was "for real" in space. From a commercial perspective, Higginbotham was excited because he believes that the U.S. responds even better to economic competition than to new security threats. [23]

From testimony before the President's Commission on Moon, Mars and Beyond, New York, May 4, 2004:

Few are against space, for the civil programs do bring a source of wonderment, pride or entertainment. But few see many of the [space] initiatives as critical to the technological and economic fabric of the nation. This arises from a fundamental lack of understanding of the deep dependencies of the nation's economy and security on the capabilities of the [space] industry...College curricula that provide an interdisciplinary education relevant to the space industry should be promoted and expanded. Such educational initiatives are critical if we are to have the talent we will need five, ten, fifteen and twenty years out, to develop the systems, manage the programs and create the businesses that will sustain the vision...technologies are not products...the skills needed to translate potential into capabilities and capabilities into functionality are as unique and sophisticated in the business world as they are in the technology world...In other words, creating commercial value from the initiative is the key to generating sustainable resources to support the initiative thus ensuring the permanent presence of humans in space.

...Our parents faced a world war, a cold war and a moon race. They accepted the risks, paid what it cost and accomplished the impossible. In short, they did their duty and in so doing ensured the very freedom that allows us the privilege of gathering today. Our children face new threats, a challenging world and a complex future. They will need the knowledge, protection and inspiration that this initiative will create for them. It is now our time to do our duty and proceed with this initiative. Go launch and...Godspeed. [24]

Leaders of Academia

Dr. John Thomas, Dean of the School of Business, La Sierra University, Riverside, California (USA)

Dean Thomas was the founder and leader of the La Sierra University Students in Free Enterprise (SIFE) team that won five years of international championships, and in this chapter he provides a critical link between the global views of those in business and those in academia:

As Ben Bova said pointedly so many years ago, "The way out is up!" The exploration of space presents limitless possibilities for humanity—economic, political, cultural, technological, scientific, and spiritual. While creating new sources of existing resources, opportunities to discover unimaginable new ones, the potential for contact with new civilizations, and the chance to resolve some of the Earth's pressing political problems, it also presents a unique and exhilarating challenge to the human spirit. Space is truly a new frontier, and as such will press us beyond ourselves, urging us on to new levels of achievement and discovery (original contribution by Dr. John Thomas).

Professor Joanne Gabrynowicz, Director, National Remote Sensing and Space Law Center, University of Mississippi School of Law (USA)

Professor Joanne Gabrynowicz is a long-time advocate for the rule of law in space. Her work provides a solid intellectual foundation for expanding human civilization into outer space. Much of her contributions are grounded philosophically in the Treaty on Principles Governing the Activities of States in the Exploration and Use of Outer Space Including the Moon and other Celestial Bodies, also known as the 1967 "Outer Space Treaty." [25] The following quote is taken from an article she published in *Space News* in 1999. Gabrynowicz skillfully disputes the claim that the Outer Space Treaty is useless simply because it is outdated in regard to issues like nuclear proliferation and the commercialization of space. She then presents her view of the larger meaning of the Outer Space Treaty in guiding the course of human events in outer space:

...the U.S. Constitution guarantees due process - a term no less ambiguous than "peaceful purposes" [term used repeatedly in the Outer Space Treaty] - but it was initially limited by the legal institution of slavery. Seventy-seven years later, when the specific crisis of slavery became obsolete, due process was again interpreted and resulted in the 14th Amendment to the Constitution. The solution was to amend the Constitution, not deem it useless. Slavery, to be sure, is very different from military space issues. But it was as significant and as difficult a national issue, involving an actual, not a potential war...the law can manage grave issues. Like the Constitution, the treaties can be amended. Whether they are depends on political will...space is, and has been for three decades, a weapons-free, peaceful, legal, political and operational environment due in large part to the rule of law and diplomatic measures used to prevent hostilities...Ultimately, humans have only two ways to resolve conflict: through agreement or by fighting. At the level of nations this means law or war. [26]

Professor David Koplow, Co-Director of the Center for Applied Legal Studies at Georgetown University Law Center, Former Special Assistant to the Director of the U.S. Arms Control and Disarmament Agency and Former Deputy General Counsel (International Affairs) at the U.S. Department of Defense

The contentious arms control negotiations process has the potential to create and reinforce deep suspicions among nation-states, in part due to the dynamics created by inevitable internal factions that may be present in any given nation-state during negotiations. Professor Koplow shares from his experience what may be useful when and if the major powers begin negotiating treaties or agreements on weapons in outer space.

Comments made at Homeplanet Defense Institute Open Seminar 1, U.S./EU/China Strategic Policies in Space: Balance of Power or International Cooperation? Arlington, Virginia, October 14, 2003:

Nobody should ever be surprised by how difficult it is to try to develop a multilateral treaty on an important topic like the eventual military uses of weapons in outer space... One formula for success, one of the few things that reliably work in this field is to negotiate and solemnly undertake the obligation not to do the things you were never going to do anyway. That was the formula that we found for the Outer Space Treaty, which for the most part does not prohibit things that the superpowers were not already committed to forego unilaterally. That is also true for just about everything else in the field of arms control... even if all a treaty does is codify a country's current expectations, plans and capabilities, that does have some beneficial long term effect. Even the Outer Space Treaty which we already have identified does not in fact change very much nonetheless contributes to a longer term stability...The law does shape expectations, it does help create conditions of legitimacy and appropriateness...It is a little bit harder to get money to do things that are illegal. It doesn't mean that it can't be done. There are plenty of instances of countries pursuing covert/illegal programs but it is harder. [27]

International Space University (ISU), Strasbourg, France

Since its founding in 1987, ISU has graduated more than 2,200 students from 87 countries, providing graduate-level training to the future leaders of an emerging global space community at locations around our world. ISU offers a unique curriculum of space science, space engineering, systems engineering, space policy and law, business and management and space and society. ISU also serves as a neutral international forum for the exchange of knowledge and ideas on challenging issues related to space, both in the public and private sectors.

ISU Credo. Signed by Peter H. Diamandis, Todd B. Hawley and Robert D. Richards, April 12, 1995:

*WE, THE FOUNDERS of the International Space University, do hereby set forth this Credo as the basis for fulfilling ISU's goals and full potential. INTERNATIONAL SPACE UNIVERSITY is an institution founded on the vision of a peaceful, prosperous and boundless future through the study, exploration and development of Space for the benefit of all humanity. ISU is an institution dedicated to international affiliations, collaboration, and open, scholarly pursuits related to outer space exploration and development. It is a place where students and faculty from all backgrounds are welcomed; where diversity of culture, philosophy, lifestyle, training and opinion are honored and nurtured. ISU is an institution which recognizes the importance of interdisciplinary studies for the successful exploration and development of space. ISU strives to promote an understanding and appreciation of the Cosmos through the constant evolution of new programs and curricula in relevant areas of study. To this end, ISU will be augmented by **an expanding base of campus facilities, networks and affiliations both on and off the Earth** [bold added]. ISU is an institution dedicated to the development of the human species, the preservation of its home planet, the increase of knowledge, the rational utilization of the vast resources of the Cosmos, and the sanctity of Life in all terrestrial and extraterrestrial manifestations. ISU is a place where students and scholars seek to understand the mysteries of the Cosmos and apply their knowledge to the betterment of the human condition. It is the objective of ISU to be an integral part of Humanity's movement into the Cosmos, and to carry forth all the principles and philosophies embodied in this Credo.*

THIS, THEN, IS THE CREDO OF ISU. For all who join ISU, we welcome you to a new and growing family. It is hoped that each of you, as leaders of industry, academia and government will work together to fulfill the goals set forth herein. Together, we shall aspire to the Stars with wisdom, vision and effort. [28]

Leaders of Faith

His Holiness the Dalai Lama, Leader of the Tibetan People (exiled in India)

His Holiness the Dalai Lama contributes to the discussion of alternative global power structures and also provides a unique view on the relationship between science and religion. The Dalai Lama offers a fresh perspective to those engaged in the emerging debate concerning the controversial "Intelligent Design" concept now being taught in some U.S. states.

I have learned a great deal from my encounters with scientists, and I see no obstacle to engaging in dialogue with them even when their perspective is one of radical materialism... My concern is rather that we are apt to over-look the limitations of science. In replacing religion as the final source of knowledge in popular estimation, science begins to look a bit like another religion itself. With this comes a similar danger on the part of some of its adherents of blind faith in its principles and, correspondingly, to intolerance of alternative views. That this sup-planting of religion has taken place is not surprising, however, given science's extraordinary achievements. Who could fail to be impressed at our ability to land people on the moon? [29]

...I believe the United Nations has a critical role to play...the United Nations is the only global institution capable of both influencing and formulating policy on behalf of the international community...One of the particular weaknesses of the United Nations as it is presently constituted is that although it provides a forum for individual governments, individual citizens cannot be heard there. It has no mechanism whereby those wishing to speak out against their governments can be heard...it would be worthwhile to establish a body whose principal task is to monitor human affairs from the perspective of ethics, an organization that might be called the World Council of the People (although no doubt a better name could be found). This would consist in a group of individuals drawn, as I imagine it, from a wide variety of backgrounds...with a common reputation for integrity and dedication to fundamental ethical and human values. Because this body would not actually be invested with political power, its pro-nouncements would not be legally binding. But by virtue of its independence - having no link with any one nation or group of nations, and no ideology - these deliberations would represent the conscience of the world. They would thus carry moral authority. [30]

When considering the matter of a governing structure for established human societies in outer space, at some point in our future the issue of *interplanetary government* will have to be broached. While many modern secular cultures around our world may justly demand firm separation of church and state in outer space, non-Western societies, which will no doubt be involved in joint human space exploration missions, may express different views. These views will need to be respected and considered by the human community.

Reverend Marc Wessels, Space Exploration and Theology Institute, Louisville, Kentucky (USA)

The Reverend and Dr. Marc Wessels offers a moderate view from a Christian perspective, applicable to a potential question of separation of church and state in human civilizations beyond Earth, in this case, Mars.

While we live in an age of continuing scientific discovery and technological marvels, as humans we must never forget that our human nature is not limited to the physical plane. Humans are by nature spiritual creatures designed by a loving Creator with a higher purpose in mind. The world's religious traditions that have emerged over thousands of years articulate the perspective that humanity is called to understand its relationship to the natural world by acknowledging the Creator and Sustainer of all of life... Those who are adventurous enough to leave the "home planet" and dedicated enough to be "bringing life to Mars - and Mars to life" will be guided by some ethical construct, some theological principles which will under-gird their personal lives and that of the Martian community which is to be established.... Now, at the commencement of this challenge to establish human habitation on Mars, we need to confess that if we do not acknowledge the importance of religious values in our Martian venture we shall build a society which will be unstable and eventually collapse on its own due to spiritual decay. This Martian venture calls for the highest of human ideals to be brought to the forefront - the best of the world's religious values in the establishment. A word of scriptural truth reminds us that "Unless the Lord builds the house, the laborers labor in vain." [31]

Dr. Lawrence T. Geraty, President, La Sierra University, Riverside, California (USA)

Dr. Geraty's life has been dedicated to scholarship, science, the Christian Faith, education and archeological research while being a citizen of the world. He is known both locally in his community and around the globe for his positive oriented leadership ability.

For hundreds of years, humans have longed to depart Earth to experience the mysteries of space. For fifty years a few fortunate people have actually begun the process. The myriad of reasons humans should, and will, migrate to space are well documented in this volume "Beyond Earth." I fully endorse the fundamental reason stated of improvement of humanity through peaceful exploration and utilization of space resources for solving the serious problems on Earth. And, though, as a Christian, I can imagine Christ's Second Coming as ultimately a mis-

sion to rescue us from ourselves, in the meantime, I do not disagree with Carl Sagan's view, stated in 1994, that we have a basic responsibility to our species to venture to other worlds because human survival is at stake.

My life dedicated to education and to the Christian faith, leads me to link the future of humans in space to the goals La Sierra University has etched in its seal and mission: "TO SEEK truth enlarging human understanding through scholarship; TO KNOW ourselves broadly educating the whole student; TO SERVE others contributing to the good of the global community."

Thinking seriously and imagining about a future human adventure civilization has never attempted opens up a vast spectrum of unknowns and possibilities for hopes and dreams humans have expressed throughout history. The one huge variable, from both a scientific and human perspectives is how we humans in the 21st century design and plan for our penetration of the universe. This book is a solid step forward. And the authors of the "Code of Ethics" Chapter 17, end with the critical statement: "The answer to the question Cain asked of God, 'Am I my brother's keeper?'" (Genesis: 4), for humans in space, needs to be "Yes" (original contribution by Dr. Lawrence T. Geraty).

Conclusion

In this chapter we have reviewed some of the *views of global leadership* from across our global society - government, business, academia and faith. All of these views reflect the idea that while military relationships dominate the political agenda on Earth today, it is possible that the lure of human space pioneering can deliver us from our dangerous propensity for self-destruction. Perhaps the most compelling idea presented in this chapter is given by the current president of India, Dr. A.P.J. Abdul Kalam, calling for an International Space Force, with conceptual precedent set by President John F. Kennedy of the United States at the United Nations in 1963. Another interesting overlap of ideas concerns the moral responsibility of planetary defense against Earth crossing orbits of asteroids and comets, as articulated in the 1999 United Nations Conference on the Exploration and Peaceful Uses of Outer Space (Vienna Declaration on Space and Human Development) and by two political representatives from the great State of California, spanning two decades and two political parties.

The views expressed by the selected leaders of business provide inspiration of what is truly possible in the near term, internationally and regionally while the selected leaders of academia and faith provide greater historical and spiritual backing for what is possible over the long term. By integrating these views into our daily approach to, or in support of space pioneering, it is our challenge and our hope that six billion together can achieve our constructive potential.

References

(1) Hobsbawm, E.J. (2000). The world unified. In Frank Lechner and John Boli (Eds.), *The globalization reader* (p.52). Malden, Massachusetts: Blackwell Publishers Ltd.

(2) Ministry of Foreign Affairs of the People's Republic of China. (2004). *Prevention of an arms race in outer space.* Beijing. Retrieved November 27, 2005, from, http://www.fmprc.gov.cn/eng/wjb/zzjg/jks/kjlc/wkdd/fzjbjs/t141326.htm.

(3) Kalam, President Dr. Abdul of India. (2003, January 4). Vision for the global space community; prosperous, happy and secure planet Earth. Address to the 90th Session of the Indian Science Congress, Indian Space Research Organization. Retrieved November 27, 2005, from, http://www.spaceref.com/news/viewsr.html?pid=7457.

(4) Foust, Jeff. (2002, March 27). Administrator O'Keefe pitches his vision for NASA. *Spaceflight Now.* Retrieved November 27, 2005, from, http://spaceflightnow.com/news/n0203/27okeefe/.

(5) President's Commission on Implementation of United States Space Exploration Policy. (2004, June 4). *A journey to inspire, innovate and discover: Moon, Mars and beyond.* Washington, DC: U.S. Government Printing Office.

(6) Bush, President George W. (2005, August 2). President calls crew members of space shuttle Discovery. Telephone call from the White House. Washington, D.C. Retrieved November 27, 2005, from, http://www.whitehouse.gov/news/releases/2005/08/20050802-1.html#.

(7) Bush, President George W. (2004, January 14). President Bush announces new vision for space exploration program. Address to the nation given at NASA Headquarters. Washington, D.C. Retrieved November 27, 2005, from, http://www.whitehouse.gov/news/releaes/2004/01/20040114-3.html.

(8) Bush, President George W. (2002). *National security strategy of the United States of America.* Retrieved November 27, 2005, from, http://www.whitehouse.gov/nsc/nss8.html.

(9) Isachenkov, Vladimir. (2004, April 12). Putin calls for demilitarization of space. *Associated Press.* Retrieved November 27, 2005, from, http://www.space.com/news/putin_space_040412.html.

(10) Ibid.

(11) Scott, William B. (2001, January 29). Wargames zero in on knotty milspace issues. *Aviation Week and Space Technology.* Retrieved November 26, 2005, from, http://www.home.datawest.net/dawog/Space/space%20war%20%28awst%29.htm.

(12) Hobbes, Thomas. (1985). Leviathan. In George Seldes (Comp.), *The great thoughts* (pp. 187-188). New York: Ballantine Books. (Original work published in 1651).

(13) Kennedy, President John F. (1962, September 12). Address at Rice University on the nation's space effort. Houston, Texas. Text and video of address retrieved November 27, 2005, from, http://www.jfklibrary.org/j091262.htm.

(14) Kennedy, President John F. (1963, September 20). Address before the 18th General Assembly of the United Nations. New York. Text and video of address retrieved November 27, 2005, from http://www.jfklibrary.org/j092063.htm.

(15) National Archives and Records Administration (2005). American originals. Retrieved November 27, 2005, from, www.archives.gov/exhibits/american_originals/apollo11.html.

(16) NASA. (2005). Nixon telephones Armstrong on the Moon [web page]. Retrieved November 27, 2005, from, http://grin.hq.nasa.gov/ABSTRACTS/GPN-2000-001672.html.

(17) United Nations. (1999). *Resolution adopted by the third United Nations Conference on the Exploration and the Peaceful Uses of Outer Space.* Retrieved November 27, 2005, from, http://www.oosa.unvienna.org/unisp-3/res/html/viennadecl.html.

(18) AIAA and The Aerospace Corporation. (2004). Planetary Defense Conference: Protecting Earth from Asteroids. Slide shows and videos retrieved January 20, 2006, from, http://www.aero.org/support/planetarydefense/index.html.

(19) AIAA (American Institute of Aeronautics and Astronautics). (1995). Responding to the potential threat of a near-Earth-object impact [position paper]. Retrieved January 20, 2006, from, http://impact.arc.nasa.gov/gov_aiaa95.cfm.

(20) David, Leonard. (2002, April 30). Pete Conrad Act tied to threatening asteroids. *space.com.* Retrieved November 27, 2005, from, http://www.space.com/scienceastronomy/astronomy/png_test_020514.html.

(21) Musk, Elon. (2003, July 24). Testimony before the Joint Hearing on Commercial Human Spaceflight, U.S. Senate Science, Commerce and Technology Subcommittee on Space; U.S. House of Representatives Subcommittee on Space and Aeronautics. Washington, D.C. *SpaceRef.com.* Retrieved November 27, 2005, from, http://www.spaceref.com/news/viewsr.html?pid=9833.

(22) Arianespace (2004), *Annual Report 2004.* Evry-Courcouronnes Cedex, France. Retrieved November 27, 2005, from, http://www.arianespace.com/site/documents/sub_main_annual.html.

(23) Higginbotham, John B. (2003, July 23). Critical issues in financing new space ventures. Address to the International Association of Space Entrepreneurs Speaker Series, McLean, Virginia.

(24) Higginbotham, John B. (2004, May 4). Testimony to the president's commission on Moon, Mars and beyond. New York. Retrieved November 27, 2005, from, http://www.spacevest.com/svdocs/news-old.html.

(25) United Nations. (1967). *Treaty on principles governing the activities of states in the exploration and use of outer space, including the Moon and other celestial bodies* ("Outer Space Treaty"). Retrieved November 27, 2005, from the United Nations Office for Outer Space Affairs (OOSA). http://www.oosa.unvienna.org/SpaceLaw/outersptxt.htm.

(26) Gabrynowicz, Joanne Irene. (1999, July 26). Space power and law power. *Space News.*

(27) Koplow, Professor David. (2003, October 14). Comments at U.S./EU/China Strategic Policies in Space: Balance of Power or International Cooperation? Arlington County Central Library, Arlington, Virginia. Hosted by the Homeplanet Defense Institute. Retrieved November 27, 2005, from, http://www.homeplanetdefense.org/Open_Seminars.htm.

(28) International Space University. (2005). International Space University - for future space leaders [web page]. Retrieved November 27, 2005, from, http://www.isunet.edu/EN/4.

(29) Dalai Lama, His Holiness the. (1999). *Ethics for the new millennium* (p.11 and pp. 214-216). New York: Riverhead Books/Penguin Putnam Inc.

(30) Ibid.

(31) Wessels, The Rev. Dr. Marc A. (2002). Religious values and the Martian venture. In Frank Crossman and Dr. Robert Zubrin of the Mars Society (Eds. and Comps.), *On to Mars: colonizing a new world* (pp. 186-191). Burlington, Ontario: CG Publishing, Apogee Space Press.

About The Author

Martin Schwab is a world security strategist and author of *Homeplanet Defense: Strategic Thought for a World in Crisis* (West Conshohocken, Pennsylvania: Infinity Publishing, 2005). He founded the Homeplanet Defense Institute, an informal global network of consultants based in northern Virginia (www.homeplanetdefense.org). Mr. Schwab is a member of the Aerospace Technology Working Group, serves on the steering committee of the International Association of Space Entrepreneurs, has been an active member of the Mars Society and is president emeritus of the Washington, D.C. chapters of both Citizens for Global Solutions and the World Future Society. Mr. Schwab has served on the board of directors of the Philosophical Society of Washington and has worked with the Space Security Working Group at the Eisenhower Institute in Washington, D.C. as well as the Federal Trust for Education and Research in London. Mr. Schwab holds a Master of Public and International Affairs degree in security and intelligence studies from the University of Pittsburgh. He lives outside of Dayton, Ohio. www.martinschwab.com.

PART II.

HUMAN FACTORS IN SPACE

The many authors of this book propose, each in their own way, that the future migration of humans to space to be characterized by Jonas Salk's concept of "Survival of the Wisest." [1]

While Darwin's "Survival of the Fittest" provides a brilliant scientific explanation of the evolution and survival of life on Earth, it does not account for the possibility that humans can play a positive role in evolution, as John Stewart proposes in Chapter 22; nor does Darwin's theory consider Norbert Wiener's prescription for the *"The Human Use of Human Beings,"* [2] that technology must free humanity rather than enslave us.

The 19 chapters of Part II explore these human and humane dimensions of space exploration, development, and settlement, dealing topics ranging from governance, leadership, law, collaboration, psychology, bacteria, biotechnology, education, the arts, mythology, and evolution itself. Among the many messages you'll find here, one stands out perhaps most clearly, which is that humanity now has the capabilities to create our future, but we can create for better or for worse. These chapters contain the many dimensions that we must master for us to achieve better.

(1) Jonas Salk. (1973). *The Survival of the Wisest*. New York: Harper and Row.
(2) Norbert Wiener. (1954*). The Human Use of Human Beings: Cybernetics and Society*.
     New York: Doubleday Anchor Books, 2nd ed. revised. Originally published
     in 1950 by Houghton Mifflin Co.

Chapter 4

The Overview Effect and the Future of Humans In Space

By Frank White

"And that whole process of what you identify with begins to shift. When you go around the Earth in an hour and a half, you begin to recognize that your identity is with that whole thing."

Apollo 9 astronaut Russell L. Schweickart in The Overview Effect, Chapter Three

Abstract

The Overview Effect is a simple yet profound concept when we consider the future of humans in space. It refers specifically to the experience that human beings have in orbit or on the moon, but it can be expanded to go far beyond that initial change in awareness.

The Overview Effect is only the beginning; it describes one of several changes in consciousness that have occurred as a result of even the limited experience of humans in space over the past 45 years. When we speak of the future of humans in space, the key is how far ahead we wish to look at how consciousness will evolve. The changes that will take place will be very different over the next 50, 100, or 1,000 years, but they are likely to be increasingly profound.

When I wrote the manuscript that became the first edition of The Overview Effect: Space Exploration and Human Evolution in 1987, my emphasis was initially on "The Overview Effect," and "Space Exploration." [1] By the time I had finished the book, however, my interest had shifted to a significant degree to the other part of the subtitle, "Human Evolution." When we begin thinking about the changes that will affect human beings and human society as we move off the home planet, into the solar system, and then into the wider universe, evolution necessarily becomes a prime topic, if not the prime topic. And it is evolution at every level: biological, philosophical, political, economic, and spiritual.

Motivations for Exploration

When human beings first began to venture onto the space frontier in the 1960s, a whole range of motivations drove their adventurous explorations. These motives included nationalistic competition, the spirit of exploration in its purest sense, and the effort to increase scientific knowledge. At the forefront of the reasons stood the struggle for supremacy on Earth between the "Free World," led by the United States, and the "Communist Bloc," led by the Soviet Union. Since the they couldn't fight a nuclear war without destroying civilization, the quest for superiority shifted to the heavens, and the "space race" became a proxy contest, the winner supposedly proving that its system for organizing society was superior. While this was, to some extent, a spurious notion, other nations watched the contest with interest that was linked to their desire to be on the winning side. It was also expected that whichever side dominated the high ground of space could also dominate the Earth militarily, and this has been proven to be valid as we see in light of the impact of satellite imagery on warfighting today.

The United States won the space race by landing a man on the moon and returning him safely to Earth before the end of the decade, as President Kennedy had vowed we would do. The Soviet moon program failed, even though they had been extremely successful in Earth orbit, and the Soviet Union collapsed some 20 years later. The demise of the Soviet empire was in large part due to the technological Overview Effect, as satellites in orbit beamed images of a better life to the suppressed peoples of Eastern Europe and other nations of the "evil empire" that secret police, Berlin walls, and gulags could not suppress.

Now a new space race may be beginning as China, the only remaining Communist superpower, plans for space programs in orbit, on the moon, and across the solar system. We can hypothesize that the recently-announced dramatic changes in US space policy, which had previously limited most effort to Low Earth Orbit but is now pointing to the moon, Mars, and beyond, and has announced intentions to weaponize the frontier, is at least in part being driven by Chinese actions. Only time will tell how this contest turns out.

But meanwhile, new and more positive motivations are now appearing on the scene as private enterprise, supplemented by the enthusiasm of wealthy entrepreneurs such as Elon Musk, Robert Branson, Paul Allen, and Jeff Bezos begin to seek profits in orbit through space tourism and space settlements.

Beyond Competition

Many of the early responses to the existence of the frontier were centrifugal, divisively separating humanity according to Earth-based national identities, rather than centripetal, i.e., drawing humanity together as a single species that is exploring the universe together. That is where the Overview Effect comes in, as it is, first and foremost, a profound realization of the unity and oneness of our planet and our species, an insight that is triggered by seeing the Earth from orbit or the moon.

This experience as been felt strongly by many astronauts, some of whom, such as Rusty Schweickert, returned to Earth to start organizations intended to help to bring unity to their home countries. The Association of Space Explorers, for example, came into being in the waning days of the Cold War, and made a modest contribution to ending it.

Advocates of space exploration have long held out the dream of human beings living in a new way on the frontier, together rather than as separate tribes, in permanent peace rather than engaged in continuous war, and the Overview Effect, represented by images of the Earth from orbit, is one of the most powerful symbols of that dream.

This dream needs substantive proposals, however, to become real; pictures alone are not enough, because as we move farther outward into the universe the forces pulling us apart will increasingly impact on our species and our emerging space-based civilization. These forces of evolution, either natural or shaped by humanity itself in the struggle to adapt to new environments, will likely create not only new and different political and economic systems, but new species as well.

This vision of the future is clearly described by Ben Finney and Eric Jones when they write:

"This advance will not be limited to the birth of one new species... By spreading into space we will embark on an adaptive radiation of hominidae that will spread intelligent life as far as technology or limits placed by any competing life forms will allow." [2]

The Human Space Program

How can human unity be maintained in the face of such mind-boggling diversity? In writing The Overview Effect, it became clear to me that the process of speciation, the process that describes the emergence of new species on Earth, will also be at work in space. There seems little doubt that the impact of radiation, low gravity, and other non-terrestrial forces will begin to shape us in new ways. Some of those on the space frontier may well choose to use the new tools of genetic engineering to accelerate the evolutionary process, and it's even possible to imagine a being able to live in free space, roaming the vacuum among the stars and planets like dolphins in terrestrial seas. Out of these processes will likely emerge at least one new species, Homo Spaciens, and perhaps many more.

In The Overview Effect, I proposed that we should create a "Human Space Program," a millennium-long commitment to exploring the universe and understanding the transformations of consciousness inherent in the exploration process. The idea of this program is to constitute a conceptual unity for our thinking as we move out into the universe and begin to evolve in different directions.

Such a program would also have to include policies and procedures for dealing with extraterrestrial cultures that are likely to be contacted, some of them more advanced than humans, some less. For example, the idea of the "Prime Directive" as invented in Star Trek is a good start, and as our descendants also become extraterrestrials in the truest sense of the word, the process will become ever more complex.

The Post-Human Space Program

Having failed to achieve unity with diversity on our home planet, what makes us believe that we can accomplish it on the larger canvas of the solar system, galaxy, and universe? The answer may be that we have little or no choice. If we do not create and evolve a broader philosophy of life than we now have, and take it out into the universe with us, the result is likely to be a repetition of the worst aspects of life on Earth as expressed in David Brin's novels about the "uplift wars."

This new "Post-Human Space Program" will need to be both broad enough and deep enough to encompass all sentient life in the universe, including humans, non-humans, post-humans, extraterrestrials and other forms of intelligence, including any advanced non-organic intelligence that might be embedded in non-biological media such as silicon.

This post-human space program also lays the foundation for a universal civilization, one that's currently beyond our limited comprehension, but as Rusty Schweickart points out, leaving the planet begins to shift our identity from national entities to "the whole thing." And as we evolve further into the universe, the nature of that "whole thing" will also evolve, from planet to solar system to galaxy to universe.

So just as we are now creating a "planetary overview system" that includes the human species, the biosphere of Earth, and the global technology infrastructure, we will eventually be part of a universal overview system, and while this book is about humans in space, a longer term view would point to the idea of intelligence in the universe as the most important concept for us to think about. The first dawning of that new intelligence begins with The Overview Effect, the view of the Earth from orbit or the moon, yet as profound as it must have seemed at the time, it was really just the beginning of a long journey. Many of us are realizing now that our journey into space is a collective "hero's journey," one in which we can all play a heroic role.

References

(1) The Overview Effect: Space Exploration and Human Evolution, Houghton Mifflin, Boston, 1987 (1st edition). Second edition published by AIAA in 1998.
(2) Excerpt from a paper by Finney and Jones delivered at a Space Studies Institute conference and quoted in The Overview Effect.

About the Author

Frank White is author or co-author of six books on space exploration and the future, including *The Overview Effect*, *The SETI Factor*, *Decision: Earth*, *Think About Space* (with Isaac Asimov), *March of the Millennia* (with Isaac Asimov), and *The Ice Chronicles* (with Paul Mayewski). He has been a space advocate for some 25 years, and is a frequent speaker at space-related conferences.

Chapter 5

Governance for a Human Future in Space

By Yehezkel Dror

Introduction

New forms of governance are essential for engaging seriously in moving humanity beyond Earth. This chapter discusses this requirement, and proposes first steps in the suggested directions. [1]

Advancing into a Radically New Epoch

Space exploration is focused on learning about space, while the "human future in space" aims at providing homo sapiens with an additional habitat, which may over time become more important than Earth. Therefore, moving from the first to the second involves a mega quantum leap.

All of human history took place on Earth. What is on the agenda is a human-created deep rupture in human history, leading to a radically new epoch in the development of homo sapiens, and perhaps of some parts of the Cosmos.

The repercussions of moving into space are largely inconceivable, and efforts to predict them on the basis of a very different past are extremely doubtful. Therefore, applying NASA's experience beyond some technologies to the problem of building a future for women and men beyond Earth offers a much too conservative perspective. Similarly, cost-benefit analysis in terms of contemporary economic realities misses the tremendously different implications of settling human beings outside Earth on all aspects of thinking and living, individually and collectively. Hence, new social structures are needed, with novel core capacities meeting the requirements of moving humanity into space. This is all the more crucial because space settlement is only one of a number of extreme changes that add up to a radically novel epoch into which humanity as a whole is inexorably moving, with tremendous potential for better and for worse.

Moving into a Radically Novel Epoch

Moving humanity into space is only one of many revolutionary alternative futures made possible thanks to evolving modern science and technology. However, science and technology is a main driver of many other radical futures, whether combined with space settlement or not. Thus, implications of biotechnology may be even more radical, perhaps enabling human movement into deep space by augmenting biological and cognitive adaptability. [See Chapter 11, *Evolutionary Psychology and its Implications for Humans in Space*.]

Not less radical in its implications is the rapidly escalating capacity of fewer and fewer fanatics to kill more and more human beings, up to the specter of a doomsday device wielded by a group that considers the collective suicide of humanity as the way to salvation. This illustrates another aspect of this novel epoch, with complex cross-impacts on space settlement: it may require a "Global Leviathan" to provide minimum security - consuming vast resources that otherwise could go to space activities, but also forcing upon humanity strong global governance able to engage in moving humanity into space.

Therefore, considering space settlement in isolation from other, and in part even more tremendous mutations awaiting humanity, is a mistake. Only by considering it within the context of homo sapiens leaping into a radically novel epoch in many central aspects of human existence can the issues of space policy be correctly understood and coped with.

However, instances of the comprehensive consideration of space settlement within the perspective of humanity moving into a multi-dimensional, radically new epoch are rare, impairing otherwise interesting and comprehensive treatments of human cosmic futures. [2] This chapter cannot make up for this omission, but it should be kept in mind.

In particular, I would like to emphasize that the requirements for radical innovations in governance discussed in terms of the needs of humanity moving beyond Earth are consistent with new features of governance that are needed to successfully guide humanity into a novel epoch as a whole, but I leave this for another occasion. [3] This is especially so in considering required governance redesigns.

Governance is Crucial

In line with contemporary fashion, some might think and hope that moving humanity into space can be achieved in the main by non-governmental actors, such as business, civil society and global communities of interested activists. While these and other actors have important and perhaps critical roles in moving humanity beyond Earth, unavoidable governance will have to serve as the pivotal actor. This is the case because of the large scale of resources required, the need to mobilize mass support, as well as the need for authoritative rules and some enforcement.

There is also a crucial normative reason for governance direction of space settlement, namely the democratic imperative: critical choices shaping the human future should be democratically legitimized. Furthermore, in terms of realpolitik, it is inconceivable that governments will let other actors dominate space settlement with all its social, economic, political, security, and ideological implications.

This does not mean that most of the job will be done by governments, or dominated by them. Civil society, grass root movements, business, scientists and technologists, public opinion leaders, and a variety of entrepreneurs will do most of the work. But without overall governance direction, support, and supervision there will be no human settlement in space.

Let me add that the hope to do with less bureaucracy is a chimera. Bureaucracies, correctly understood as big formal organizations, are essential for large scale human action, whether by business or governance. The correct question, therefore, is not how to reduce the roles of bureaucracy, but how to change core features of bureaucracy to make them fit the requirements of a new epoch that includes human movement into space.

Required Governance Characteristics

For governance to be willing and able to engage in human settlement of space in the context of a new epoch in the making requires at least seven main characteristics that are very different from main features of contemporary governance.

1. Global

We are speaking on an endeavor of humanity as a whole. No single country, including the USA, has the material, scientific and technological, human, or political resources sufficient for moving humanity into space. Furthermore, unless it is undertaken by a strong global governance, the human movement into space is sure to involve conflict, including militarization, with dismal consequences.

This does not mean that consensual global governance is necessary, however desirable it is in principle. De facto global governance dominated by the USA [4] in cooperation with other willing major states may do the job, while being more feasible in the foreseeable future. But, however structured, without strong global governance human settlement of space has a very low probability, because of lack of adequate resources and the likelihood of conflictive activities. There is scope for division of labor and innovation-stimulating competition between countries in space activities, but these need regulation by a central authority to assure synergism and prevent misuses.

2. Inspirational

The case can be made that human settlement has long term economic advantages and may even be necessary for long term human survival. However, this case is based on extrapolating present trends into a long term future, in line with classical "limits of growth" speculations. Such an extrapolation of current dynamics into a different epoch is very doubtful. [5] Alternative scenarios that show humanity thriving are not less valid, such as stabilized populations, restrained consumerism, and disciplined energy use, together with new technologies that cope with pollution and green house effects, and value changes that provide new meanings of "the good life."

Therefore, human settlement in space needs additional justifications, which are in part at least "inspirational." These include, for instance, seeking knowledge, searching for "the new," the human desire to be challenged and, very importantly, a spirit of "adventure." [6] It is up to cultural leaders to provide most of the inspiration for human settlement of space, but mundane governance preoccupied with current pressures will inhibit even consideration of such a mega-endeavor, and very likely make it impossible. Therefore, governance for space settlement must be inspirational, with top politicians, for instance, being very active as educators of the public instead of creeping behind opinion polls and occupying themselves with media spinning.

3. Long term perspectives and persistence

Despite the optimism of many space settlement enthusiasts, my guesstimate is that serious steps towards making parts of space a human habitat will take at a number of generations, say, until between the middle and the end of the twenty-second century.

Taking into account the many other pressing issues that are sure to accompany the move into a radically new epoch, which will include many serious crises, disagreements, scarcities and also bloodshed, it will be very difficult for governance to adopt the long term perspective and engage in the persistent policies essential for humanity moving into space. No contemporary governance system is capable of doing so. Therefore, radically different governance structures and rulers are essential.

In particular, while being democratic as far as possible and as fits different human cultures, governance will have to overcome a main problem of democracy, namely the tendency to discount future generations because they do not vote now, never mind the large impact of present decisions on them. Longer terms in office illustrate possibilities to address this problem.

4. Large scale mega-project resources and management

Human movement into space cannot be achieved by an upgraded and upscaled NASA-type organization. It is a mega-project that will not only take generations, but which requires as a minimum critical mass very large resources, including both economic and political, as well as highly qualified human ones. Also essential are unprecedented mega-project management systems, though here much can be learned from NASA and other large scale technology projects.

These, again, are far beyond the present capacities of governance, requiring new ways of financing, such as a global "space settlement" tax, as well as uncertainty-coping and constantly-learning project management modalities.

5. Will and enforcement tools

Space settlement is unlikely to be a harmonious and smooth activity. Fed by nationalistic traditions, fundamentalist ideologies, economic egotism, and the "tyranny of the status quo" [7] as a whole, there will be many resistances, misuses and abuses, including dangerous ones. Therefore, a strong will together with effective enforcement tools are essential requirements, taking the form of democratic but "strong" global governance.

6. Outstanding cognitive capacities

Governance should not and cannot engage in most of the activities of moving humanity into space. But governance most direct and overview these activities. This requires a deep understanding of the many and complex issues involved in this mega-project, combined with an open mind and rapid learning. Such essential cognitive capacities are absent in all known governance systems, and scarce in human structures as a whole.

7. Raison d'Humanité values

All the governance redesign directions mentioned above can add up to very powerful evil governance doing grievous harm to humanity, including misusing space settlement for the worst. [8] Therefore, new values focused on the long-term good of humanity, within pluralistic normative systems, are needed. This goes far beyond a code of ethics for space settlement, however important, [9] involving human values as a whole.

Needed is what I call Raison d'Humanité [10] values displacing, in part at least, Raison d'Etat, and also going beyond the propensity of countries to regard what is good for them as good for humanity as a whole. Developing Raison d'Humanité is a sorely neglected task for value creators and moral philosophers. Present efforts in this direction are often very narrow in scope, doubtful in terms of serious moral reasoning, and not fitting the nature of human settlement of space, [11] which necessarily will be "tough" in many respects.

Developing Raison d'Humanité and increasingly being guided by it is an urgent necessity for the new epoch into which humanity is moving as a whole, but particularly so for space settlement with all the difficult moral dilemmas and tragic value judgments it involves, such as taking high risks with human lives for essential but dangerous experimental and explorative activities. Also, unless governance is committed to pluralistic versions of Raison d'Humanité as its main value compass, the danger of misuses of human settlement of space are very serious, such as building of "evil" empires. Indeed, a strong case can be made that large scale space settlement should wait until human values achieve a quantum leap. But, though this is not assured, engaging in human space settlement may serve as a catalyst for bringing about a much-needed mutation in humanity, including its governance values.

Redesign Steps

Building up governance so as to meet the required characteristics will be a slow and difficult process. But some first steps can be taken on the basis of existing institutions and through relatively feasible innovations, within a selective radical reform strategy focusing on critical components to be redesigned significantly. Five initial directions are recommended:

1. Structures

The first steps may well include building up a network of NASA-type organizations around the world, putting space settlement on the agenda of such organizations and of visionary politicians, and starting to develop what can metaphorically be called "a central brain of space-settling governance" - perhaps best in the form of a Think Tank working in depth on main issues of moving humanity into space. Various forms of working groups, such as ATWG, can be of much help, but more is required. Essential is a critical mass of full time, interdisciplinary, high-quality professionals, together with staff experienced in politics, policy making, handling of public opinion and large scale management working as teams on major space settlement issues within the context of humanity moving into a radically new epoch. Let me add that historians and social scientists are an essential component of the needed interdisciplinary staff, space settlement being much too complex, multi-dimensional and significant to be left only to physical scientists. In short, what is needed is a "RAND-type" Think Tank, though smaller in scope and focused on the space settlement mega-project.

2. Rulers

Another chapter in this book deals with required global leadership. [12] What I would like to add is an emphasis on "rulers," senior politicians in primary decision making positions and those who are candidates for such positions. Whether we like it or not, a very small number of top decision makers will make the crucial decisions concerning whether and how to move into space. True, such rulers are conditioned by culture and situations, and constrained by political and economic realities. Still, they have a lot of leeway - and without their support governance will not be adjusted to the needs of space settlement, and no serious steps to move beyond Earth in desirable ways will be taken.

Improvement of rulers is required for even more urgent reasons related to the need for guidance in the movement of humanity into a new epoch, but it is critical for space settlement. However, improvement of rulers is a very difficult endeavor, but some measures can be taken while waiting for overall adjustment of rulership to a new epoch. Recommended initial steps that are quite feasible today include: organizing workshops on space settlement issues for interested senior politicians and young promising ones; injecting the discussion of space settlement into existing leadership programs; and setting up a "Global Leadership Academy" [13] where movement into space will be one of the main subjects.

3. Professionals

With due respect to the many outstanding and original thinkers, scientists, engineers, etc. thinking and working on various aspects of space exploration and space settlement, a more encompassing new profession is needed that would consider in depth, plan, and later implement the movement of humanity beyond Earth. This is a multi-disciplinary endeavor for which knowledge in both social and physical sciences is needed within broad historic thinking on one hand, and future-oriented epoch-guiding policy planning on the other.

Much of the work can be done by multidisciplinary and interdisciplinary teams within the "Central Brain" (think tank) suggested above. But, for implementing space settlement, a new "space settlement profession" is required. [14]

Designing the curricula for such a profession, or for sub-specialization in existing professions, to be offered at universities and special training activities is therefore recommended as an urgent step.

From Dreams to Action

Dreaming about the human settlement of space that leads to realistic visions increasingly shared by humanity as a whole, and by influential elites in particular, is essential for making human movement beyond Earth a reality. Systematic and realistic thinking on how to accomplish such realistic visions is a next essential step, to be followed by modular implementation. On all these levels much attention needs to be given to governance, because without restructuring governance, the movement of humanity into space will remain a dream or, even worse, may take the form of nightmares becoming a dismal reality.

References

(1) This chapter in part applies to human settlement in space the ideas developed in my books *The Capacity to Govern: A Report to the Club of Rome* (London: Frank Cass), 2002).

(2) E.g., Nikos Prantzos, *Our Cosmic Future: Humanity's Fate in the Universe* (Cambridge, UK: Cambridge University Press), 2000 (first published in French in 1998).

(3) Some central aspects will be discussed in Yehezkel Dror, *Superior Ruler: Mirrors for Epoch-Guiding Political Leaders* (in preparation).

(4) As proposed, for instance, by Michael Mandelbaum, *The Case for Goliath: How America Acts as the World Government in the 21st Century* (New York: Public Affairs), 2005.

(5) For a more sophisticated approach see Robert J. Lempert, Steven W. Popper, Steven C. Banes, *Shaping the Next One Hundred Years: New Methods for Quantitative Long-Term Policy Analysis*. (Santa Monica: The RAND Corporation), 2003.

(6) See Michael Nerlich, *Ideology of Adventure in Modern Consciousness, 1100-1750* (Minnesota, MI: University of Minnesota Press), 1987; and his more recent book available only in German, *Abenteuer, oder, Das verlorene Selbstverständnis der Moderne: Von der Unaufhebbarkeit experimentalen Handelns* (München: Gerling Akademie), 1997.

(7) See Milton and Rose Friedman, *The Tyranny of the Status Quo* (New York: Harcourt Brace), 1984.

(8) Still, it is a moot question what is worse, no human settlement in space or such settlement misused in its initial stages. An interesting historic analogue is the sending of criminals into newly settled territories, such as Australia, which in the longer run proved very beneficial.

(9) See Chapter 17, *A Code of Ethics for Humans in Space*.

(10) See *The Capacity to Govern*, ibid., chapter 9.

(11) A case in point is "animal rights" in their more extreme forms, for instance Peter Singer, *Animal Liberation* (New York: HarperPerennial), 2001.

(12) See Chapter 8, Stage Three Leadership: From Good Ideas to Unified Action

(13) See Yehezkel Dror, "School for Rulers", in: Keyon B. De Greene, ed., *A Systems-Based Approach to Policymaking* (New York: Kluwen), 1993, pp. 139-174.

(14) An interesting comparison is with a profession of applied anthropologists developed in The Netherlands for the settlement of land reclaimed from the sea.

About the Author

Yehezkel Dror is Professor of Political Science and Public Administration, Emeritus, at the Hebrew University of Jerusalem. Since February 2002 he has served as Founding President of the Jewish People Policy Planning Institute. He has filled senior positions in Israeli governments, including two years as full time Senior Policy Planning Advisor in the Office of the Minister of Defense, consultancy for a number of Prime Ministers, advisor of the Israeli Cabinet Office, chairman and member of public commissions dealing with various policy issues.

Professor Dror has served as policy planner, strategic consultant and senior professor in many countries, including two years with the RAND Corporation in Santa Monica, California and two years working on European Union policy issues at the European Institute of Public Administration in Maastricht. He was a Fellow at major Institutes of Advanced Study, including in Berlin, Palo Alto, New York and Washington D.C. Former member of the Club of Rome and the International Institute of Strategic Studies.

He has published many articles and fifteen books in seven languages, including: *Public Policymaking Reexamined; Crazy States: A Counterconventional Security Problem; A Grand-Strategy for Israel* (in Hebrew); and, most recently, *The Capacity to Govern: A Report to the Club of Rome;* and *Epistle to an Israeli Jewish-Zionist Leader* (in Hebrew).

He is a Fellow of the World Academy of Art and Science and of the European Academy of Sciences and Arts, and has received a number of awards from the Policy Studies Association. In 1999 he received the Israeli Anniversary Arthur Ruppin Prize from Haifa Municipality for his original contributions to policy making; in 2002 the annual Landau Prize for outstanding contributions to the social sciences; and in 2005 the Israel Prize in administrative sciences for his contributions to the theory and practice of strategic planning and policy making.

Chapter 6

Space Law in the 21st Century... and Beyond

Dr. George S. Robinson

"Space law is an interdisciplinary journey; not an isolated profession."

Commercial Space Law

Although many of the current genre of space lawyers practicing both in domestic and international arenas are establishing the early legal foundation for the private sector's entrance into identification and commercialization of space resources, it is the coming generation that will be addressing an explosion of legal issues regarding those that are potentially exploitable. Lawyers and jurisprudents will face the tangle of activities and related policies flowing from the economic potential of asteroids, comets, the lunar and Martian resources, and other celestial body resources barely on the horizon of our awareness. The most difficult issues to be faced by lawyers trying to make sense of, and implement decisions by, policy makers and legislators will be the "soft" issues...human relations and the characteristics of related decision-making by politicians, statesmen, high-level public, military, and industrial administrators, and...even law-makers and lawyers themselves.

The not altogether future space and space-related resources available to the public and private sectors to exploit commercially will include the development of activities, services, and products designed to help sustain human life in the synthetic and alien life support environment of space. Regardless of the multitude of evolving justifications, whether serving civilian and/or military requirements, extraordinary efforts will be pursued to assure that humankind has a reasonable chance to survive off-Earth. The relevant commercially-oriented efforts include biomedicine, directed human genetic intervention, the very difficult pursuit of gene sequencing that could take from fifty to a hundred years to achieve, and genetic manipulation for specific individual and generational survival requirements, on Earth and in long duration and permanent space habitats.

Another developing area of important and lucrative research relates to human biotechnological integration - both gross technology and also nanotechnology - for survival and evolution in space. Even more in the not-so-soft aspects of space-related research is the need to identify the psychopathological workings and expressions of human and humankind, astronauts, and the future spacekind, or *Homo alterios spatialis*. Put a bit differently, defining the empirical characteristics of "human nature," or the "essence" of being human, with some certainty in order to understand just how our evolving technology is changing representatives of our species will provide commercial opportunities for our scientists, engineers, ethicists, and lawyers. Developing laws and regulations to control and implement related policy decisions regarding these undertakings will be mind-boggling, not only for the practitioner of the implementing laws, but also for those identifying and formulating the applicable and very likely necessarily unique jurisprudence, or underlying legal theory.

In addition to amendment of the existing five basic United Nations space treaties, unique concepts of public and private international law will continue being developed. Space lawyers will focus on the proliferating private international agreements relating to the global funding for exploration, development, use, and settlement of near and deep space. Practicing space lawyers will require increasing expertise in the related hard sciences, i.e., there must be a significant ability of lawyers to asses the laws reflecting the physical characteristics and realties of space-related resources and the technologies available for recovering them, and the problems associated with developing them commercially. One of the most pressing domestic and/or international legal issues to be resolved is whether private ownership of space resources will be permitted, and whether sovereign appropriation of those resources will be necessary to establish private ownership. Without these issues being resolved legally, or settled politically and diplomatically, significant private capital investment in space resource exploration will remain with the small-change investor.

The implementing characteristics of private space "business" will result in significant part from revisiting the histories of demographic expansionism and applying certain of the motivating or underlying principles finding their roots in economic, military, and cultural imperialism. For example, some of today's principles and practices of space exploration, economics, and politics can be compared with the imperialism and exploration carried out by England in the late sixteenth century. Queen Elizabeth, by the end of the Sixteenth Century, started issuing what was called "monopoly charters" to explore, exploit, and trade around the globe as a means of encouraging, in part, both private and public economic interests. In various respects, these charter companies might serve in part as models for how near and deep space, and the known and yet to be discovered useable resources, can, perhaps should, or indeed should not, be secured, developed, and exploited. The practice of space law in the 21st Century will involve innovative legal variations of some of the principles, good and bad, established for and by the charter companies and used to open up remote areas of Earth to commercial development.

Finally, some of the legal regimes ripe for application and development to evolving private/public commercial, military, and public interest space activities during at least the next fifty years, include antitrust laws in the United States, as well as foreign laws prohibiting certain activities that are clearly designed to restrain trade; laws relating to economic globalization, including the increasing reliance on global participation to fund various space undertakings, such as technology transfer laws and export/import laws, a variety of fiscal laws, securities laws, contract law, corporate and tax laws; international trade agreements and attendant national implementing legislation and regulation; private and public placement laws; insurance laws and other evolving risk management principles and policies; health care laws and quarantine protocols; transportation laws (land, water, air, and space tourism, etc.); domestic and foreign employment laws; human and humankind rights, policies, and laws; domestic, international, and interplanetary communications laws; and all aspects of intellectual property rights, domestic as well as international and global.

Some of the other frequently ill-considered or totally ignored areas of space law that will be addressed in the Twenty-First Century include:

* Changes in legal education curriculum that will allow examination and formulation of employment laws necessary to meet radically changing characteristics of human capital with specific requirements needed for conducting space activities, both technically and in the context of fiscal globalism and transnational corporations. Education policies and implementing laws must recognize the consequent requirements for drastic changes in education objectives and methodologies.
* Communications law will have to keep pace with the rapidly changing technical improvements that emphasize the importance of "community of knowledge" unique to exploring, commercially exploiting, and settling near and deep space...in furtherance of domestic and international policies based on reasonably informed transglobal public consent.
* Space laws will stress and protect "inclusiveness" where appropriate in decision- making, domestic and global, rather than an either-or decision-making methodology.
* Space jurisprudence will incorporate an international recognition that creating a viable space civilization is critical to the survival of humankind and/or its essence(s).

Laws Relating To Forward, Back, And Cross-Contamination of Planetary Bodies

... States Parties to the Treaty shall pursue studies of outer space, including the moon and other celestial bodies, and conduct exploration of them so as to avoid their harmful contamination and also adverse changes in the environment of the Earth resulting from the introduction of extraterrestrial matter and, where necessary, shall adopt appropriate measures for this purpose.

So provides, in part, Article IX of the 1967 United Nations Treaty on Principles Governing the Activities of States in the Exploration and Use of Outer Space, Including the Moon and other Celestial Bodies, known by its short title as the Outer Space Treaty of 1967. This Article reflects the serious concerns of the various United Nations member states involved in negotiating the Outer Space Treaty regarding the unknown dangers of sending launch vehicles and scientific probes/payloads to explore near and deep interstitial space as well as celestial bodies. Given the comparative paucity of the scientific knowns during the Treaty negotiations with respect to the potential for existence of extraterrestrial life forms, carbon-based or not, the primary initial concerns of the negotiators was with potentially harmful contamination of Earth's biotic ecosystem by alien matter returned in or on recoverable manned and unmanned missions.

Despite being cloaked in such precautionary language, the underlying question not really addressed fully...and certainly not publicly...might well be "Why do we need to look for extraterrestrial life?" Does it help in the search to find out about the past and potential future options of *Homo sapiens sapiens*? Is the search for evidence of extant or paleo-extraterrestrial life and life precursors related to the migratory survival requirements for our species...or evolving specieskind? How do philosophical and theological considerations fit into the decision to search for and locate extraterrestrial life or evidence of it? How does President George W. Bush define "curiosity" of humankind when he says it is the driving force behind his February 2004 policy statement that the United States will establish a manned presence once again on our moon, and ultimately on Mars? How carefully, thoroughly, and effectively have these questions been addressed by the global public as critical components of governmental policies relating to the search for potential extraterrestrial life? Not very thoroughly.

As observed in an October 11, 2005 communication to the author, Dr. John Rummel, Director of NASA's Office of Planetary Protection, NASA controlled for biological contamination prior to the Space Treaty, so it is nice but not necessary that the Space Treaty exists for this purpose. Dr. Rummel also observed that "biological contamination is top priority because its existence could destroy the science that could be done by the mission(s) that we send. Nevertheless, Dr. Rummel doesn't think that Article IX of the Outer Space Treaty "was even intended to be limited to biotic contamination. It could equally be applied to other contaminants...if necessary."

In addition to protecting the integrity of our experiments to detect past and present extraterrestrial life forms or life precursors, it is imperative to determine whether we as a species have a right to interfere with the evolution of any non-Earth indigent species, regardless of level of evolution; indeed, whether we have any biological, moral, or ethical "right" to interfere at all even with non-biotic components of celestial bodies. Nevertheless, whether biotic or abiotic, astrobiology offers a very unique and marvelous opportunity to involve, much more directly and effectively than has been encouraged in the past, the broad general global public in policy and programmatic decisions relating to planetary exploration and, perhaps, settlement. This type of involvement is essential to developing a broad and informed dialogue about the future of *Homo sapiens sapiens* and, indeed, the more ephemeral aspects of our sentient nature, i.e., the future of "humanity" in space.

In the process of accomplishing this undertaking of involving a broad sector of the general international citizenry in astrobiological research, mutual respect and a meaningful dialogue must be encouraged...insisted upon...among the broad variety of interests involved and expressed. In other words, it does not mean a public information lecture by various national and multinational space agencies conducting current, as well as planning, future astrobiological research programs and projects. Because of the complex multidisciplinary nature of astrobiology and its profound implications, NASA and appropriate agencies of foreign spacefaring governments and space related international organizations must be attuned to the global public's informed input...and consequent consent or rejection of proposed exploratory, migratory, and/or settlement activities. Controversy should be rapidly and fully engaged with the public. Unfortunately, scientists are not necessarily the best or most effective communicators of their research results and research plans. Effective public dialogue leading toward an informed public consent to the research objectives and methodologies ultimately selected for implementation requires establishment of an equally as effective legal infrastructure to assure that kind of transglobal consent. At present, such an effective infrastructure is not available; and certainly not internationally. Nevertheless, our rapidly evolving communications capabilities through the internet and incredible near-term advances should go a long way to encouraging and assuring a high quality of informed public consent. The search for extraterrestrial life and the reasons for doing it constitute a biological and spiritual, a secular and a humanist, survival imperative...and we had better get as many of the lay public as possible on board as soon as possible to determine whether, if not how, to go about doing it.

Finally, although the forward contamination control effort to date has focused primarily, but not exclusively, on primitive life forms and precursors to life, we cannot exclude the prospect of contacting what we assume might be extraterrestrial intelligent life forms...or at least as we currently define "intelligent" in an anthropocentric context. Here, we are dealing with issues that relate to biocultural interaction and/or contamination with those life forms on other celestial bodies, i.e., forward contamination (although it is possible, of course, that such cultural interactions might take place on Earth, in which event we may be facing issues of back contamination creating adverse changes on Earth pursuant to the concerns addressed in Article IX of the 1967 Outer Space Treaty). From one perspective, however, since we also presently are developing biotechnologically transhumanistic characteristics in representatives of our species, such representatives surviving in long-duration or permanent space habitats could well be our own grandsons and granddaughters during the next 100 years. They, in fact, may well be our first tangible contact with a true extraterrestrial intelligence and culture. Toward this end, we might well...and should...find ourselves revisiting the set of eleven basic metalaws (i.e., interactive infrastructure rules between and among humans, humankind, and those long duration and/or permanent inhabitants of off-Earth space referred to as spacekind) suggested in 1970 by Austrian jurist and legal writer, Dr. Ernst Fasan. Using a few basic, and occasionally debatable assumptions, Dr. Fasan asserted seven fundamental metalaws, which he assumed possessed the characteristics of universal applicability and validity. In descending order of importance, they are:

No partner of Metalaw may demand an impossibility.

* No rule of Metalaw must be complied with when compliance would result in the practical suicide of an obligated race [perhaps an embarrassingly anachronistic and inappropriate characterization of alien life form].
* All intelligent races of the universe have in principal equal rights and values.
* Every partner of Metalaw has the right of self-determination.
* Any act which causes harm to another race must be avoided [i.e., pioneering space lawyer Andrew.G. Haley's posit of an "Interstellar Golden Rule"].
* Every race is entitled to its own living space.
* Every race has the right to defend itself against any harmful act performed by another race.
* The principle of preserving one race has priority over the development of another race.
* In case of damage, the damager must restore the integrity of the damaged party.
* Metalegal agreements and treaties must be kept.
* To help the other race by one's own activities is not a legal but a basic ethical principle.

As noted by attorney Robert A. Frietas, Jr., "Fasan's metalaws are the beginning of a new era of metalegal development. Such principles of conduct may soon become a matter of survival; our thinking must be guided by metalegal precepts in astropolitical contexts...When intelligent extraterrestrial life is discovered, mankind must be prepared, for in all of human history there will be but one first contact," and that is likely to be with those springing from our own loins and minds, our sons and daughters, our transhuman grandsons and granddaughters. The demands of creative jurisprudence are immense for dealing with biological and cultural forward contamination issues over the next twenty-five, fifty...a hundred years.

Intelligent Autonomous Biorobotics, Telepresence, Virtual Presence, Avatars, And Transhumans In Space Environments

With the rapid and far-reaching advances being made in human biotechnological integration capabilities, an entirely unique jurisprudence must be formulated to assure individual accountability of these entities...separate from the basic research scientists, engineers, programmers, and the like who create, program, and manipulate them at the outset...under some uniquely responsive regime of law applicable only to highly intelligent and continuously evolving autonomous biorobotics, telepresences, teleported entities, avatars, transhumans, and the like. This unique accountability jurisprudence also must recognize and relate to the establishment of a metalaw regime transitioning the broad sociopolitical, economic, and military relationships between and among humans on Earth (i.e., Earthkind) and those having been altered to possess significantly different biocultural survival characteristics while inhabiting near and deep space long duration and permanently, and who or which may be referred to as Spacekind or *Homo alterios spatialis*.

Humans unenhanced by appropriate biotechnological integration are extraordinarily fragile in terms of surviving the alien environmental extremes off Earth. For this reason, alone, relying on robotics to explore and settle space may be an end in itself...at least at the outset of establishing a permanent human or humankind presence on the Moon and Mars. Nevertheless, a succeeding intermediary step relying on telepresence technology, or even robotics enhanced by telepresence capabilities, may well precede truly intelligent autonomous robots and human biotechnologically enhanced humankind. "Telepresence" might be defined as the projection to a distant environment of an operator's or user's sensory, cognitive, and motor capabilities; or, the distant environment can be recreated from distant actuators (say, on the lunar or martian surfaces) virtually at the location of the user/operator. In both circumstances, virtual interactive environments, i.e., virtual realities, are created. At some point, these types of entities can and will attain a level of independence that incorporates the embodiment of independent "personhood" requiring a new or even unique jurisprudential way of holding these uniquely independent entities accountable under legal regimes separate from those finding their genesis in Natural Law from which the so-called immutable rights of *Homo sapiens sapiens* derive.

These immutable human rights were said to be discovered through the use of reason, and, most important-ly, represented the progress of knowledge. This understanding allows for a rational transition from the unknown (spiritual?) to the evolving "known" through advancing scientific data and the beliefs of the secularists. It allows, ultimately, for the creation of "new" principles embodied in or flowing from natural order and finding temporary jurisprudential expression in Natural Law theory, that is, it allows for new principles applicable to humankind and not those applicable only to unenhanced or partially enhanced humans. In certain circumstances, the latter have been referred to as "transhumans."

For the above reasons among others, one of the principal undertakings for the secularists, humanists, and secular- humanists during the early years of the twenty-first century will be the determination of whether humankind have their own inalienable rights, both those shared with and those separate and distinct from *Homo sapiens sapiens*. And in the process, will "faith" continue to diminish in the light of evolving reason...or will that evolving reason lead to ever greater realms of the secular unknown and, hence, create greater realms for expand-ing faith-based beliefs? The initial issue to be addressed, it seems, is whether immutable Natural Law Rights are really that immutable when referring to the characteristics of human or humankind nature; and whether the corre-sponding jurisprudences and implementing positive laws are all that immutable as well. In certain respects, cur-rent formal legal education does not seem to spend too much time on the jurisprudential efficacy of a non-evolv-ing Natural Law Theory.

Jurisprudence has been described as more a formal than a material science, and has no direct concern with issues and questions of moral or political policy, which fall under the province of ethics and legislation. In this con-text, philosophy (as in the philosophy of law) has been defined as a discipline comprising as its core logic, aesthet-ics, ethics, metaphysics, and epistemology...the pursuit of wisdom and the search for a general understanding of values and reality by chiefly speculative rather than observational means. Clearly, the operative terms used to define "philosophy" are inconsistent and confusing at best...as are the variations in terminology and conceptual-izations used to define "jurisprudence." Perhaps an equally as confusing, but substantively more accurate defini-tion of jurisprudence, or "The Law," would be "the psychoneurophysiological interpretation of external and inter-nal bio-ecological influences and dictates affecting and shaping the motivational characterizations of individual and collective representatives of humankind life forms."

One of the basic questions to be addressed in formulating both a transitional regime of "metalaw" and in cre-ating a unique jurisprudential infrastructure for transhumans in a space environment would be whether *Homo sapi-ens sapiens*, in the biotechnological integration process of becoming transhumanistic, would no longer respond effectively to a jurisprudence embracing basic, traditional principles of Natural Law; particularly since they are not apt to be applicable and survival effective to a significantly altered human, say, *Homo sapiens alterios*, or *Homo alterios spatialis* functioning in a completely synthetic life support environment of a space habitat. Must jurispru-dence for unaltered humans first define human "spirit," or "soul," or "essence" in a secular sense before attempt-ing to determine at what point a non-human entity can be ascribed the characteristics of individual personhood...as having some independent accountability under some regime of law? Is the nature, or soul, or essence of being human a critical component of personhood and what we want to convey to space as our human envoys? Can a tran-shuman convey that aspect of *Homo sapiens sapiens* into near and deep space so as to serve as a true "envoy" of humans? What is the principal difference between these envoys and the unaltered human species? Is it the capac-ity for abstract perception? Of being "sentient?" If the individual human is genetically societal, is the "whole" of the individual greater than the sum of its "parts?" Just what is it of ourselves that we want to put into space...that we want to survive?

Since the Renaissance, intellectual pursuits relating to the defining and assessing of human nature have been divided into two major camps...often locked in ongoing political and/or legal combat: For example, the issues sub-tending the concept of Intelligent Design, which certain humanists are attempting to insert into public school cur-ricula, and wending its way through the courts of Pennsylvania and other jurisdictions as this chapter is being writ-ten. The evolutionist/creationist controversy, particularly in the United States, of the last two decades or so is but a relatively recent manifestation of arguments in theological antiquity. One camp consists of the humanists who frequently are accused of being little or ill-informed regarding the knowledge of scientific and technological progress. On the other hand, conservative humanists frequently accuse the secularists of being uncultivated about, and even disinterested in, the classical studies embracing culture, philosophy, and theology. They may use some-what similar analytical methodologies in their respective disciplines, but they rarely communicate them socially or meaningfully and in depth...outside a courtroom.

In the late 1950s, as the space age was just really beginning to show distinctive activities and a civil complexion of its own not altogether reflecting its military genesis, C.P. Snow observed in The Two Cultures and the Scientific Revolution (Oxford Univ. Press, 1959) that societies were being separated into two distinct cultures. Instead of characterizing the humanists (often a non-exclusive characterization ascribed to Christians, and particularly Christian Fundamentalists) and the secular humanists as representing two different "cultures," British psychologist Liam Hudson referred to the two camps, respectively, as "divergent" and "convergent" thinkers. To him, the approach of "divergents" is to work on the whole being greater than the sum of its parts...a kind of open-ended thinking...while the preference of the "convergents" is to unfold empirical data as the basis of problem solving, i.e., they seek closure to the question, issue, or problem and use it as the foundation for the next set of inquiries. And these characterizations are misleadingly simplistic explanations for extraordinarily complex and integrated disciplines.

It might be said, then, that the "essence" of being an individual human is the biological ability to create a separate "out of body" self-image; a kind of "cyber-essence" or "cyber-spirit." In other words, a "cyber-soul." The next question, then, might be "how is the collective essence or spirit of the species created, and who or how does one see or perceive it?" And then, the most difficult question of all..."for what purpose?" Are survival and evolution of the spirit tied to survival and evolution of human biology, i.e., *Homo sapiens sapiens* as a biological or even biotechnological entity (individually and/or collectively)? In this context, is the "soul" a reflection of the formula that the whole is greater than the sum of its parts? How, then, do we define "sum?" Does the soul evolve and change in character in proportion to or context of bio- and biotechnological evolution? Can the soul, once defined, be projected or otherwise transmitted by directed energy fields, much like telecommunications images? Can souls, and/or the biological components and carbon-based systems characterizing the individual and which give form to the soul, be held separately accountable under specific regimes of law? Can a soul reflect or characterize a societal autonomy of cooperating/interacting individual biotechnological components?

These are questions and issues that traditionally have been addressed solely from anthropocentric perspectives...even when observing and assessing the highly evolved societal cetaceans (e.g., whales and porpoises) and certain land mammals previously considered to be of lower taxonomic orders than humans (e.g., elephants). Again, traditionally, humans have been unable to approach these questions and issues from any perspective but human, i.e., "we posit our own intelligence, our behavior, emotions, and language skills, as the norm." But what if we ultimately accept that lower orders of animals do have "standing to sue," do have "Natural Rights," even if it requires an intermediary interpreter to assure expression and safeguarding of the "rights" sought through such a lawsuit. How much more difficult would it be to give the same rights and needs, temporal and spiritual, to humankind...including human biorobotics? *Homo alterios spatialis*?

In many respects, Earth's current civilizations do not seem quite ready to recognize the technologically and human bioevolutionary imperative to avoid dragging frequently irrelevant Natural Law principles and corresponding expressions of legal constructs into space simply because we are familiar and comfortable with them. We must be pressing with much greater urgency during the Twenty-First Century to catch up with our rapidly evolving space technology in terms of philosophical, theological, and biocultural constructs necessary for establishing a civilization that embodies a framework of values reflecting realities that characterize a new and perhaps unique civilization, and not just another colony. Toward this end, a space age "constitutional convention" must be undertaken to formulate a type of metalaw-based Migratory Manifesto for humans and their humankind or biotechnologically enhanced descendants, their own sons and daughters, grandsons and granddaughters, that will allow them to evolve according to their own unique survival requirements in space.

Summary

In the Twenty-First Century, legal regimes relating to private and governmental activities in space of a commercial nature will reflect both old principles of economic imperialism experienced on Earth, and economic and fiscal principles of a steadily evolving and unifying transglobalism...unless the recidivism of ethnic, religious, economic, political/military, and cultural parochialisms are the survivors. If the latter occurs, an entirely different scenario will evolve relating to space exploration, migration, commercial exploitation of space resources, and settlement in the Twenty-First Century. Whatever the directions, and whatever the implementing policies of the decision-makers, the applicable regimes of law will evolve hand-in-glove to reflect those policies...or new regimes of law, often unique, will share the stage of space jurisprudence. Space law in this century will embrace and respond to new and disparate concepts relating to educational requirements for the changing natures of the private and public sectors workforce, military-private commercial ventures, transglobal trade arrangements, and recognizing the independent trading nature of long-duration and permanent space communities. No area of domestic and international regimes of commercial law will be left untouched in their application to humankind activities in space.

With the expanding global and interplanetary communications capabilities, decision-making by government bodies and officials will become increasingly subject to direct input and effective influence over those decisions by the international or global citizenry, and how those decisions are, or are not, implemented. Interplanetary research and non-Earth celestial body intervention by humans and humankind will experience tighter prior influence, if not control, by the global population, and the resulting law will incorporate an evolving space ethos heretofore pretty much left to the discretion of governmental requirements and officialdom. In short, environmental awareness will expand its ecotone of interest beyond Earth's atmosphere, and will be forced to accommodate an extraordinarily broad array of public interests and demands. Both domestic and international laws will have to evolve that protect not only those concerns and interests, but more importantly, encourage and protect those responsibilities of the general public to help lay a foundation of relevant ethics and philosophic constructs for exploring and intervening with the natural state of other celestial bodies.

Finally, the Twenty-First Century may well see an extraordinarily complex jurisprudential assessment of humankind, or transhumanists, regarding issues of intelligent biorobotic individuality and independent personhood, and separate accountability under regimes of law in a fashion only slightly being addressed curently by lawyers and legal philosophers in the context of cyberspace. Intelligent biorobotics, teleportation, and telepresence very likely will become defined in terms of independent accountability under unique legal regimes, if not jurisprudence, in the ambience of a multitude of space activities. In the process, *Homo sapiens sapiens* will be forced to attend in a pragmatic and secular fashion the basic question filling endless theological, philosophical, and scientific tomes, i.e., what is the nature, the soul, the very essence of being *Homo sapiens sapiens*? Toward that end, one of the first all-encompassing legal efforts will be attempting to articulate the underlying values of a space constitutional convention, a Migratory Manifesto incorporating the metalaw ethic that will allow spacekind, whatever its biological or biotechnological genesis, to evolve according to its own needs in a unique space civilization without dragging along unnecessary and unresponsive laws formulated in the immediate physical environment of Earth's surface.

About the Author

Dr. George S. Robinson is a graduate of Bowdoin College in Maine, where he majored in biology and chemistry. He holds an LL.B. from the University of Virginia School of Law, and an LL.M. and the first Doctor of Civil Laws degrees from McGill University's Graduate Law Faculty, Institute of Air and Space Law, Montreal, Canada. For twenty-five years, Dr. Robinson served as legal counsel at the Smithsonian Institution in Washington, D.C. Before that, he served as an international relations specialist at the National Aeronautics and Space Administration. He also served as legal counsel at the Federal Aviation Administration.

Dr. Robinson has authored over 100 articles and books on a broad range of subjects. Upon leaving the Smithsonian Institution in 1995, Dr. Robinson established with his sons the firm of Robinson & Associates Law Offices, P.C. While emphasizing commercial space law, the practice also has addressed issues of law relating to corporations and non-profit organizations. Dr. Robinson has taught and lectured in law and business relating to space commerce at numerous universities in the United States and abroad. He has served on the boards of directors of various science research facilities, foundations, and hospitals. Dr. Robinson currently is working on his next book addressing the establishment of a unique jurisprudence to accommodate individual personhood of transhumans functioning in space.

Chapter 7

Creating the First City on the Moon

By Thomas F. Rogers

Doing So Should Grow To Become
The Major United States Civil Space Activity
Of the Space Age's Second Half-Century

"We have the power to shape the civilization that we want. ...those who came to this land sought...a new country. ...[we must] make their vision our reality. So let us from this moment begin our work so that in the future men will look back and say: it was then, after a long and weary way, that man turned the exploits of his genius to the full enrichment of life"

Lyndon B. Johnson. His "Great Society" speech, 1964, The World Is Flat - A Brief History of The Twenty-First Century, Thomas Friedman Farrar, Straus and Giroux; New York; 2005; page 276

"... The assumption that because America's economy has dominated the world for more than a century, it will and must always be that way, is as dangerous an illusion today as the illusion that America would always dominate in science and technology was back in 1950. But this is not going to be easy. Getting our society up to speed for a flat world is going to be extremely painstaking. We are going to have to start doing a lot of things differently."

Thomas Friedman; , above, page 278

Real Estate Developer James Rouse once observed that creating new cities can provide *"...a new, creative, thrust that will not only produce new communities but will release among the people in them the potential for the noblest civilization the world has ever seen."*

James Rouse, "Better Places, Better Lives - A Biography of James Rouse" Joshua Olsen; The Urban Land Institute; Washington D.C.; 2003; page 135

Introduction

The publicly funded United States civil space program came into being in response to the unexpected launch of the world's first man-made satellite into Earth orbit in 1957, and later satellite and cosmonaut launches, by the then Soviet Union. As the U.S.S.R. continued to demonstrate prowess in space, Cold War considerations led the U.S. to demonstrate its superior technological capability to the world by deciding to be the first Country to send astronauts to the Moon and back.

By the end of 1972 the US had demonstrated the continued ability for two astronauts to take such a trip and stay there for three days.

Thereafter, the role of people in space became less clear, and the most important space activity became the conduct of scientific research, primarily in the natural sciences, using telescopes and robotic probes. Several attempts since then to extend human activities beyond Low Earth Orbit failed to generate widespread national support.

Now, however, the Federal civil space program is being fundamentally changed to institutionalize human expansion throughout the Solar System and the Cosmos. NASA has been drawing up plans for what some see as the commencement of the next human evolutionary advance.

At the outset, NASA says it will spend just over $100 billion (2005 dollars) to send four astronauts travel to/from the Moon's surface in 2018, and remain there for up to a week.

But considering its earlier actual civil space program cost experiences, and Federal fiscal circumstances, the cost could be as much as roughly $250 billion, and the trip date could be delayed to 2024-5. (See below for details.) "House Science Committee Chairman Sherwood Boehlert said [on] Nov. 3 [2005] that 'There is simply not enough money in NASA's budget to...maintain the current schedule' with the relatively flat NASA budgets forecast for the year's ahead." (Space News; November 7; 2005; page 6.)

Thus, we could spend some one-quarter trillion dollars and 20 years to merely double the capability we demonstrated in the early 1970s. That is, over a period of more than 50 years since the last Apollo flight, doubling of capability will have achieved a paltry average rate of less than 1-1/2% improvement per year. And there is little NASA or public discussion about how this increased space capability is to demonstrate its value.

In addition:

1. There is no public discussion about the U.S. public civil space effort being broadened to include other Federal offices than NASA in what would become a truly great human activity requiring a very broad and sophisticated national involvement;

2. The American public is not to be directly involved in the effort. The 1991 report "America's Space Exploration Initiative" should be recalled in this regard, as after stating that the then described human Moon-Mars program should "Involve the public in the adventure of exploration" (page 26), nothing further was said about doing so;

3. Private U.S. commercial-industrial participation is not addressed. Again, the 1991 "Initiative" report noted that "Industry effort should be limited to studying elements of the [program] architectures." (Page 7); and

4. Except for some potentially significant space transportation vehicle development efforts, private investment is hardly addressed; the tax-paying public is expected to pay for almost all of the program's financial cost.

Finally, what is the fundamental purpose of the initial human Moon activity that is planned? Is it to be conducted primarily, and perhaps only, to demonstrate that extra-Earth human trips and stays could also involve Mars? And, later, would astronauts visit Mars primarily to demonstrate that the technological-operational capability to visit Pluto? And continuing, then to have them visit ... ?

The United States can, and should, be doing more and better than this in non-military space.

The U.S. has great experience and capability in civil space, and it faces grave national and international safety, security and economic circumstances. Potentially, the Moon offers safety, security, economic, and cultural potentials, but we are not moving to realize these potentials.

This chapter suggests that the Moon program that it is now commencing could be, and should be, modified and enlarged in a fundamental manner. Doing so would see us involving a broad cross section of U.S. professional, business, and financial interests and capabilities to advance civilization's most fundamental interests and aspirations.

✳✳✳

In thinking about past civil space advances (here "civil" includes all non-military, Federal government and private sector initiatives), it should be kept in mind that, measured from Sputnik, 2007 will mark the end of the first half-century of the civil space age.

Today, the U.S. Federal civil space program is focusing upon implementing the "space exploration" vision advanced by President George W. Bush in January 2004. An initial program has been set forth by NASA. (See, for instance, the contemporary summary: "Reaching for the Moon", Nature; 6 October, 2005; page 789.)

The space community expected the vision to call for astronauts to visit Mars, and as Wendell Mendel observed "Most people seem to think 'been there, done that', about the moon, which appears to most to have nothing to offer." (In "Making Space Happen", Paula Berinstein; Medford Press, Plexus Publishers, Inc.; 2002; page 193.)

This is expected to be the first step in a continuing United States program that would see robots and people engaged in physical discovery activities throughout the Solar System and, indeed, the Cosmos, for as far into the future as can be imagined. But the present NASA Administrator sees much more - - he has recently emphasized that NASA would be involved in "extending the range of human habitat[ion] out from Earth into the solar system [because] a single-planet species will not survive." ("Ground This Mission", The Washington Post; October 2, 2005; page B6.) And Nature (see above) notes that this activity would be "... not just flags and footprints [as in Apollo] but the beginning of humanity's permanent expansion into the Solar System."

This view implies that we must consider not just Columbus-like, or Lewis and Clark-like astronauts as explorers of space, but their immediate pioneer followers as well, people concerned not only with exploring, mapping, and geology, but with extending civilization itself beyond the Earth.

If not just a few individuals visit the Moon, but large numbers of people migrating to, bodies far from the Earth, it will be an extraordinary, novel, and complex undertaking for today's space professionals to carry out. Indeed, by themselves, they cannot do so - - it is simply too large and complex a task for them. ("The U.S. civil space sector; alternate futures", Courtney Stadd and Jeff Bingham; Space Policy; November, 2004; pages 241-252.)

Therefore, in considering this initial program's goals, cost and schedule, it's instructive to compare it with another "super-exploration" program, the one that, 500 years ago, immediately after Columbus' arrival in the New World, saw mass migration by Europeans.

The first human visits to the Moon extended over an interval of nearly 3.5 years, but there have been none since December 1972, so the Moon has been "uninhabited" for nearly 33 years. If the next trip happens in 2018, which would be almost exactly 50 years following the first one.

In comparison Santo Domingo, capital of the Dominican Republic "was founded in 1496 by Bartolomeo Columbus, brother of Christopher Columbus, as capital of the first Spanish colony in the New World. It became the starting point of most of the Spanish expeditions of exploration... of the other West Indian islands and the adjacent [North and South American] mainland." ("The New Encyclopedia Brittanica, Volume 10, Micropedia, 15th Edition; Encyclopedia Brittanica, Inc., Chicago; 2002.) That is, it took only 10 years after Colombus' original voyage to establish a permanent presence, and not just a "base" or "outpost," but a city. Today Santo Domingo is the industrial, commercial, and financial center of the country with a population of about 2 million, about double the population of Washington D.C.

Of course, the physical circumstances on the Moon are markedly different from those faced in the West Indies, but in principle the difficulties that they present should be offset to some considerable extent today by the financial and political wealth that United States can call upon. In addition, our scientific knowledge, technological assets and experience, and our economic capabilities are far greater. And a voyage to the Moon takes only days, while it took Europeans months to travel to the Caribbean.

Why Should The United States Lead Human Immigration To The Moon ?

There are several important reasons why we should return to the Moon, and other than the first one, they are not yet given much attention by the space world or the general public.

A. Safety:

Fundamental Support for the Future of Civilization
The past several decades have seen increasing concern expressed by experienced scientists and others about the increasing peril that our civilization faces due to the growth in natural and manmade threats.

For example, Martin Rees, Royal Society Professor at Cambridge University, Fellow of King's College, and England's Astronomer Royal, has recently declared that "...terror, error and environmental disaster threatens humankind's future," and suggests that the odds are no better than fifty-fifty that humankind will survive to the end of the twenty first century.

If Rees is correct then many people already living today will face an awful fate, and their children could well be among the Earth's last. But if enough people carry human civilization with them a quarter-million miles away to the Moon, as they did across the Earth's surface from southern Africa starting some 50,000 years ago, many/most Earthly risks thereto should be eliminated for a long time to come.

B. Economics:

A Fundamental Response to "Economic Globalization"
A major change is underway in international economics. Over the past two decades, nations such as China have made rapid progress and have begun to join the United States, Canada, and Western Europe in business capability. They look forward, confidently, to equaling or surpassing the performance of today's economic leaders.

Billions of people in China, India, Russia, Eastern Europe, Latin America, and Central Asia are becoming capitalists. "The scale of the global community that is soon going to be able to participate in all sorts of [economy-related] discovery and innovation is something the world has simply never seen before." (Friedman; pages 181-2.) And this is raising grave problems for the United States and other Countries that have global developed economies.

For instance, Thomas Friedman has observed that "China's real long-term strategy is to outrace America and the E[uropean] U[nion] countries to the [economic] top, and the Chinese are off to a good start." (Friedman, page 118.)

The National Academy of Sciences recently released a report prepared in response to a request from Congress to study economic globalization and its likely influence on the U.S. economy. (The Internet: www.nationalacademies.org; The New York Times; October 13, 2005; page A16; The Washington Post; October 13, 2005; page D5.)

"'America must act now to preserve its strategic and economic security,' panel Chairman Norman R. Augustine said. 'The building blocks of our economic leadership are wearing away. The challenges that America faces are immense.' ... The underlying goal, the panel said, is to create high-quality jobs by developing new industries...based on the bright ideas of scientists and engineers." (The Times) Augustine also observed that, "The U.S. is not competing well in this new world." (Aviation Week and Space Technology; October 31, 2005; page 70.)

Section 102 (c) of the U.S. Federal Space Act of 1958, as Amended, states (in unusually strong language) that "The Congress declares that the general welfare of the United States requires that the National Aeronautics and Space Administration [should] seek and encourage to the maximum extent possible the fullest commercial use of space."

But to date, civil space public expenditures have not produced significant private sector economic gain, but now is the time to think about the use of Moon-related assets and activities in an imaginative and determined fashion to advance America's economic interests at a time when we are threatened by economic globalization.

Learning how to move people to the Moon and have them reside and work there would be a large-scale imaginative advance that would take advantage of our past half-century's civil space scientific research, technology development, and space operations efforts, and could lead to enormous economic gain for the US.

C. Peace:

Seizing the Opportunity to Avoid War
The Moon is nearly a quarter-million-miles from the Earth, a distance that should offer protection to its inhabitants from any natural disaster that occurs on Earth other than a large meteor strike. However, in considering the formation of an Earth-Moon civilization, two generally- unappreciated Lunar characteristics stand out:

While the Moon is billions of years old and has some four times the surface area of the United States or Europe, there are no people there!

And, except for a near-two week total interval in the some 150,000 years that modern humans have existed, no one has been there, so the Moon has essentially no human history!

In human terms the Moon is essentially a tabula rasa - - a blank space. There are no lingering hatreds, or even recollections of war, slavery, genocide, or religious persecution. There are no political or economic alliances or political boundaries, no land ownership, no weapons of any kind, especially weapons of mass destruction.

Imaginative and sensitive advantage could be taken of such an extraordinary civilization-related circumstance not to be found here on Earth. Indeed, political, economic, legal, and other governance-related steps could be undertaken on the Moon that can hardly be thought of in many parts of today's Earth based world.

For we now have the extraordinary, encouraging circumstance of being able in effect to restart civilization under the best circumstances possible, learning and demonstrating how the evolving New World could become a peaceful one for all of its inhabitants.

Is this possible?

In the opening section of "Wilson's Ghost - - Reducing the Risk of Conflict, Killing, and Catastrophe in the 21st Century", (Public Affairs, New York, Perseus Books Group; 2001) Robert S. McNamara and James G. Blight remind us of the awful experience that the world's people have had with war over the past blood-drenched century alone.

Many would say that because human nature is what it is, civilization will always be faced with the prospect of terrible wars, that such aggression "is in our genes." But this may well be an incorrect understanding of the human condition. Archaeologists are now concluding "...that warfare, despite its malignant hold on modern life, has not always been part of the human condition. The global archaeological record... shows that warfare is largely a development of the past 10,000 years." ("The Birth of War", R. Brian Ferguson; Natural History; July/August, 2003; pages 28-35.) Hence, for about 150,000 years of human existence it is seen as quite possible that there was no such thing as war.

This suggests that we could use human development of the Moon to learn how to change our culture. To achieve peace is one of civilization's most important goals and activities, one of the greatest in history and pre-history. It would certainly be the outstanding civil space accomplishment of the second half-century of the civil space age!

D. Diplomacy:

"For the Benefit of All Mankind", and Public Diplomacy.
The Space Act states that U.S. "...activities in space should be devoted to peaceful purposes for the benefit of all mankind."

Certainly learning how to live and work on the Moon in a manner that would formally seek out how to do so peacefully would fulfill the spirit and letter of this law, and would benefit all people throughout the World!

And it would be of particular importance to the U.S. today. For unlike the wide plaudits it received for its victorious World War II efforts followed by the Marshall Plan, its work in helping to create the United Nations organization and the defeat of Communism followed by the destruction of the Berlin Wall, recent years have seen the U.S. lose the approval of many of the World's peoples and governments: "Washington's image and influence [are] at a low ebb." (The Wall Street Journal; November 3, 2005; page A10.)

The extraordinary and worldwide understanding and appreciation of the Apollo program was a kind of "soft power" used by the U.S. in a manner that now could be described by the term "public diplomacy." Thoughtful pursuit of peaceful immigration to the Moon, in which it would lead the international community in seeking ways by which safety, economic gain, and peace would be sought for all people there and on the Earth could generate widespread approval for the U.S. again - - approval which could extend over decades to come. "We could certainly help our case by boasting about our benevolence less and proving it more...by acting, that is, in ways that seem worthy of a great democracy." ("Their Highbrow Hatred of Us", James Traub; The New York Times; Sunday Magazine Section; October 30, 2005; pages 15-16.)

But Why A Lunar City ?

The Moon's surface area is nearly as large as North and South America combined. Under proper circumstances, well over one billion people could reside there. Even if the number in residence were only to replicate that of Alaska, the U.S. State with the smallest population density, it would exceed 700,000.

While NASA plans to spend over $100 billion to see four astronauts reach the Moon's surface in 2018 and stay there for a week, a study of NASA technology development-operational programs by the Congressional Budget Office suggests that it could well cost some 45% more than this. ("A Budgetary Analysis of NASA's New Vision for Space Exploration", The Congress of the United States, Congressional Budget Office, Washington, D.C.; September, 2004.") i.e., some $150 billion.

Since we can expect to run annual Federal budget deficits of many $100s of billions per year for decades ahead, this discretionary program sum will have to be borrowed, so it may cost roughly $250 billion in public funding to see this seventh astronaut Moon stay come about. Additional trips and extended surface stays will add to this cost.

Whether this is an acceptable sum to spend for the acquisition and demonstration of this particular Federal civil space capability does not depend upon its technological-operational success alone, but upon how it will be employed. To date, the most that is the government has said that it expects to do is to see that, in time, astronauts will "...live for six months at a time in Antarctic-style outposts." (Nature)

Certainly we can do much better than this! And if this activity is being undertaken to demonstrate "...that humanity is poised to take its next evolutionary step...the beginning of humanity's permanent expansion into the Solar system" (Nature), we must do more, much more, than this!

What needs to be done is to advance beyond government space explorers visiting Moon outposts for relatively brief intervals, to seeing large numbers of a wide cross section of the general public settled there as pioneers undertaking economic, social and cultural activities. We must see our government and private sector enter into an extraordinary undertaking together.

This undertaking must deal with very novel and complex circumstances, and therefore a very broad and most sophisticated approach is required. It must provide appropriate Earth-Moon transportation that is not only safe and reliable, but at a unit cost well below anything yet attained in spite of decades of attempts by NASA to do so. Food, water, electrical energy, communications, local transportation, housing, medical care, educational services must all be provided in a way that accommodates the Moon's great and rapid changes in temperature, its absence of an atmosphere, meteoric and radiation influxes, a much lower gravity, etc. Matters of governance and private investment practices must be created in a legal setting with which the World has had no experience. And, in principle, the varied interests of the Earth's very nearly 200 countries should be accommodated.

Several professional studies address having people on the Moon to make economic use of its physical resources have certainly been instructive and stimulating, but they have been primarily theoretical and have not dealt practically with such real-life matters. For the most part they conclude by implying what kinds of additional specific and detailed studies, analytical and experimental, are called for. (See, for instance, two examples: (a) "America At The Threshold - - the Report of the Synthesis Group on America's Space Exploration Initiative", Superintendent of Documents, U.S. Government Printing Office, Washington, D.C., 20402; 1991; this Group effort was Chaired by Thomas P. Stafford, Lt. Gen. USAF (Ret), a former astronaut; and (b) "The Moon - - Resources, Future Development and Colonization", David G. Schrunk, Burton L. Sharpe, Bonnie L. Cooper, and Madhue Thangavelu; John Wiley & Sons - Praxis Publishing, New York; 1999.)

There is one clear and early way to begin to bring this situation into useful professional, business, legal, and political focus. It calls for a novel public-private, technological-operational-economic, national-international, effort to begin to design the first Moon city in order to conceive and gain adequate acceptance of specific plans for seeing the city actually financed, designed, and created, permanently occupied, and widely and increasingly used. And to prepare to mount follow-on city investment, construction, and settlement programs.

In beginning to consider designing the first such city, the views of James Rouse, the creator of the modern "new city" of Columbia, Maryland could be relevant. He thought in terms of an "... entire metropolitan region: 'a series of small communities separated by topography, highways, public institutions [and/] or greenbelts...united by a center that provide[s] cultural, educational, [and] recreational facilities for many small towns around it.' " (Olsen, page 135.)

Private-public studies should be made of the activities that would involve the city's residents: government scientists engaged in scientific research in geology, astronomy, gravity-influenced human physiology, space tourism; low-gravity-influenced sports and entertainment, especially dance; mining; electrical energy generation and service provision; communications and position-fixing services... .

Fortunately, many countries have had very instructive experience with the creation of "New Cities" since the end of World War II. In the U.S., Reston, Virginia, and Columbia, Maryland, are the outstanding examples. (Re Columbia see Olsen; also "The Columbia Process - The Potential For New Towns", Morton Hoppenfeld; The Architects Year Book, The Garden City Press Limited, Pixmore Avenue, Letchworth, Hertfordshire, SG61JS, U.K.)

Those engaged in the conduct of space science, technology and operations activities are quite used to dealing with such professionals as physicists, astronomers, geologists, mathematicians, with engineers, computer specialists, test operators, with astronauts and radio communicators. But in planning to design and build a city and see it operate successfully, quite different disciplines must be called upon to see that the worlds of financial investment, construction, local law, and social organization all receive balanced attention. Professions such as urban planners and designers, mortgage bankers, real estate developers, architects, landscape architects, economists, marketers and schedulers, experts on nursery schools and day care, housing, recreation, and religious centers will be involved. It is quite a different professional world, and present civil space-related government offices and aerospace industry offices must be markedly broadened and enlarged to deal with them effectively. (Hoppenfeld)

Modifying and extending the new Federal human "space exploration" program that is suggested here should result in much better understanding and higher value by the general public of the United States and throughout the world. Inasmuch as this fundamental civil space program, in political terms, will be very costly and will take a long time, its prospects will be strengthened by obtaining wide and continuing support.

Perhaps the Moon city could be named "Lunar Columbia", after (a) the great explorer Christopher Columbus, (b) the Columbia river in the present State of Washington that took so much of Lewis and Clark's interest, and (c) the successful U.S. "New City" in Maryland that, even today, is actively studying how it can become more widely useful and appreciated.

And perhaps the U.S. government post-Apollo Lunar city creation program might be named "Project Odysseus", after another Greek hero - - the one who, over time, became involved in just about everything.

Some Specific Suggestions

A. Wide Federal and Private Sector Involvement
As the "Report of the President's Commission on Implementation of United States Space Exploration Policy" (2004) emphasized, both the United States public and private sectors must be involved in providing goods and services required to establish "Lunar City" and see it function appropriately, and both must be involved in providing the financing required to do so.

While NASA and its related aerospace industry would have important roles to play in seeing the U.S. create the Lunar city, a much wider approach is seen to be required than they alone can provide, as opening the Moon to immigration must also be seen as an extraordinary financial-economic investment opportunity.

Therefore, a US public plus private investment-business plan should be developed that would see the Lunar city established and operated. At a minimum, the Departments of Commerce, Energy, Interior, Housing and Urban Development, State, and Transportation should be involved, as should the United States Chamber of Commerce and countless companies. They should all become active advisors to NASA to assist in seeing the Moon initiative become a large-scale economic engine that provides a significant U.S. response to the growth in "economic globalization".

And the United States, under the aegis of the Department of State, and perhaps with the Council on Foreign Relations, The Carnegie Endowment for International Peace, the United States Institute of Peace, and the United Nations, should lead in seeing a national-international plan developed for city governance, with international plans for the related governance of localities throughout the entire Moon, and a plan or plans that would deal with Earth-Moon matters.

A research and demonstration program dealing with the avoidance of physical conflict, Moon governance, and Moon-Earth interrelation should be planned and conducted by NASA and others.

B. Interim Use of the Shuttle Fleet for Low Earth Orbit-Lunar Orbit Transportation

Earth-Moon transportation of people and cargo must be safe, reliable, and of large capacity and low unit cost (i.e., dollars per person / per pound).

While NASA has announced a space transportation vehicle development program that would allow astronauts to return to the Moon in 2018, but if the CBO "cost overrun" judgment is correct and NASA's annual appropriation is not increased to meet it, the return would not take place until as late as 2024-5. A Space News editorial has just observed (November 7, 2005; page 18) that NASA "...is looking at flat budgets for [at least] the rest of this decade." If so, its purchasing power could come down from today's by 25 - 33% or more.

And were the annual growth rates of the $ multi-trillion gross domestic products (GDPs) of the U.S. and China continue at the same pace that each has averaged over the past two decades (roughly 3% per annum U.S., and 9% China), by 2018 China's GDP could equal that of the U.S., and by 2024-5 China's could be some 50% greater than that of the U.S.!

These fundamentally negative financial-political considerations call for (a) our learning how to reduce the unit cost of Moon-related space transportation as far as possible, and (b) seeing our return to the Moon and its economic development made as soon as possible.

Public discussions of the astronaut Moon-return plan often note that it will take important advantage of Shuttle vehicle technology and operational experience. Therefore, careful and imaginative attention should be given to extending this thinking to considering the use of the present Shuttle fleet to provide Earth LEO-Moon Orbit cargo- and people-carrying transportation service for 10-15 years after 2010. Doing so would make full use of already paid-for vehicles and already experienced operations and maintenance professionals. They could be fueled by private sector Earth surface-LEO vehicles now being developed to support the International Space Station (ISS) and to meet the space tourism market - - vehicles that, in serving large, continuing, markets can expect to achieve much lower unit costs than the Shuttle and today's ELVs. Inasmuch as they need not travel through the Earth's atmosphere, they could carry much greater payloads, and they should require less maintenance attention. An initial private study of this approach reached a positive conclusion.

A relatively small vehicle to be used for Lunar orbit - Lunar surface transport would have to be used as well. Perhaps the model now located at the Smithsonian Institution's National Air and Space Museum in Washington, D.C., could be regained and refurbished; and NASA is moving to support private sector development of such a vehicle capability.

(In passing, it should be noted that the U.S. Air Force continues to plan for keeping its early 1960s B-52 bombers in service for as long as 100 years.)

C. Changing the Way that Basic "Space Exploration" is Publicly Funded

While it is not often spoken to, the President's "space exploration" vision must expect to see NASA and other Federal offices dealing with infinities - - infinities of places to be explored, of their distances from here, of travel time, and thereof, overall, of cost. Already the cost and time involved in astronaut return to the Moon have been seen to be so high-long that any human Mars trip program activities have had to bee essentially deferred.

If it will cost the U.S. tax-paying public more than $200 billion to see four astronauts make a week's return to the Moon's surface, what will it cost to replicate doing so on Mars, at least 140X as far away? And what of other Solar System bodies out as far as Pluto, some 3.5 billion miles away? Several of the Solar System's planets have forbidding physical characteristics, and Pluto, "...once seen as...at the [solar] system's edge...has...been revealed as just one of a growing swarm of dwarf planets reaching far out toward the depths of interstellar space." (From a review of the book "The Planets", Dava Sobel; Viking; 2005; in The Wall Street Journal; October 28, 2005; page W8.)

Given the anticipated Federal fiscal circumstances and the other priorities besides civil space that call for Federal funding, it should not be surprising to see the "space exploration" program's level of appropriations come under wide and serious political fire - - especially if most of its attention is essentially culturally focused, as it is today. (The Washington Post)

Under these circumstances, a fundamentally different way of paying for "space exploration" activities is called for. The following steps are suggested:

* Starting in a few years so as to allow NASA to adjust to a new way of fiscal life, each year the President and the Congress would see that its basic C.P.I.-adjusted appropriation is reduced by a fixed percentage - - say (arbitrarily, for discussion here) by 2% per year. This would result in its basic public expenditure level being reduced by nearly 67% by 2060. Doing this so would certainly encourage NASA to focus upon cost containment.

* Each year the budget would be adjusted upward to reflect taxes collected from all new private sector businesses that involved people in economically attractive Moon activities.

What might reasonably be expected to result can be seen from considering two existing civil space business areas: (a) The Department of Transportation has determined that space transportation supported just under $100 billion of business three years ago. ("The Economic Impact of Commercial Space Transportation on the U.S. Economy: 2002 Results and Outlook for 2010", Associate Administrator for Commercial Space Transportation, Federal Aviation Administration (FAA); Washington, D.C.; March, 2004); and (b) "...by some estimates, commercial revenues from satellite navigation exceeded $12 billion in 2002, growing at more than 20 percent annually." [...which could reach some $35 billion per year by the end of 2005.]. ("Space Diplomacy", David Braunschvig, Richard L. Garwin, and Jeremy C. Marwell; Foreign Affairs; July/August, 2003; page 158.)

The satellite communications business began to produce income in 1964, and satellite navigation in about 1978. Taken together, these two new space-related businesses have been growing at an average rate of about $2 billion per year. Inasmuch as the Internal Revenue Service obtains 16-20% of the U.S. Gross Domestic Product (GDP) in taxes, this suggests that today the Federal government's tax revenue from these two space businesses could approximate $25 billion per year.

If these early space business / tax results were replicated in the Moon city economy, by 2060 the Federal government could expect to be receiving nearly $20 billion per year in taxes. Such an amount would be used to more than restore the NASA space exploration 2/3rd appropriation reduction that would amount to some $8 billion per year by then (all in 2005 dollars).

So NASA's space exploration program could continue to maintain its purchasing power, and therefore its program ambitions and schedule, and perhaps even see it increase significantly even though its original appropriation rate was sharply reduced. Of course, this would see NASA doing something that it has not done since the 1950s-60s, when it worked closely with the private sector (AT&T, IT&T, ...) to create the large and unusually useful satellite communications business.

Adopting such a plan for sharply constraining the need for public financing of "space exploration" would certainly gain NASA wide and prompt appreciation for leading the Federal government in such a novel fiscal direction.

Summary

The United States is still the world's professional leader in human space flight and related human Moon analytical and experimental operations and studies. It is now starting a large, human, Moon-focused activity as the initial stage of a large, long term, overall "space exploration" program designed to realize President Bush's 2004 civil space vision. In addition, it has obtained very useful experience in the design, creation, and operation of "new cities," experiences that are actively broadening. And it still continues to be the world's economic leader, generating roughly 30% of global gross national product (GNP).

Given this fundamental national context, for several national and international security reasons - - reasons of physical safety, security, economics and diplomacy - - the U.S. now should enlarge and modify its Moon program so as to begin to lead the world in commencing human immigration there, to seeing pioneers settling and working there and learning how to advance civilization's prospects. It would marshal its public and private sector capabilities and mount a sophisticated and early effort to do so.

Specifically: it should now commence planning to create a unique "new city," humanity's first city in space, on the Moon.

The city would not be a Lunar outpost to be visited only by astronaut space explorers; a program to obtain and utilize such a capability is already underway. Rather, it would draw upon this experience in the creation of a true city - - a growing one to be lived in permanently by a broad and growing cross-section of the general public. Its residents would settle there primarily for economic reasons: to make widespread commercial-industrial use of the Moon's unusual physical characteristics and potential resources.

Doing so will take imagination, energy, financial investment and perseverance in technology development, city design and construction, space and Moon operations, and Moon-related economic development. Unusually wide and continuing Federal Government involvement and support will be required, as well as active and widespread national-international private investment and effective modern governance.

This will be a large, novel, complex, and therefore unusually difficult activity - - perhaps as much so as any non-military one carried out in the history of the world.

But much has already been accomplished in space and in modern city building since the end of World War II, and continues today. And there is extraordinary security and economic value to be obtained by opening up the Moon to imaginative civilization; the realization of these values justify proceeding, and some of their features justify speed in doing so.

It is now time to begin involving not only natural scientists and professional space explorers in Moon activities, but a wide array of people with broad experience and education, with varied public and private U.S. and international community and investment interests. We must move beyond the idea of astronauts making relatively short visits to the Moon and involve others as pioneers, commencing immigration there, to settlement and working there; broadly speaking, to "living there," striving to create an expanded and improved civilization created in and for a two-body, Earth plus Moon, New World! A New World that is safe, peaceful, secure and economically sound.

It is time that we learn and demonstrate that we can move to the Moon now as rapidly as we did in the Apollo era, with new ways of financing such an effort, in the interest of broad national security (not military security) and "for the benefit of all mankind."

❋❋❋

In 2007, the 50th anniversary of the commencement of the civil space age will take place. So will the 400th anniversary of the founding of Jamestown, the first permanent "New World" English colony in America in 1607. 2007 will commence the second half-century of the civil space age, and it could also mark the beginning of a program to create the first city on the Moon and the initiation of the permanent expansion of the World's human civilization beyond the Earth.

In commenting upon the opening stage of the new "Space Exploration" civil spaceprogram, NASA administrator Mike Griffin said that, "the United States is... building the space transportation systems needed for traveling [between] the Earth [and] the Moon... a project he likened to the creation of America's... interstate highway system more than 50 years ago [, and] that NASA fully intends '...to figure out what we can ...accomplish at the exit ramps.' " Space News; November 7, 2005.

"On the Moon, we can learn if mankind has what it takes to settle the solar system."
Paul D. Spudis *"The Once and Future Moon"*
Smithsonian Institution Press; 1996; page 215.

About the Author

Thomas F. Rogers has B.Sc. and M.A. degrees in physics, was a research associate in the Harvard University Radio Research Laboratory, and worked in, formed, and headed Air Force and Massachusetts Institute of Technology (MIT) R&D laboratories. He has been engaged in military and civil space matters since the late 1940s. He was a member of the Steering Committee of the MIT Lincoln Laboratory, a Deputy Director of Research and Engineering in the Office of the Secretary of Defense, the first Director of Research in the Office of the Secretary of Housing and Urban Development, and Vice President for Urban Affairs in the Mitre Corporation. He directed the Congressional Space Station study in its Office of Technology Assessment (OTA).

He was a senior member of the DOD-NASA Aeronautics and Astronautics Coordinating Board (AACB), the NASA Administrator's Space Program Advisory Council (SPAC), and the National Academy of Sciences (NAS)-National Research Council (NRC)'s Space Application's Board (SAB). He led the group that first relayed television signals via orbiting satellite, was responsible for the basic design of the first global satellite communications system and oversaw its acquisition by Defense; conceived of, and commenced Defense R&D on, the use of orbiting satellites for precise and prompt three-dimensional navigation / position-fixing that led to the Global Positioning System (GPS).

He is Chairman of the Sophron Foundation; a Fellow of the Institute of Electrical and Electronics Engineering (IEEE) and a member of its Transportation and Aerospace Policy Committee; a member of the Cosmos Club. He was awarded medals by both the Secretary of Defense and the NASA Administrator.

Chapter 8

Stage Three Leadership:
From Good Ideas to Unified Action

By Charles E. Smith

As we move toward space, new standards for leaders and leadership are necessary: cooperation and coordination at a level currently thought impossible will become essential. "Stage Three Leadership" is a blueprint for developing humane, competent leaders who can fulfill the possibility of Unified Action in the unforgiving environment of space, and on Earth.

Unified Action means sharing the same boat while traveling one's own path to a common destination. You don't have to like or agree with the other voyagers, but your mutual dependence is absolute. To live in a condition of Unified Action is like being a trapeze artist — you had better catch the other person, and they had better catch you. You need not agree on style, but intention and timing are critical.

In practice, Unified Action demonstrates concern for the needs and interests of everyone involved. This is immediately obvious when it comes to your own children, job or close friends. It's less obvious, but equally real, when it comes to the future of humans in space. Moving beyond the planet of our origin is literally a global issue. As with issues such as weapons of mass murder and the destruction of irreplaceable ecosytems - the fate of humanity as a whole is involved. Only a cooperative response beyond anything we have previously achieved as a species can enable us to successfully address these issues.

Yehezkel Dror calls this collective response "Raison d' Humanité," and declares it a normative imperative. This imperative is the cornerstone for Unified Action for the future of humans both on Earth and in space.

The Talmud says that there are thirty-six righteous people in hiding who, if they can find each other, will transform the world. This is a suitable metaphor for the quest to find, develop, and support Stage Three Leaders.

These rare individuals embrace Yehezkel Dror's normative imperative at a global level and are also competent in achieving technical, administrative, and financial bottom lines. They think on a scale of generations while acting in real time.

Developing Stage Three Leaders is akin to nudging evolution, like teaching fish to crawl from the sea to the land.

The Stages of Leadership

Through thirty-five years of helping private and public organizations manage change, we have distinguished three discrete developmental stages of leadership excellence. Stage One Leaders are successful by conventional measures, and produce results. Yet they often undermine the creative process of moving good ideas to Unified Action by devoting excessive attention to immediate transactions, instrumental relationships, and bottom lines of order and control. We are all familiar with Stage One Leadership; a mindset in which leaders and groups usually react to circumstances regarding constituents, strategy, structure, process and people. They do what is necessary to meet the requirements of the moment, but seldom look further ahead.

Stage Two Leaders are also successful, and they also produce results. They are more aware of the need for coherent and managed approaches to Unified Action and change. They are more willing than their Stage One counterparts to try such approaches, particularly in times of crisis.

However, these innovative approaches usually give way to 'normal' prior practices after the crisis has passed. Results-focused Stage Two Leaders often do not persist in "walking their talk" with respect to Unified Action, but continue to be moved by circumstance and 'business as usual' ideology.

Stage Three Leaders creatively bring people and organizations together. Given human nature, and the almost inescapable pull for conformity from institutional culture, Stage Three Leadership is a tall order. It calls for uncharacteristic ways of seeing the world, and turning some commonly held beliefs and assumptions on their heads. What's needed is to embrace a new set of beliefs and practices, while replacing old paradigms without compromise, leadership challenge that in no way diminishes the absolute requirements for competence, judgment, and experience.

Mahatma Gandhi said, "Be the change you want to see in the world." For Stage Three Leaders, "being the change" means thinking and speaking the language of possibility. It means personally and always standing for the imperative of Raison d' Humanité — to be humanitarian on the widest scale while producing both tangible results and Unified Action. "Being the Change" also means moving ideas from the abstract realm of concepts to the arena of coordinated action, often in partnership with discordant individuals and groups. It demands that you confront your own attitudes, behaviors, and self-limiting beliefs, to get yourself out of your way. It means becoming an ambassador from the future you want to see and intend to create.

Many of our basic social institutions fail to meet this standard. People who control the most valued resources and carry senior accountability in large organizations often act as if they care more for order and control than for the rights and wellbeing of the people whose efforts sustain those systems. In schools this means that interests such as state test scores become more important than individual children's learning. In corporations this means the company's quarterly earnings outweigh its long-standing employee pension commitments. For the future of humans in space, this means that parochial bureaucratic, scientific, commercial, and national interests collide in ways that impede success. Short-sightedness reigns precisely where the need for leadership vision and Unified Action is greatest.

Processes and prescriptions for Unified Action abound. Methods for constructive conflict resolution, searching for common ground, and coordinated action are easy to find. What we lack are leaders committed to using them. Institutional cultures are merciless in their gravity and grip on people to conform. This usually stops people from "helicoptering" and seeing events from perspectives above and beyond what is already sanctioned by the culture. It is difficult to apply new knowledge, however productive it may potentially be, when it runs counter to deeply embedded beliefs.

The majority of leaders and organizations I have known operate at Stage One or Stage Two. Most of these are fine men and women, but they're stuck in human and institutional paradigms which guarantee that their futures will look pretty much like their past.

For Stage One and Stage Two leaders and organizations, their personal and cultural point of view is more important than anything else. Their very sense of survival is tied to their already existing beliefs about what's possible and what's right or wrong with technology, politics, identity, and economics. Anything that threatens this point of view is endured, ignored, suppressed, or attacked. This identity-driven mental defense mechanism, which we all share, prevents good ideas from turning into Unified Action. It prevents both leaders and organizations from taking risks and seeing possibilities with others who have different points of view. Without evolving leaders who are able to transcend their own point of view, Homo Sapiens will arrive at a developmental dead end.

Courageous, unified, and coordinated action is widely recognized as important in public, private, international, interdisciplinary, and inter-govern-mental situations. Too few players are willing to step up to this personal and global challenge. Uncommon commitment is missing. From greed to self-interest, to the immovability of existing structures, there are many reasons why Unified Action seems impossible and the exercise of power and position is the best leadership we get.

Stage Three Leadership Assumptions

For developing Stage Three Leaders on and in space, there are hypotheses to test:

If a small group stands for the emergence of Stage Three Leadership for the future of humans in space, it will propagate broadly over time despite entrenched disagreement and resistance. Theories, political philosophies, models, and intelligent proposals open the conversation and point the way. What will move these from concept to reality is what people stand for, and what they are irrevocably committed to.

Stage Three Leadership is imperative for the future of humans in space, and on Earth, to prevent the catastrophic negatives that have occurred internationally and institutionally throughout history.

Stage Three Leadership needs to be demonstrated now, in preparing for a future in space, if we expect to achieve it there. We can't get there from here. We need to put the organization of the future in place now in order to create the organization of the future. Flexibility cannot be invented from bureaucracy.

We can expect little agreement for a new paradigm of Stage Three Leadership or for the Future of Space, and in the absence of broad agreement, the development of Stage Three Leaders calls for training and mentoring in certain uncommon assumptions and sensitivities.

These assumptions include:

Unified Action is the new bottom line. A new bottom line looks crazy to others with different bottom lines. Unified Action will thus find little agreement as the new bottom line in a world where money, power, technology and territoriality are primary. Unified Action as the new bottom line is not simply a matter of process but a matter of context, process, and outcome together. A new bottom line of Unified Action requires an emotional and spiritual context of relationship, a physical process of coordination, and a commitment to outcomes that recognize the basic human needs for freedom, money, technology and social order. Institutional and social forms we will create from this new bottom line cannot be known at the outset. With a new bottom line, all we have is conviction, uncertainty, and faith.

Leadership is plural. It takes two to see one. Without some dedicated 'other' committed to our commitments, our own blind spots keep getting in the way of our higher intentions. An effective coach is the bridge across the gap between who we are and who we can become. Without dedicated mentors and coaches to help us achieve 'escape velocity,' it's difficult to sustain Stage Three Leadership individually or Unified Action together.

Thought is largely collective. Physicist David Bohm offers evidence that thought is largely a collective phenomenon, even though it seems to us that we are each individually doing the thinking. It is only through open-ended dialogue that shared meaning and unification can occur. Debate, opinion, and sharing information are not dialogue, and rarely result in shared meaning.

Reality is constructed of possibilities. Quantum physics shows us that the universe, at its most fundamental level, consists of an infinite set of possibilities, literally evoked by the questions that we ask. Creating new possibilities for moving good ideas to Unified Action requires the commitment of leaders and groups dedicated to inventing new questions that evoke and precipitate energy and action simply by posing the unpredictable possibility.

Power comes from holding opposites in your mind. Victor Sanchez, a teacher of ancient Toltec Indian philosophy and practices, says that the source of Toltec success as a harmonious society was the practice of "Kinam" — the ability to hold opposites in their minds without contradiction. To achieve large scale Unified Action, our Aristotelean "either/or" thinking must be displaced by a more inclusive mode of understanding. One pioneering investigation into the essence of creativity studied the milestone contributions of fifty-eight famous scientists and artists, including Einstein, Picasso and Mozart. They shared a common pattern: all breakthroughs occurred when two or more opposites were conceived simultaneously, existing side by side as equally valid, operative and true. In an apparent defiance of logic or physical possibility, the creative person consciously embraced antithetical elements and developed these into integrated entities and creations.

Problems are often good. Michael Reid, a gifted trainer and coach, said that most problems are not as bad as is normally considered. Rather, if problems are used to reveal underlying commitments, and these commitments are used as a basis for action to produce breakthroughs, a world of opportunities emerges.

Stage Three Leadership Sensitivities

Stage Three Leadership also calls for sensitivities for reliably moving good ideas to Unified Action. Out of many such sensitivities, the following are primary:

Energy Sensitivity. Systems with the most focused energy will prevail. In fostering Unified Action, energy is the decisive context. In any moment, energy, vitality and inspiration are either strong or weak. Stage Three

Leadership creates an energy envelope in which others can operate. While everyone knows this, few leaders act as if energy makes the critical difference. Red Auerbach said that as coach of the Boston Celtics his main job was to assure 'team spirit.' While the truth of this is broadly accepted, it's usually not acknowledged as the leader's main job, especially if they don't think they can control the parties involved.

Alignment Sensitivity. Alignment sensitivity is seeing the chaos or distortion caused by individuals and groups. Higher levels of strategic alignment, coordinated action, and cooperation are often a function of a specific shift in a leader's point of view about what s/he believes is possible. Even when the call for alignment comes from a shift in external conditions, competition, or new technology, the response may be limited or non-existent if there is no accompanying change in point of view.

Breakthrough Sensitivity. Breakthrough sensitivity is seeing and embracing opportunities for discontinuous outcomes. Most people will not promise themselves or others what they cannot predict. Unified Action in complex systems is not predictable, so it is rarely promised. Since commitment—not reasoning—does the mental sorting to produce breakthrough results, readiness to throw one's hat over the wall and then follow it is necessary. For example, our government promised to go to the moon within ten years, before planning how it would be done.

"Thou" Sensitivity. Thou sensitivity is the recognition that all people are human beings, not objects to be manipulated, and to behave consistently with that awareness. Most public political and commercial communication is "I-It" based, often concealed in caring or solicitous language. When people feel they are being treated as objects, the opportunity for willing, Unified Action disappears. Treating other people as full human beings with an inner life and perspectives as legitimate as our own is a challenge in any relationship, and may be especially difficult for an impatient visionary leader. Coercion and manipulation are tempting shortcuts, but they foreclose the possibilities of Unified Action. Objects can be used, or refuse to be used, but they can never choose to partner with us.

Possibility Sensitivity. Possibility Sensitivity is seeing the world in terms of adjacent possibilities. This gives the ability to invent new possibilities in the face of apparent impossibility, and to give that which is chosen such compelling importance that it can't be resisted.

Developing Stage Three Leaders

In the past fifty years, powerful methods have been adapted from Marine, SAS, and Navy Seal training, and from intensive personal transformation programs, to provide people with powerful experiences in which they see for themselves the possibilities that arise from changing their points of view. Methodologies now exist to transform how leaders experience the world in ways that will increase the likelihood of Unified Action. These methodologies give them new choices for transcending their former points of view.

As in an effective boot camp, a moment of truth arrives sooner or later, a moment in which leaders come to appreciate the terrible cost of their current segmented and self-centered world view. This moment is followed by seeing what is really possible as a result of claiming their own courage and leadership. They then have a choice to move the best ideas toward Unified Action for the greater good.

In developing Stage Three Leaders for the future of space and other critical challenges, there are examples to follow. A former marine told me that his drill instructor's motto was, "Let no man's ghost say my training did not do its job." The same level of commitment and persistence is necessary in training leaders to invent desirable futures and unify intractable boundaries and conflicts.

Steve Jobs, founder of Apple Computer, remarked in his 2005 Stanford University Commencement address that, ".., it was impossible to connect the dots looking forward when I was in college, but it was very, very clear looking backwards, ten years later. If we can't connect the dots looking forward ... only looking backwards ... then we have to trust that the dots will somehow connect in the future. We have to trust in something - our 'guts,' destiny, life, karma, whatever belief systems we've grown up in or chosen. Believing that the dots will connect down the road gives us the confidence to follow our hearts, even when it leads us off the well-worn path ... and that makes all the difference."

The choice to stay where we are or to turn away from the challenge ahead is always with us. Ultimately, the crossroads we face is whether to stay comfortable with and 'right' about our particular point of view, or to work together in service of our common humanity.

The major challenge in formulating the training and development of Stage Three Leaders involves defining areas of agreed and limited goals, and assuring the imperative of Raison d' Humanité. It's not just a matter of implementation, but rather "to stay on a path of "experimentation and inspiration ... to let go and follow the course of what is emerging."

Living in Space

When America was founded in the late 18th century, the dominant paradigm of the natural world was one of independent particles, mirrored by a society of individuals pursuing their own ends and only related externally through contracts. Today our world no longer exists in that way. Einstein and theories of quantum physics remind us that the world is inextricably intertwined - on micro and macro-levels.

Living in space, everything will be interdependent to an extreme degree that is hard to imagine now. Without Stage Three Leadership and a prime directive of "humanity in space," we can expect people bent on power, position, and greed to duplicate the violence, inequity, and lack of cooperation that often diminishes effectiveness and the quality of life on Earth. On our home planet we could (at least until very recently), rebound from the consequences of this behavior without overstressing the capacity of our environment to renew itself. In the habitats of space our closed biological systems will not withstand a crash in our social systems.

In the face of these vastly higher levels of complexity and interdependence, an atomistic theory of humanity is no longer relevant. Whether the subject is the future of humans in space, or controlling nuclear weapons, or reorganizing a government agency, our nation and the world urgently need more Stage Three Leaders to generate unified and committed collective action. This is a personal invitation to join in a collective hero's journey, and to stand for their selection, training, and support.

References

Dror, Yehezkel. The Capacity to Govern: A Report to the Club of Rome, 2002. Frank Cass Publishers.
Manji, Irshad. "Why Tolerate the Hate?", August 9, 2005. New York Times Op Ed.
Sherman, Howard. Open Boundaries, 1998. Perseus Books Group. Personal conversations, 2001.
David McCullough, 1776, 2005. Simon and Schuster.
Berne, Eric. What do you say after you say hello?, 1975. Corgi.
Bohm, David. Wholeness and the Implicate Order, 1980. Routledge Classics.
Senge, Peter M., with C. Otto Scharmer, Joseph Jaworski, & Betty Sue Flowers. Presence: An Exploration of Profound Change in People, Organizations, and Society, 2004. Society for Organizational Learning.

About the Author

Charlie Eliot Smith, Ph.D., author and organizational leadership consultant, has been working with executives and their employees in corporations, associations and government agencies throughout the United States, Canada and Europe for the past 35 years His work aims to help leaders and groups discover how they personally hold themselves back as a platform for unparalleled success.

Dr. Smith has a Bachelor's degree in Social Relations from Harvard College, a Master's from the Harvard Business School, and a Ph.D. in Organizational Behavior from Case Western Reserve University. He also has a certificate in Gestalt Methods from the Gestalt Institute of Cleveland. He was a Visiting Associate Professor of Organizational Behavior at Sir George Williams University in Montreal and taught at the McGill University Centre for Management Education. He was on the Boards of the National Peace Academy Foundation and The Foundation for Mid East Communication. He is currently on the advisory board of Mighty Mojo Studios. His book, The Merlin Factor: Keys to the Corporate Kingdom, is jointly published in Great Britain and in the U.S. He has also written numbers of other articles on personal and organization development, including "The Merlin Factor: Creating Ambassadors from the Future", "Explainers Anonymous", "Mentoring Leaders with Vision", "Board Dialogue as a Path to Competitive Advantage" and "Who Stole My Synergy?"

Chapter 9

Tennis Time and the Mental Clock

By Howard Bloom

"I long for an experiment that would examine, by means of electrodes attached to a human head, exactly how much of one's life a person devotes to the present, how much to memories, and how much to the future. This would let us know who a man really is in relation to his time. What human time really is. And we could surely define three basic types of human being, depending on which variety of time was dominant."
Milan Kundera

"A turning of our states of consciousness toward the future...[makes] our ideas and sensations succeed one another with greater rapidity; our movements no longer cost us the same effort."
Henri Bergson

"If man's reach does not exceed his grasp then what's a heaven for?"
Robert. Browning

We visited a peculiar creature: the spade-foot toad of Arizona. During long dry spells, the toad nursed the stores of moisture and food packed away in its cellular structure by crawling under the sand, shutting down its metabolic systems, and slipping into slumber. Lethargy was the life-saver that allowed the snoozing beast to go for months—perhaps even years—without a sip of water.

On the other hand, when an infrequent shower soaked the desert floor, the toad shook off its torpor, wriggled to the surface, cried out for company, listened for the croaked sounds of a gathering crowd, then headed for the nearest puddle. There, it leaped into a frenzy of action, wooing females at a rapid rate, then grappled with them in sexual ecstasy. Its manic eroticism was as much a survival mechanism as its former inertia. For only by coupling quickly could the toad sire a new generation that might grow to maturity before the pools of moisture could be sucked away by the desert sun.

The habits of the toad show up in one form or another over and over again. They appear in the hibernation of squirrels and bears, the seasonally fluctuating fat deposits of woodchucks, and in a wide variety of human annual rhythms. [1] They even show up in the moodswings of societies. When there's little to be gained, nature slows an organism down. When opportunity arrives, she speeds it up. One consequence is the strange gyration of your mental clock.

See if this sounds at all familiar. On a work day when I'm under extreme pressure, I rush through one task, hurry on to the next, then move quickly from that job to the one after it. I see my allotted hours as time in which I can easily accomplish a lot. But on a day when I have less work than usual, my mental clock readjusts. Suddenly, instead of seeing the day as a period in which I can easily perform a plethora of tasks, I get the sense that it'll be a struggle to finish anything at all. My mind grows sluggish. It's the phenomenon behind the common sense expression, "Work expands to fit the time available to it."

In a person with little to do, the mental clock slows down. In a person with a great deal to accomplish—or a person excited about what she's doing—it speeds up. Take, for example, the athlete who sees every eighteenth of a second [2] of a tennis ball's motion and calculates in a wink exactly where the ball is going to be when she attempts to swat it. For her, every micro-instant is filled with meaning. But for the person lying on a beach catching some rays, a whole morning can go by without a single meaningful moment. [3]

For the athlete under high stimulation, there is more time. Her world is richer. Far more data is processed by her brain.

One difference between a society on the rise and a society in decline may be that the rising society is on the fast clock. It sees each impediment as a challenge, absorbs information quickly, and finds new ways to overcome its obstacles. It operates on tennis time.

But the society that has peaked has moved to the slow clock. It has ceased to absorb data rapidly. It is on beach time. Tennis time is the clock of the newly-emerged toad, spending energy in a frenzied burst. Beach time is the clock of the dormant toad, hoarding every gram of substance on his bones.

Superorganisms on the trail of growth gravitate toward chemicals that speed the system up to tennis time. The British, when their Empire was enthusiastically seizing new possibilities, were fueled by a new import called coffee. The English commercial conquest of the world was planned in the coffee houses of London in the late 1600s and early 1700s. [4] The Chinese, during the roaringly successful years of the T'ang Dynasty (AD 618-907), filled their lives with another beverage that set their mental clock to "fast." They expanded their empire under the influence of tea. [5] In the 1980s, when the Japanese were the fastest-charging competitors in the world, they showed the same predilection for chemicals that turbo-charge the system. The leading drug problem in Tokyo's nightlife neighborhood during the late 1980s was not heroin or marijuana—the drugs that slow you down. It was amphetamine. [6]

With perceptual shutdown, we put on our eye-protectors and crawl into the stupor of beach time. But exploding societies like those in today's India and China may well be racing on tennis time, a clock that allows them to outrun us as we sit in front of our television sets, cradling a can of beer in our hands, cozy in the low-stimulus, low-challenge life.

How can America put itself back on tennis time? By focusing on the trigger that moves the toad from torpor into overdrive—opportunity. Nearly 100 years ago, the historian Frederick Jackson Turner proposed his influential Frontier Hypothesis. [7] The existence of the American frontier in the 19th century, he said, had invigorated the American mind. The possibility of unending resources just over the horizon had filled Americans with zest, imagination and exuberance.

America was not the only nation thrown into high gear by the presence of a new frontier. England was a puny and somewhat pathetic power up to the time of Henry VIII. Nonetheless, the country had its dreams of glory, and those dreams were associated with the notion of expansion. The single form of expansion the English could imagine, however, was conquering some of the only world they knew—Europe. The British vainly bumped their heads against a brick wall, attempting to saw off pieces of France in the futile bloodbath known as the Hundred Years War. They had their share of victories, suffered humiliations at the hands of folks like Joan of Arc, and were utterly thwarted in their efforts, eventually losing even the one scrap of territory they'd managed to cut out for themselves—Calais. Meanwhile, they mauled five generations of French peasants innocently trying to plant the next year's crops. [8]

Historian A. L. Rowse, a leading British expert on the Elizabethan age, considers Henry VIII's final failure to conquer the French one of the luckiest embarrassments England ever endured. [9] It forced the English to turn their attentions away from the Continent and made them focus on a sphere in which England would eventually make a fortune—the New World.

The Old World England reluctantly turned its back on was a land of little opportunity. Hungry Italians were reduced to eating songbirds off the trees. Tourists in southern Europe 200 years later would be puzzled by the eerie silence of the countryside. Melodious warblers like robins, titmice and wrens had disappeared into the cooking pots of the area's humbler citizens. [10] Meanwhile, the average French peasant lived so close to starvation that in the fairy tales he recited to his children, the hero was rewarded—not with a pot of gold, but with a decent meal. [11]

At first, the New World looked equally unrewarding. Christopher Columbus was bitterly disappointed by this hulking mass of landscape. He had set off to find the riches of China and ended up in a territory no one had ever heard of. The poor sailor insisted for years that this had to be some previously unreported part of the Chinese Empire. [12]

But Columbus' disappointment became England's New Frontier. The British missed out on the easy pickings. The Spanish beat them to the Aztec and Inca lands, where a few hundred Europeans armed with steel swords, muskets and cannon could overwhelm ten thousand Indians wielding wooden paddles and make off with a king's ransom in gold. [13] But Englishmen settled a land that seemed barren and forbidding, a land that had killed its first British settlers in the Roanoke Colony of 1583. [14] The British saw opportunity in catastrophe, settled the North American continent, planted cotton in its southern regions, and set up sugar plantations in the Caribbean Islands, bringing back a bounty that boggled the mind.

The new economic horizons utterly transformed England. In the 1400's, the country had been a barbaric backwater, producing almost nothing in the way of literature, art or science. But charged with the energy unleashed by a whiff of fresh resources, Britain became a cultural dynamo. The era in which she awoke was The Age of Elizabeth, and it gave us, among many other things, the plays of Shakespeare. [15]

America has a host of new challenges available to it—biotechnology, nanotechnology (the construction of microscopic machinery from units a few hundred molecules in size), expansion of the cybersphere, renewable resources, next-generation energy, and the construction of intelligent self-replicating devices [16] to name a few. But in the long run, another more daunting frontier awaits us...one that looks as bleak at first glance as our land once seemed to Columbus. Like Britain under Henry VIII, the country that sucks its energy from that next frontier will not necessarily be the one that manages to conquer some small swatch of the world we know. Instead, it may be the one that first finds a way to mine mineral-rich asteroids the size of Manhattan. It may be the country that first "terraforms" another planet—turning its atmosphere into breathable gas and its surface into a place where humans unencumbered by pressurized suits could take a pleasant stroll. It may be the country that relinquishes dreams of conquest in lands where the birds no longer sing...and turns its eyes toward space.

Space could provide a new rain of resources...or it could bankrupt us. But its habitation does offer two other advantages.

The first: international cooperation. No single nation can afford the price tag for full-scale extraterrestrial development. To turn the wastelands of asteroids and planets into lands of plenty would involve consortia that pull together the European Union, Russia, Japan, India (which has been developing space technology since 1969), [17] and China. Some of those partnerships are already underway thanks to the clumsy mistake known as the International Space Station. But when it comes to establishing a manned presence on the moon, building moon processing plants capable of turning lunar water into fuel, erecting moon construction centers able to turn lunar dust into interplanetary vehicles, and heading for Mars and beyond, we are still trying to go it alone. [18]

The second, and perhaps more important advantage of following in the footsteps of Captain Kirk is this. Man has as yet invented no way to prevent war. We have found no method for shaking the consequences of our biological curse, our animal brain's addiction to violence. We cannot free ourselves from our nature as cells in a super-organismic beast constantly driven to pecking order tournaments with its neighbors. We have found no technique for evading the fact that those competitions are all too often deadly.

Many optimists feel that the mere threat of nuclear annihilation will weld us together as one human race. If only the great communicators can shrill at us loudly enough about the threat of holocaust, all nations will see themselves as brothers, realizing their common stake in the survival of the species. Unfortunately, these peace advocates—much like you and me—have been known to quibble harshly with others who share their goals but differ in beliefs. Even the peace-makers cannot entirely restrain the urge for battle.

Nor can human beings as a species stop their inexorable itch for war. We're like a teenager in the days before the sexual revolution of the 1960s who has been told that masturbation will drive him insane. His guilt makes him feel nearly suicidal, but he still can't stop himself from the unspeakable act. We've found ways to halt illnesses, we've invented means to leapfrog continents in hours, and someday we will find a way to stop war. But only if we survive long enough. Until then our task is to outlast our own impulses. Our task is to outwit The Lucifer Principle.

You could think of us as a species trapped in a car hurtling out of control toward a tree, the steering locked, the brakes frozen. We could sit behind the wheel and pretend that if we felt enough guilt the tree would disappear. Or we could throw ourselves out of the auto's door and live. For us the equivalent of hurling ourselves to safety is moving a few humans off this planet. It's putting enough of our kind into colonies in space so that if the rest of us down here on earth disappear, those left in the rotating habitations above could keep the species going. Hopefully, the survivors could carry on the knowledge we've acquired so far, and with that wisdom and their own fresh discoveries, someday learn to overcome what we could not.

We need a new horizon, a new sense of purpose, a new set of goals, a new frontier to move once again with might and majesty, with a sense of zest that makes life worth living, through the world in which we live. One of the most challenging frontiers left to us hangs above our heads.

References and Notes

(1) David McFarland, ed., The Oxford Companion to Animal Behavior, pp. 479480. Joseph Altman, Organic Foundations of Animal Behavior, p. 425.

(2) Mihaly Csikszentmihalyi, "Memes Vs. Genes: Notes From the Culture Wars," in The Reality Club, John Brockman, p. 117. For the fraction of a second of light an eye can discern as a discrete flicker, see J.A. Deutch and D. Deutch, Physiological Psychology, p. 350.

(3) A few rare athletes can actually glean a message from a mere hundredth of a second of a ball's trajectory. Baseball player Ted Williams at the age of 50 demonstrated that he could register exactly where the seams were on the ball as it smacked into his bat at eighty miles an hour. Williams smeared pine tar on the bat's barrel and called out the part of the ball he'd hit. A sample call: "one quarter of an inch above the seam." When the ball was checked to see where the tar had left its mark, Williams was right five out of seven times. (Dr. Arthur Seiderman and Steven Schneider, The Athletic Eye, Hearst Books, New York, 1983, pp. 1718, 91.)

(4) "In these coffee houses you could borrow money, lend it, invest it, or spend it." In fact, one coffee house owner actually began selling insurance to his merchantcapitalist clientele. In time, the insurance venture proved more lucrative than serving cups of Java. The coffeehouse proprietor was Edward Lloyd, as in Lloyd's of London. (James Burke, Connections, pp. 193194. See also Fernand Braudel, The Structures of Everyday Life: Civilization & Capitalism, 15th18th Century, Vol. 1, trans. Sian Reynolds, Perennial Library, Harper & Row, New York, 1981, pp 254260; Mitchell Stephens, A History of the News: From the Drum to the Satellite, Viking, New York, 1988, pp. 4143.)

(5) Wolfram Eberhard, A History of China, pp. 169, 196. E.N. Anderson, The Food of China, pp. 5556. Philip D. Curtin, CrossCultural Trade in World History, pp. 104105.

(6) Robert Christopher, The Japanese Mind, Fawcett Columbine, New York, 1983, p. 163.

(7). Frederick Jackson Turner first presented his thesis, "The Significance of the Frontier in American History," in 1893. The concept wasn't published in book form until Turner put out his The Frontier in American History (Henry Holt, New York, 1920). For a modern variation on the frontier hypothesis, see Daniel Boorstin, Hidden History: Exploring Our Secret Past, pp. ixxxv.

(8) For the Hundred Years War, see: Barbara Tuchman, A Distant Mirror, pp. 48594; and G.M. Trevelyan, A Shortened History of England, Penguin Books, Harmondsworth, Middlesex, England, 1959 (originally published 1942), pp. 181188. For the loss of Calais in 1558, see: James A. Williamson, The Evolution of England: A Commentary On the Facts, Oxford University Press, Oxford, England, 1944, p. 179; and The New Encyclopaedia Britannica, Vol. 2, p. 731.

(9) A.L. Rowse, The Expansion of Elizabethan England. Cambridge University historian Eric Walker agrees with Rowse's assessment. (Eric A. Walker, The British Empire: Its Structure and Spirit, 14971953, Harvard University Press, Cambridge, Massachusetts, 1956, p. 2.) For the futile campaigns against the French with which Henry nearly bankrupted his government, see: J.J. Scarisbrick, Henry VIII, University of California Press, Berkeley, California, 1968, pp. 3335, 453456, and virtually the entire rest of the book; J.D. Mackie, The Oxford History of England: The Earlier Tudors, 14851558, Oxford University Press, London, 1962, pp. 410412; Kenneth O. Morgan, ed., The Oxford Illustrated History of Britain, Oxford University Press, New York, 1984, p. 256; and The New Encyclopedia Britannica, Vol. 5, pp. 840841.

(10) Keith Thomas, Man And The Natural World: A History of the Modern Sensibility, pp. 116117.

(11) When granted anything they wished, the heroes of these tales chose such items as "a bun, a sausage, and as much wine as he can drink," "white bread and chicken," or "crude wine and a bowl of potatoes in milk." Robert Darnton, The Great Cat Massacre and Other Episodes in French Cultural History, Vintage Books, New York, 1985, pp. 22, 2434.

(12) Daniel Boorstin, The Discoverers, pp. 236244.

(13) On March 12, 1519, Cortes landed at Tabasco and overwhelmed an Aztec army that outnumbered his tiny force 300 to one. (Hammond Innes, The Conquistadors, Alfred A. Knopf, New York, 1969, pp. 4252.)

(14) # Page name: Roanoke Island, # Author: Wikipedia contributors, # Publisher: Wikipedia, The Free Encyclopedia., # Date of last revision: 21 December 2005 00:17 UTC, # Date retrieved: 21 December 2005 00:28 UTC, # Permanent link: http://en.wikipedia.org/w/index.php?title=Roanoke_Island&oldid=32171948, # Page Version ID: 32171948

(15) G.M. Trevelyan, A Shortened History of England, p. 206.

(16) See Steven Levy, Artificial Life: a report from the frontier where computers meet biology, New York, Vintage Books, 1993, 35-42.

(17) Department of Space: Indian Space Research Organisation. November 10, 2005. Retrieved March 5, 2006, from the World Wide Web http://www.isro.org/

(18) For NASA's latest go-it-alone plan, see: NASA. Exploration Systems Mission Directorate. The Vision for Space Exploration. Retrieved March 5, 2006, from the World Wide Web http://www.nasa.gov/missions/solarsystem/explore_main.html

About the Author

A recent visiting scholar in the Graduate Psychology Department of New York University and a Core Faculty Member at The Graduate Institute, **Howard Bloom** is the author of two books: The Lucifer Principle: A Scientific Expedition Into the Forces of History ("mesmerizing"-The Washington Post) and Global Brain: The Evolution of Mass Mind From The Big Bang to the 21st Century ("reassuring and sobering"-The New Yorker). Bloom is the founder of two new fields-paleopsychology and mass behavior. Christopher Boehm, the director of the Jane Goodall Research Institute says, "Howard Bloom should be taking notes on what he's doing every minute of the day. He is single-handedly creating a scientific revolution." Gear Magazine has said Bloom is "....the next Stephen Hawking. But he's not just interested in science. He's interested in the human soul." Britain's Channel 4 TV has dubbed Bloom "...the Darwin, Einstein, Newton, and Freud of the 21st Century". And Joseph Chilton Pearce, author of Evolution's End and The Crack in the Cosmic Egg, says, "I have finished Howard Bloom's two books, The Lucifer Principle and Global Brain, in that order, and am seriously awed, near overwhelmed by the magnitude of what he has done. I never expected to see, in any form, from any sector, such an accomplishment. I doubt there is a stronger intellect than Bloom's on the planet."

Chapter10

Cooperative, Worldwide Space Collaboration Epiphany: A Turning Point, or Else

by Michael Hannon

The implicate desire of all life, without exception, is to live long, prosper, and reproduce.

These are my impressions and interpretations of my perceptions:

Obstacles to Worldwide Space Collaboration:

Success for international space cooperation over the long term future requires an idealized space community in which no nation, or bloc of nations, dominates the others. Out of respect, trust, and understanding, a new international "supernation" of enthusiastic, cooperative, and collaborative nations would work together to bring forth the best results for all. The participants would include all nations presently involved in space, as well as those who aspire to participate, to achieve both a strong sense of unity and positive outcomes unprecedented in space endeavors.

This would be a single group with shared aims, rather than competition involving one-upmanship. A shared sense of identity would be foremost among all members regardless of other affiliations such as nationality, race, education, or other influences that would lead to separatism. Cooperation can be achieved by the open sharing of common goals, priorities, and results so that a strong sense of global identity and team spirit pervade as the foremost influence in decision making.

In doing so, these nations will experience equality and fairness, sharing both burdens and rewards which they could not otherwise. An inherent sense of balanced responsibility would preclude what might otherwise turn into unnecessary conflict and competition. Each member goes about its tasks in the certainty that all are doing their fair share - a strong sense of team and group effort pervades the general atmosphere, and non-productive, competitive, self-seeking attitudes give way to a shared identity with the entire group as a whole.

In contrast, the current state of space exploration and research is one of competition between agencies and nations, with only limited cooperation between them. Although the concept of worldwide space cooperation appears to be a wonderful idea, acting against the implementation of this noble enterprise and making it seem virtually impossible are issues including radically disparate national policies, and growing mistrust of efforts coming from the major world powers on anything involving international cooperation in the aftermath of the invasion of Iraq.

Despite the significant economic advantages of working together, the world is far from ready to unite in space efforts, and shall remain so as long as governments demonstrate inconsistency between their actions and their stated intents, and a general lack of open shared honesty. A stigma of unilateralism has been established so powerfully that any sudden shift to authentic internationalist goals has every appearance of disingenuousness and lack of authenticity, regardless of how sincere the intent.

Thus, a realistic outlook for a sudden shift to large scale cooperation in 2006 is neither foreseeable nor entirely possible, but able leaders worldwide in both the public and private sectors must be willing to break the ice in a sincere commitment to worldwide space cooperation research and exploration, the results of which shall be shared openly for all mankind's benefit.

Looking ahead for the next ten years, a major push needs to be made toward a mutual building of cooperation and trust so that healing to bring about serious collaboration in space can commence.

In the hostile environment of space, every move must be well-considered beforehand so that mistakes are not compounded and do not act in concert. Those involved must trust each other and themselves at a level that transcends the demands of daily life on Earth. Proven methods of nurturing and preserving trust are necessary.

It frequently happens that the best solutions do not reach fruition because of lack of shared trust and intention. Economic pressures from corporate and other influences take priority, centered not in the best effort, but in self-serving interests seated in financial gain, nationalism, cultural priorities, and all those things that have kept people separated and at odds for millennia.

Thus, the goal of maximum immediate profit must give way to meeting the needs of humanity via creation and preservation of a highly cooperative international community. The economics of such an effort must be dictated by a sense of practicality that only wisdom can nurture. It requires that we consider the entire human race as we would one human being - that humanity as a single organism exists, has needs and goals, and that the means exist for its maximum benefit, health, and well-being.

Our most common concepts of power in this respect must be expanded to include all human beings, as the limited focus only on select individuals and groups has exposed their extreme degree of negligence, waste, and environmental destruction. This poor and unwise use of power has magnified into dilemmas without known solutions, including societal, economic, financial, environmental, health, and climatic crises.

Thus, the world cannot continue as it is indefinitely without suffering dire consequences. We must choose wisely, or forever pay the price for ringing a bell that we shall discover afterwards we cannot un-ring, no matter how we try. So if there ever was a time to join together in long term visionary efforts worldwide aimed at space and space projects to benefit Earth, it is now.

The time of selfish, irresponsible gain must give way to a new era of contribution and transcendent wisdom. 2006 is the year to begin, as otherwise it will be yet another year lost to history with old problems compounded. It may very well mark the year that the war against poverty was finally undertaken in earnest at the international level, and the potential for success in the noble pursuits could create an even greater front of authentic, sincere innovation and quest toward a planetary and space renaissance. We would look back on this as the beginning, when the Earth's peoples transcended themselves to a higher level of existence rooted in potential yet dormant in their hearts and minds, as they would will to become more than they have ever been in their rightful and inevitable evolution toward their celestial future.

Let us hope and pray that it shall be so.

A Future Psychology for International Cooperation:

Human behavior has now reached a point, due to advancements in weaponry and the capacity for destruction, that it has become absolutely necessary that a new psychology and methodology be developed. The capacity of humanity to continue to inflict harm and even to destroy itself must be tempered by insight and wisdom into human nature. A profound sense of purpose must emerge from within the framework of possibilities for all peoples.

If, for example, the first word children worldwide learn to read and write is "we," then child by child the world could become a much better place for all of us to enjoy life and liberty, and pursue happiness.

While spending vast resources over centuries and becoming more and more sophisticated at such methods as psychological warfare, we have spent disproportionately little energy on needs such as psychological welfare. The price of this is self-evident. If we continue dealing with the realities facing us now in a pathological manner, we shall indeed reap the fruits of our misuse of what has been given to us. Rather than acting as the true custodians of our planet and its surroundings, we shall perish as delinquent caretakers who, out of ignorance of our own self-inflicted perceptions and actions stemming from them, could never fathom why the world we created became what it has.

Thus, we descend into the same patterns suffered by previous civilizations that also failed to survive, and for the same reasons. Fixation on destructive tendencies has brought us to this, and the only way to stop this pathological state is to change the way we think, feel, and deal with ourselves and the world around us. Proven methods exist by which we can do this when we choose to.

We have in our languages words that show us the way out of the apparent dilemma in which we have placed ourselves, and we must recognize those words as belonging to us rather than simply as ideas that we have grown incapable of manifesting. Truth, honesty, honor, compassion, love, wisdom, understanding, responsibility, conscience, decency, knowledge, being, custodianship, care, and so many others have authentic meanings that have been exhibited by many of our ancestors, yet for some reason we veer ever farther away from the realities which those words describe and name. The noble dreams of our youth are clouded by icons of selfishness and greed, the thunder of war fills their ears, while among us globally those in need find neither audience, nor aid, nor solace.

We have lost our way, and we must restore it by actively pursuing world harmony through concerted efforts to overcome our disunity from millennia of negative associations, even at the subconscious level, which have plagued humanity and shall continue to do so unless we find a new way. We must do everything we can to evolve psychologically beyond where we are now, and we have the means available to do so. We simply need to make the choice and do it.

The costs borne by individual nations without international cooperation are becoming more difficult to continue, let alone expand their budgets and programs into even more serious and expansive space goals such as the Mars missions and even beyond. We are approaching the limit of what individual nations can shoulder to finance future space goals, particularly as the Earth's climate change may well saddle us with huge natural disaster recovery burdens as never before seen in history.

It thus behooves all nations to realize that what may have worked in the past is fast becoming obsolete in terms of our common wealth, and to actively seek a fundamental shift in their cultural perceptions. This will require serious effort to transform old patterns of thinking and behavior to favor shared programs instead of unilateral efforts focusing on maximum benefit for a much narrower spectrum of interests. For centuries there have been various traditional disciplines which could help us change those patterns and bring us out of this dilemma, and we are compelled by our realization of the truth to use them to do so now as never before.

As Albert Einstein expressed it, problems created at one level cannot be solved by thinking at the same level. Thus, much of what plagues mankind today, and which prevents the sort of cooperation the world indeed needs today, cannot be resolved through the same patterns of thought and action which created them in the first place. It is simply impossible to do so, and we must admit this to ourselves or continue to suffer.

The necessity of peaceful, multinational cooperation has become a major priority.

Worldwide reports of imminent climate changes, evidenced by historically unprecedented shrinking of the Arctic ice [1], and the unstoppable melting of the vast, methane-rich permafrost of western Siberia [2], which holds 70 billion tons of until-now-frozen methane and has begun its first melting in more than 10,000 years, clearly suggest that the hurricanes which struck the Gulf Coast in 2005 were a portent of worse weather coming, which will cost the United States and other nations dearly. Budgeting for space must of necessity give way to coping with such disasters.

A shift in thinking toward a general sense of "we" among all nations is needed as never before in history, or few of us may survive what is coming. It is our destiny to thus change our minds both for our own benefit and that of all life.

Thus, as our own perceptive patterns change, science itself could very well change dramatically, and in so doing stretch the possibilities before us, and the requirements to meet them.

In a world that is arguably dying from the over-use of fossil fuels and other accepted methods of energy production, we must consider any inventions not founded in conventional thought. All claims need serious review and testing in order that we all may survive on Earth and in space, where it is entirely possible that inventions such as Tom Bearden's Motionless Electromagnetic Generator [3] may provide us with long awaited safe solutions to our vast energy needs.

We cannot afford to leave any stone unturned when it comes to investigating possible energy sources, just as we cannot afford to continue thinking as we have for so long about each other. Our lives and those of our children depend on our rapidly gaining the wisdom to recognize the destructive path that we have trod so far and forever change it for our mutual benefit.

Expectations of Collaboration:

It's not that there is no value in protecting what we have - we are of course absolutely obliged to protect it. But we have taken a road which, by forcing others to submit to our will, we fail to protect. We have no intrinsic need to continue to rely on such force, particularly when its cost is so high when we can ill afford it, and when new disasters appear at our doorstep that cost us so dearly. War is increasingly losing its value as a viable option when natural disasters affect the lives of so many, and shall continue to do so because we ourselves are generating the very causes of them. Shall we fight with each other as the ship upon which we stand slowly sinks, and we all drown? It is our choice to make.

Ultimately, if all nations remain on the same path, the world will find itself in an international conflict as never experienced before in mankind's history, a conflict over conventional sources of energy and the territories that hold them. This continuing conflict will drain the resources of every nation involved, weakening them to the point of potential extinction, while any dreams we have held about space exploration and research other than for the military use will find no home, and the entire world will suffer from the insanity of conflict as never before.

We do not need to do this - any of us. We can work with each other around the world to restore our planet as a safe home for all of us and our children, and pursue unprecedented projects together internationally on Earth and in space to demonstrate those better, creative, adventurous, courageous, inventive, nourishing, enthusiastic, giving, embracing, inquisitive, loving, bold qualities in ourselves which are so capable and willing to perform, given the opportunities to do so. We can indeed make our dreams of the world and space real, and do so peacefully.

It is certain that we cannot continue with the patterns we have been manifesting for so long; we must realize how mistaken those patterns are for our own lives and those of our children, and change how we live and treat each other worldwide. Such a change opens the possibility for the fulfillment of a destiny in space that will just as certainly leave us breathless in its wonders and rewards. We can share such dreams because they are not outside the realm of possibility, and we can choose to experience them in reality and pursue them with all our hearts, minds, and bodies. We must endure in peace and make it so.

It is our choice to make with the greatest sense of conscience, responsibility, and love toward each other which each of us can muster, because the future depends on our doing so. The stars do indeed await us for the time when we have earned the right to them by our own thoughts, words, and deeds, when we at last share our destiny as members of the Divine Cosmos. Look into the heavens on a clear night and you will see our heritage. Trust in this and make no mistake about it.

Are we to take the challenge with the clarity and courage to carry on into space, other planets their moons, and the stars, or are we to toss our heritage aside while bickering our way to oblivion? Are we incapable of cooperating with each other to our mutual benefit, because we cannot break free from our deeply ingrained historical patterns of pettiness, mistrust, and lack of vision that fall upon us? I find it difficult to pose the question because the answer to this is so self-evident, yet it must certainly be asked in the context of the world as it exists today.

✳✳✳

Other questions requiring serious research for our future:

What aspects of the human psyche pose the greatest barriers to creating a successful worldwide society based on mutual responsibility, enthusiasm, cooperation, and commitment to the general welfare of all participants?

How do we successfully prevent them from continuing to prevent a fully functional worldwide society capable of cooperative existence?

How can we end the destructive cycle of war that has been draining the world of its resources for millennia, so that we can simply live to achieve those things we naturally desire as part of our heritage in the Divine Cosmos?

How do we best support each other in dealing immediately with the imminent climatic changes that could devastate significant portions of Earth's lands and populations, so that we can fulfill our heritage in peaceful international cooperative collaboration?

Postscript

The Dr. Buzz Aldrin and Thomas F. Rogers *"Space Trips for Peace"* concept:

Planners for the future of humans in space have the responsibility to give their plans the best potential for solving serious Earth problems. Peace has always been humanity's most important and illusive goal. Dr. Buzz Aldrin, the Gemini XII and Apollo XI astronaut who shared the first human landings on the Moon with Neil Armstrong, during his astronaut missions conceived the idea of "Space Trips for Peace."

Tom Rogers has held leadership positions in military and civil space matters since the late 1940s. His primary contributions have been in satellite communications and navigation, space transportation, space housing, space solar power, and general public space travel. Tom joined Buzz in informal discussions, as the Third Millennium was approaching, to promote the "Space Trips for Peace" idea.

Space Trips for Peace would create crews composed of members from nations marginally friendly, hostile, or even at war with each other. Space, new to civilization and without territorial boundaries and national sovereignties, would be the ideal frontier for demonstrating that people of all cultural beliefs and religious backgrounds are able to set aside differences and work harmoniously for goals mutually considered good. The space environment would give humankind an opportunity to establish new precedents and create defining moments in the quest for worldwide peace. The selected space crew and travelers would be fully trained and active participants during these highly publicized missions.

References:

(1) http://news.bbc.co.uk/2/hi/science/nature/42903040.stm, http://www.nsidc.org/news/press/20050928_trendscontinue.html
(2) Full "New Scientist" article : http://www.newscientist.com/article.ns?id=mg18725124.500
(3) http://www.cheniere.org

About the Author:

Michael Hannon is a graduate in electronics (analog and digital) and inventor of devices for mundane to advanced applications, including computer control of textile looms, automotive fuels systems, solar arrays, and power tools. He is also an artist, musician, and combat-disabled Vietnam veteran, and has engaged in extensive study of numerous esoteric disciplines.

Chapter 11

Evolutionary Psychology and Its Implications for Humans in Space

By Sherry E. Bell and Dawn L. Strongin

As the human habitat changes, so will its control center—the human brain.

Our view of the future of humans is optimistic. Humans have evolved and will continue to do so, and humans in space are likely to evolve in ways that differ from those who are Earth-based. This chapter examines the underpinnings and extrapolates some expected evolutionary changes. It is meant as a thought-provoking piece.

Innovations in physics and engineering have made space travel a possibility. The relative time and energy allotted to propelling out of Earth's atmosphere has far outweighed emphasis placed on studying the effects of prolonged weightlessness and hypergravity on the human traveler. With the exception of moon travel, space flight studies have been limited to low Earth orbit, and those not exceeding a little over six months. Thus, our views on the psychological and physiological effects of long-duration orbital flights and non-returning spacefarers are purely conjecture at this point.

Introduction to Evolutionary Psychology

The basic premise of evolutionary psychology is that evolutionary pressures effect change in all living individuals. Larsen & Buss, 2005 [9] describe this as: "The basic elements of the evolutionary perspective apply to all forms of life on earth, from slime molds to people" (p. 237). The research goal is to discover and understand the design of the human mind by understanding the evolved properties of the nervous system, especially those of humans, which are most malleable to internal and external environmental stressors (Cosmides & Tooby, 1997, [4]). Evolutionary psychologists attempt to find the "functional mesh" between adaptive problems posed by the environment and inherited architectural changes within organisms designed to solve these problems (Cosmides, Tooby & Barkow, 1992 [5]; Cosmides & Tooby, 1997 [4]).

Because all living tissue is functionally organized and is a product of natural selection, a major assumption of evolutionary psychology is that the nervous system, too, is functionally organized to serve survival and reproductive needs, and is best understood from an evolutionary perspective (Cosmides, Tooby & Barkow, 1992 [5]). Thus, a major lesson of evolutionary psychology is that if you want to understand the brain, look deeply at the environment of our ancestors as focused through the lens of reproduction, and if the presumptions of evolutionary psychology are correct, the structure of our brains should closely reflect the reproductive ecology of our ancestors (Buss, 1999 [2]).

One way to think about the evolution of the human brain is by examining its increasing level of power over the environment. The primitive nervous system evolved over millions of years, permitting humans to become the most adaptable organism on Earth – adapting to, improving, and enhancing a tremendous variety of living conditions from wind-swept deserts to icy snow caps. In the process, "human beings have caused greater changes on earth in 10,000 years than all other living things in 3 billion years. This remarkable dominance is related to the development of the brain from the minute cerebrum of simple animals to the complex organ of about 1350 grams in man" (Sarnat & Netsky, 1981, p. 279 [13]).

Our hominid ancestors had smaller brains than present day humans. Relative brain size has increased over generations, and changes in its structure are a result of adaptation to the environment. Specifically, the neocortex has grown exponentially compared to its origins, and in fact this bulging mass almost completely occludes the more primitive brainstem. The ability to plan and to think abstractly relies on the development of the neocortex, and complexities in environment can be addressed by complex thought and variability in behavior.

The most sophisticated devices our ancestors made are primitive by our standards. They had stone knives and carving tools, and later spears, but we, in contrast, have developed telephones, computers, automobiles, airplanes, rockets, satellites, and a space station, to mention just a few. And of course, humans have traveled to the moon and back. Hence, our abilities to imagine, to reason, and to learn are quite different in degree of those same abilities in our hominid ancestors. Although they could do many of these things, they could not do them with the level of complexity that we now can.

Thus, our present day cultures and institutions, our societies, our very way of life would have been inconceivable for our ancestors, and in much the same manners it is likely that our progeny will be dramatically different from us. Many, for example, will live in space stations, some will live in extraterrestrial space settlements, while yet others will be spacefaring, so we can state that the evolution of the brain's inner space has made outer space accessible.

Humans in Space

On October 4, 1957 the Soviet Union precipitated the space age by sending Sputnik I into orbit. This artificial satellite, no bigger than a basketball and weighing as much as an average man (183 lbs), launched dreams into the reality of space exploration. Four years later, on April 12, 1961, the Soviets sent cosmonaut Yuri Gagarin into space, and virtually every year since then, Soviets or Americans sent humans into space. While the first missions were short-lived, subsequent missions were longer in duration, ranging from a day to over a year.

In August 1961 Russian cosmonaut Titov was the first to spend a full day in space, while the following year American astronaut, John Glenn, was the first to orbit the earth. During the Mercury Program (1961-63), the first six American astronauts in space accumulated less than 54 collective mission hours. In June 1970, Russia's Nilolayev and Sevastyanov spent a record 18 days in space, and in 1971 three Russian cosmonauts worked aboard a space station for 24 days.

During 1972, the longest Apollo mission, Apollo 17, lasted 12 days, 13 hours and 52 minutes. In 1973, American astronauts spent 84 days in space aboard the Skylab. In 1978, two Soviet cosmonauts set a record of 136 days in space, but a year later that endurance record was bested for a total of 175 days in space. The record was broken again in 1982 by Soviets working for seven months on a space station. The Expedition 7 crew lived on the International Space Station for over six months. The longest duration in space was by the cosmonaut Valeri Polyakov, who stayed aboard the Russian space station Mir for 438 consecutive days.

Over the decades, we have learned much from the experiences of cosmonauts and astronauts about the physiological influence of space on human beings. Experiments conducted on Earth and on the International Space Station have provided insights into the effects of weightlessness on the human body, information that is vital to human spaceflight and long-term space travel, as "plans for colonization of the Moon and Mars will move the duration into years and, perhaps one day, into generations" (Griffin, 2005 [7]).

Effects of Gravity

Long-term space travel and multi-generational colonization of planets within and outside our solar system is steadily moving towards becoming a dream materialized, but concerns regarding the psychological and physiological effects of prolonged exposure to microgravity and hypergravity are consequently getting increased attention. As described by Cohen, "gravity shapes life...It defines the character of the nervous system, our reflexes and our bones" (Malcolm Cohen as cited in Shwartz, 2002, para. 7 [14])

Researchers on Earth and the International Space Station have provided a peek into the mechanisms behind these effects, and while the gravitational force on Earth obviously has inconvenient consequences, such as laboring uphill or sagging over-the-hill flesh, the evolution of human physiology has naturally reflected the Earth's gravitational effects.

Microgravity

Orbital spaceflight is not a significant distance from Earth's center, and thus gravity is minimally decreased, creating microgravity. However the astronaut is virtually weightless due to free fall. In other words, the scale used to measure weight and the weight to be measured (the astronaut) are falling (due to gravity) around the Earth's orbit at an identical rate (Turner, 2000 [15]). Gravity remains present but weight is markedly decreased. Therefore the physiological effects studied on the space stations are more accurately described in terms of weightlessness versus lack of gravitational force.

A few days or weeks in microgravity may be uncomfortable, but it is not damaging to the health of astronauts. However, on shorter missions, astronauts have reported severe motion sickness. The vestibular system relies on gravity, so as the head tilts, hair cells in the inner ear are displaced, creating a signal to the brain regarding the head position and balance. When these signals from the vestibular system and the brain are incongruent, an individual feels nausea. However these symptoms abate over time, and astronauts on long-term missions adapt quite readily.

On longer missions, astronauts endure somewhat detrimental effects, such as blood and body fluid shifts, muscular atrophy and bone density loss. Blood pooling is a concern when astronauts have been in microgravity for prolonged periods. Without resistance to gravity, fluids shift from lower extremities to the upper body and face, creating discomfort. When a person stands on Earth, 300 to 800 mL of blood pools to the lower extremities. Standing, or mechanical load, causes muscles in legs to contract, and veins in the abdomen to restrict, forcing blood to circulate back up. Blood pooling is avoided when veins and arteries constrict while the heart rate increases blood flow through one-way valves. However, without gravitational opposition, blood pressure decreases, causing the potential of cerebral ischemia (i.e., blood loss in the brain) and neuronal (i.e., brain cell) death (Pinel, 2005 [10]).

Although rigorous exercise in space prevents lost muscle mass, sustaining bone density requires additional gravitational resistance. Without the appropriate signals precipitated from gravitational force, space-induced bone loss develops. On Earth, bones grow and change shape based on the weight put on them. Resistance to gravitational pull signals gene expression of new bone formation and resorption. Approximately 10% of bone mass is removed and replaced every year for the maintenance of healthy bone. The balance between growth and removal of bone is tipped over a lifetime. Men experience an average of 15% bone loss while women experience an average of 30% loss, mostly after menopause (Turner, 2000 [15]).

Without the signal created from gravitational resistance, bone maintenance is altered. Gene expression signaling the production of bone formation is dramatically reduced with an increase in bone resorption, creating fast bone loss (Wang, 1999 [17]). Astronauts studied under conditions of microgravity lost approximately 1-2% in weight-bearing bones each month in areas such as pelvic bones, lumbar vertebrae and femoral neck (Carmeliet & Boullon, 1999 [3]; Grigoriev et al., 1998 [8]). Because bone grows slowly, replacing it takes a long time, creating a serious health risk for those returning to the gravity of Earth (Bloomfield, 2001 [1]). Notably, the upper skeletal regions were either not significantly changed, or there was a positive trend. Thus as the pelvis and legs begin to atrophy, the upper skeletal regions remain intact or increase in density. As long as the space traveler does not return to gravitational force where bone and muscle integrity are required for gravitational resistance, such changes are unimportant.

That said, many space travelers will settle on other planets or return to Earth. Osteoporosis and muscular atrophy resulting from long-term weightlessness will be debilitating when spacefarers are required to build extraterrestrial settlements having gravity that differs from that on earth, or when they return to Earth's gravitational pull. Artificial gravity, whether close to Earth's 1-*g* or greater, will be an effective countermeasure.

Hypergravity

Hypergravity is gravitational force *greater* than Earth's. Most of us have experienced the sensation of hypergravity. At every county fair can be found a ride in which individuals stand against the interior wall of a circular contraption. As the wall spins faster and faster, the floor gives way, but the individuals are pinned against the wall and do not fall down. This ride is a centrifuge, producing a gravitational pull three times that of Earth. Astronauts currently experience hypergravity at launch (up to 3.2-*g*) and at reentry (-1.4-*g*). Due to the short duration of this force, astronauts have had short-term difficulty adapting to increased gravitation, and have experienced loss of consciousness and nausea.

Because body fluid is heavier during hypergravity, the heart must pump faster and harder to push blood to the brain (Vince, 2002 [16]). An experiment conducted by Arthur Hamilton Smith in 1970 (as cited in Regis, 1991) sheds light on this issue. Smith used the Chronic Acceleration Research Laboratory at UC Davis to study the effects of hypergravity on chickens living in 2.5-g. Chickens, like humans, are bipeds with non-bearing limbs (wings). Hundreds of chickens were put into two 18-foot centrifuges for from three to six months, and during this period 23 generations were bred. The findings showed that at the end of the experiment their bones and muscles were bigger and stronger, while their hearts circulated more oxygen than before.

Artificial hypergravity may serve to counteract permanent losses, and may provide a particular advantage for space settlers who require strength and endurance. Conditioning training before launch will provide the time the brain and body need to adapt to the additional gravitational force before living conditions actually change.

As we noted earlier, long-duration exposure to microgravity creates circulatory problems as well as muscular and skeletal atrophy. Managing the effects of hypergravity will be important for those who will return to Earth after a long stay in microgravity, and for those living and working in space settlements on planets with gravity greater than that on Earth. Under those conditions, both the skeletal and muscular systems will need to alter the way they function.

NASA is currently exploring human adaptability to artificial hypergravity by means of exposing humans to centrifuges for prolonged periods of time. It is likely that astronauts living in conditions of hypergravity will develop larger bones and muscles, and will present with increased oxygenation.

The Plastic Brain

In order for we humans to experience sensations, perceptions, thoughts, movement and basic life functions, neurons, or brain cells, communicate with each other. A network of communication develops as pathways become established. The human brain is remarkably malleable, or plastic. Communication among the neurons can develop new patterns based on the brain's experiences, as the brain is shaped by the characteristics of its environment. These changes can take from hours to years.

The developing brain requires specific input to establish appropriate connections and those connections are based on the need to adapt the environment. For example, a child born blind may over-develop other sensory modalities, such as touch and sound. The visual neurons are not needed, so they may be recruited to become part of the network for processing touch and/or sound information, or they may be recruited by both. The consequence of blindness therefore, can be either a more finely tuned tactile or auditory function, or both.

The primary cortex, a region in the brain involved in voluntary movements, is also capable of modification. Neuroimaging techniques show considerable plasticity of the motor cortex. For example, without the use of weight-bearing limbs, the neurons motor originally devoted to lower body motor control are recruited to support the intact upper muscular and skeletal systems.

Muriel Ross at NASA-Ames Research Center (1996) found rapid neuroplasticity in her study of the vestibular system of rats (who are mammals) living in prolonged microgravity on NASA's Space Life Sciences Mission. Her studies included two experiments ranging in duration from nine days to two weeks. As noted earlier, one of the effects of microgravity is the loss of a sense of body orientation and balance due to the gravitational silence in the inner ear. Within 7 to 14 days, the rats regained balance and orientation. Close examination of the rats' brains indicated neuroplasticity, most evidenced by increased synaptic density among the neurons in the associated regions of the brain. Dr. Ross postulated that the resulting formation of new synapses was an attempt by the system to return to an output which was more adaptive in the absence of gravity.

Neuronal plasticity is thus a mechanism which can be used to compensate for lost or reduced function, or to maximize remaining function in the event of brain injury, or to adapt to dramatic environmental changes. The brain can literally modify itself in response to its environment. In their effort to enhance efficiency, neurons can change their basic structure and their connections to other neurons. This neuronal plasticity does not require generations to accomplish, and in fact, changes readily occur within a short time frame. Such changes will occur in humans as well, for, as the human habitat changes, so will its control center—the human brain.

Conclusion

During relatively extended excursions under microgravity, astronauts and cosmonauts began to adapt. Motion sickness abated over time, but detrimental effects such as muscular deterioration, bone density loss, and cardiac insufficiency were pronounced upon returning to Earth's gravity.

Long-duration, and to a lesser extent, shorter exposure to microgravity induce changes in brain organization, and thus sensory and motor function, skeletal muscle and bone maintenance, and cardiovascular efficiency. These changes are appropriate to the immediate environmental requirements and demonstrate human adaptability. In particular, the adaptability of the nervous system is an important principle underlying human evolution, and studies have shown that the brain has undergone neuronal reorganizations in relatively short periods of time as the brain literally restructures itself to meet the demands of the environment.

Dramatic changes in the environment alter the organization of the brain and function of the body.

The morphology of the brain and body literally changes to accommodate the immediate environmental needs. To accommodate environmental changes individuals who live in a permanently weightless environment will develop changes in their cerebral cortex's subserving sensory and motor functions. The vestibular system, no longer receiving gravitational signals necessary for the brain to decipher body position and balance, will alter. Neurons from other sensory modalities may take over these areas. The resultant brain reorganization may therefore increase sensitivity and enhance function of other systems, for examples, the tactile and visual systems might be among those which are altered.

In addition, recall that when the brain does not receive sufficient gravitational information, the gene expression for the preservation of bone density is altered. The brain assumes that bone formation is no longer necessary in the weight-bearing lower extremities, and thus this function is ceased. Astronauts in space have commented that their legs were not necessary, and were in fact, in their way. Taken together, this could mean that the no-longer-necessary weight-bearing limbs, i.e., pelvis and legs, might atrophy over time. Perhaps selection pressure will thus act to produce generations of humans who have no legs.

Additional changes are also likely to occur. Weightlessness will also cause changes in genetic expression and neuronal connections in the skeletal and muscular systems. Spaceflight induces changes in synaptic organization of the developing neocortex. Unborn and newly born mammals show the greatest increase in brain plasticity. Without gravitational resistance, the circulatory system is impaired in its ability to pump blood from the lower extremities back to the upper body and brain. Just as one would expect, neonates are particularly sensitive to the effects of microgravity during spaceflight, exhibiting such characteristics as underdevelopment of lower limbs (DeFelipe et al. 2002). These underdeveloped limbs may become vestigial, or degenerate over time, thereby resolving the issue of cardiac insufficiency. Hearts that do not need to pump blood into lower extremities need not expend energy doing so, and instead will devote their energies into more adaptive functions. Generations born into the weightless world will thus reflect their environment in their genetic, anatomical and functional evolution.

Under the effects of hypergravity bones and muscles grow bigger and stronger, and hearts circulate more oxygen than before. Humans who live in hypergravity will adapt into beings whose bones, muscles and cardiac systems are larger and stronger than today's humans.

What this means for humans in space is that first-time settlers trained in hypergravity via centrifuge both on Earth, en-route and in the extraterrestrial space settlement, will benefit by maintaining their own conditioning and that of their offspring until such time as evolution adapts the bodies of their progeny to hypergravity. However, for many who do not intend to return to Earth, but remain in extraterrestrial space settlements, or who remain spacefarers, these issues may be unimportant, and their biological systems will have ample opportunity to restructure and reorganize themselves under either microgravity or hypergravity.

Our work to understand the psychological and physiological alterations to humans in space is ongoing. We will continue to explore and write about the changes that are likely to occur and we very much welcome input.

References

(1) Bloomfield, S. A. (2001). Optimizing bone health: Impact of nutrition, exercise, and hormones. Sports Science Exchange 82(14), 3. Retrieved December 20, 2005, from http://www.gssiweb.com/reflib/refs/288/sse82.cfm?pid=60&CFID=2914708&CFTOKEN=17231148
(2) Buss, D. M. (1999). Evolutionary psychology: The new science of the mind. Needham Heights, MA: Allyn & Bacon.
(3) Carmeliet, G., & Bouillon, R. (1999). The effect of microgravity on morphology and gene expression of osteoblasts in vitro. The FASEB Journal 13, 129-134.
(4) Cosmides L., & Tooby, J. (1997). Evolutionary psychology: A primer. University of California, Santa Barbara. Retrieved December 14, 2005 from http://www.psych.ucsb.edu/research/cep/primer.html

(5) Cosmides, L., Tooby, J., & Barkow, J. H. (1992). Introduction: Evolutionary psychology and conceptual integration. In J. H. Barkow, L. Cosmides, and J. Tooby (Eds.), The adapted mind. New York: Oxford University Press.

(6) De Felipe, J., Arellano, J. I., Merchan-Perez, A., Gonzalez-Albo, M. C., Walton, K., & Llinas, R. (2002). Spaceflight induces changes in the synaptic circuitry of the postnatal developing neocortex. Cerebral Cortex,12(8), 883-891.

(7) Griffin, M. (2005). To extend life to there. Unknown original source.

(8) Grigoriev, A. I., Oganov, V. S., Bakulin, A. V., Poliakov, V. V., Voronin, L. I.,Morgun, V. V., et al. (1998). Clinical and physiological evaluation of bone changes among astronauts after long-term space flights. Aviaksom Ekolog Med, 32, 21-25.

(9) Larsen, R. J., & Buss, D. M. (2005). Personality psychology: Domains of knowledge about human nature (2nd ed.). New York: McGraw-Hill.

(10) Pinel, J. P. J. (2005). Biopsychology (6th ed.). Needham Heights, MA: Allyn & Bacon.

(11) Regis, E. (1991). Great mambo chicken and the transhuman condition: Science slightly over the edge. Reading, MA: Addison-Wesley Publishing.

(12) Ross, M. (1996). Gravity sensor plasticity in the space environment [Abstract]. NASA-Ames Research Center. Retrieved December 14, 2005 from http://astrobiology.arc.nasa.gov/workshops/1996/astrobiology/speakers/ross/ross_abstract.html

(13) Sarnat, H. B., & Netsky, M. G. (1981). Evolution of the nervous system (2nd ed.). New York: Oxford University Press.

(14) Shwartz, M. (2002, March 19). Human biology professor seeks participants for NASA hypergravity experiment. Stanford Report. Retrieved January 18, 2006, from http://news-service.stanford.edu/news/2002/march20/centrifuge-320.html

(15) Turner, R. (2000). Physiology of microgravity environment: What do we know about the effects of spaceflight on bone? Journal of Applied Physiology, 89, 840-847.

(16) Vince, G. (2002, March). Hypergravity experiment in a spin. New Scientist 19(27). Retrieved January 5, 2006, from http://www.newscientist.com/article.ns?id=dn2076

(17) Wang, E. (1999). Age-dependent atrophy and microgravity travel: What do they have in common? FASEB Journal 13, 167-174.

About the Authors

Sherry E. Bell, Ph.D. received her Doctor of Philosophy degree in Industrial/Organizational Psychology from Capella University and her Master of Science degree in Clinical Psychology from Capella University. She specializes in Leadership Development and is an affiliate of Peak Strategies, an organization dedicated to Leadership Development, Executive Coaching, Individual and Team Performance Enhancement, Personality Assessment, and Consulting. She is certified in the use of the Myers-Briggs Type Indicator and is currently in the process of gaining Executive Coach certification. Dr. Bell is a member of the following organizations: Aerospace Technology Working Group, The Planetary Society, American Psychological Association, Society for Industrial Organizational Psychology, Society of Consulting Psychology, Golden Key National Honor Society (Lifetime Member), and Psi Chi - the National Honor Society in Psychology (Lifetime Member). One of her aspirations is to become a writer of futuristic humankind. Dr. Bell's email address is DrSherryBell@aol.com

Dawn L. Strongin, Ph.D., is an assistant professor at California State University at Stanislaus, specializing in cognitive neuroscience. She earned an M.S. in Human Neuropsychology and a Ph.D. in Applied Statistics and Research Methods with an emphasis in Neuropsychology and a minor in Educational Psychology at the University of Northern Colorado. Her areas of research include neurotoxic effects of organophosphates and psychometric development. When not teaching, researching and traveling, she is dreaming...of life on distant rocks.

Chapter 12

Space as a Popular National Goal

By David Livingston

Introduction

Space is not the popular national goal that it once was. President Kennedy's pledge to place an American on the moon and return him safely to Earth was a source of national pride and a strong motivator for American space policy, but after that goal was met, the American public's interest in a national space policy waned. Those of us in the space community know that the space program has proven to be a strong force for world peace and stability, as evidenced by United Nations treaties and agreements assuring cooperation and restricting weapons in space that were enacted by the super-powers and most other countries. Today, former enemies are partners in space development, pursuing space commerce and development together.

Those of us in the space community and those involved in the latest advances of technology also know that the American space program, under the auspices of the National Aeronautic and Space Administration (NASA), has produced extraordinary benefits for people all over the world, not just the wealthy or those associated with space enterprises. These benefits extend to the fields of health, medicine, public safety, energy, the environment, resource management, art, recreation, computers, automation, transportation, manufacturing, construction, and many more.

Unfortunately, the many benefits of space are not often understood by average Americans, and only when regular citizens recognize the far-reaching humanitarian advantages, or can personally experience the technological advantages of the space program will a national space policy have broad support.

How space agencies communicate with the public and how the public understands space policy are issues that I examine on a weekly basis through my radio talk show, The Space Show(r). For almost five years, in nearly 450 interviews, I have been talking with the general public and the informed space community about how best to become spacefaring, to expand and enhance space commerce, and to develop an economic space infrastructure. In this chapter, I will share with you some of what I have learned, and I will propose solutions for restoring space to its rightful place as a popular national goal.

Communications 101

The Space Show(r) has taught me how to talk to the public about space. For our society to devote a great deal of attention and resources to space, we as space advocates must be able to communicate with the population at large, not just with those on the inside of the space community or with those who have romantic notions of space travel. Obtaining general support for space as a national goal is not easy, especially during a time of war, federal budget deficits, and political scandals. Still, if we can present a strong, practical argument that space enterprises and explorations are in the best interests of our nation and its citizens, then a popular space policy will surely follow.

Typically, when space is being discussed or promoted, several reasons are given for the need to be in space and to spend public money on space development. Among the reasons most often cited for a continued and expanding space presence are scientific knowledge; environmental monitoring; possible extraction of extraterrestrial resources, including vast supplies of energy; planetary protection from a large meteor or asteroid on course to strike the Earth; settlements that could save humanity if the Earth is somehow destroyed; jobs; and just plain fun. While these are valid, even exceptional, rationales for spending public money on space development, they appear to be distant and not easy for people to assimilate. For example, to say that we need to go to Mars to save humanity in case the Earth is destroyed by an incoming asteroid is not an easy sell to large numbers of people. When this topic comes up on the show, a listener invariably will e-mail or call in to say that it only makes sense if his family is the one chosen to be saved. This objection is mostly tongue-in-cheek, but the point is that it is difficult to get motivated about something when you know at the outset you will likely be the loser. As I have also discovered by hosting the show, not everybody who has an interest in space cares about future generations of humanity.

If the space community wants to make space a popular goal for our country, then we need to begin an open discussion about space. For starters, a productive exchange of ideas is more easily facilitated when representatives of the space industry communicate in plain language to other people or groups. The purpose is not to dazzle or impress others with technical jargon, but to include those outside the space community in the discussion of why it is important for humanity to be in space. Rather than present our ideas as the only important ones in the discussion, we need to both respect and welcome the ideas of others into the dialogue; then we need to listen to what is being told to us, even if it is anti-space. At the same time, we must let people know that benefits from space accrue to everyone in society, not just selected communities or individuals. We need to respond reasonably and appropriately to specific concerns. Our arguments for space will probably not be their arguments for space. We need to remember that the public did not enroll in our lecture series. Other people may have strange reasons for liking or disliking space, and we must be prepared.

I would like to cite a recent Space Show program as an example of a breakdown in communication between a guest and a listener. The guest for this particular show was a well-respected scientist, engineer, and proponent of manned missions to Mars. A listener asked the guest why we should spend taxpayer money to go to Mars. The listener said that there was no economic justification for putting tax dollars into a manned Mars mission right now because Mars would still be there in a hundred years. The listener wanted justification for the hurry-up attitude to get there. The listener further reasoned that if Low Earth Orbit (LEO) is not even cost effective, why not focus on very low-cost space access rather than an expensive project having no economic merit, like a manned Mars mission.

Without thinking, the guest began rattling off the usual reasons for wanting to go Mars, including exploration, adventure, scientific knowledge, and the establishing of a "second home" for humanity should the Earth become destroyed or uninhabitable. The listener was not buying any of it. The exchange between the guest and the listener continued for several minutes. The listener talked about wasting tax dollars on Mars, which, he asserted, would produce little or no economic benefit for the country. The guest continued to cite his preferred reasons and his agenda, all of which the listener could care less about. I was disappointed that neither party really listened to the other, that neither party addressed the issues and concerns of the other. They were talking at each other. Sometimes I think we would rather hear ourselves talk as an expert than listen and learn-and this applies as much to the guests as to the listeners. If space is to become a popular national objective, then those nonexperts outside the space community must be heard.

After pondering this exchange for several weeks, I began asking other guests and listeners how they would have handled this line of questioning regarding public financing of a manned Mars mission. Perhaps the best response I received was from another guest, who suggested that, like the rain forest today, Mars may very well prove to be a productive place for the discovery of new medicines and pharmaceuticals. He said that it might be possible to use what is unique on Mars to significantly improve our health and well-being on Earth, as long as we are extremely careful about returning Martian samples to Earth. This guest then logically added that sickness and disease reduce economic productivity. For a fraction of what is cumulatively spent on these conditions on Earth, if we as a nation would invest some of our wealth in going to Mars, we might well create a national and even global economic boost, let alone realize significant benefits for all mankind. What this guest did was take the listener's taxpayer issue and modify it to show that going to Mars may be a better public investment than not going to Mars. He then talked about the risk-reward ratio of a Mars mission. He said he would ask the listener if the potential reward were worth the risk of a mission failure. Even if the risk were not worthwhile to the listener, the two would have had a useful discussion. They would not have talked at cross purposes, and maybe the listener would have had his mind opened a bit, or at least the issue might have been diffused.

Logistical Barriers Are Not the Real Hindrance

Industry advocates, policy makers, and other adherents of space development frequently cite scientific, financial, and political barriers that prevent the realization of space objectives. While these barriers are formidable, they do not always justify the relatively slow pace of space development. Those who claim that enterprises in space are too risky and that our taxpayer dollars could be spent more wisely create the real drag on our progress in space.

Another recent Space Show interview highlighted this critical perspective on future space projects. The guest for this program was San Francisco Chronicle science writer Keay Davidson, an articulate and well-educated journalist with a strong background in space technology. A former L5 supporter and member, Mr. Davidson has reported on the space community for a long time, both the government program with NASA and the fledgling entrepreneurial, alternative space community. As he suggested in his interview (which can be heard at www.thespaceshow.com for November 22, 2005), he was an early advocate of human endeavors in space-that we would be on the Moon and well on our way to Mars and beyond, performing useful research and making beneficial discoveries in space that would enhance our lives. His focus was, and still is, the betterment of humanity. As time went on, however, a struggle ensued between his fascination with space and his core political values. What happened to Mr. Davidson has actually happened to millions of people across this country, and during his interview he placed the issue square in the face of the space community. He lost his conviction that all that cool space stuff was worthwhile. In other words, space cities and the potential for technical innovation in space once seemed fascinating to him; but now these priorities seem to pale in comparison to basic human needs on Earth. To be clear, Keay Davidson did not say that space cities were worthless. He believes that space cities will probably exist some day, probably long after we're dead. But then he asks, what's the rush? Mr. Davidson further claimed that there was no need to develop space cities as fast as the L5ers (how young we were, then!) once advocated, suggesting that we were all young and idealistic back in the days of L5.

In our after-show e-mail exchanges, Mr. Davidson told me that for two decades he tended to view much of the hype over space exploration as a modern form of bread and circuses, i.e., a distraction from the more pressing issues that face humanity. He confirmed that not just he but others across our nation have disconnected from the space program, viewing it as mostly driven by corporate and political interests who are vying for billions in NASA dollars, especially in electoral vote-rich states like Florida, Texas, California, Louisiana, and Alabama. He pointed out that space has become a type of entertainment that distracts Americans from serious and challenging issues. Nonetheless, he remains interested in and enthralled by space exploration, specifically the robotic space missions. He would send a million robots to Mars before sending humans (again, sound familiar, anyone?), though to him and others sharing his views, space exploration must always take a back seat to what he and others see as the more pressing needs of humanity.

Toward the end of the show I asked Mr. Davidson what it would take to convince him, as well as millions of other Americans, to support space exploration and research as a predominant national objective. His reply was short, saying he wanted proof of how space can improve humanity. He was not interested in more rhetoric or promises. He and millions like him want to see results. They want to see and experience on a personal, national, and global level the benefits of space that we talk about, promise, and say are possible, but are not yet available to us. He wants to see better prospects for human life on Earth, not just read about it or hear it promised by government, space, and business officials.

Keay Davidson was not talking about complex and challenging technological space barriers, economic issues concerning the cost to orbit, or the rigidity of the laws of physics. He was talking about political and human failures that dampened his enthusiasm for space development. For those in the space community, both public and private, if you are willing to listen and learn from Keay Davidson, you will certainly change the way you advocate space as a national priority.

We all know the barriers to space commerce and development. We talk about them and discuss the solutions among ourselves all the time. We even address the problems Mr. Davidson addresses. But for the public, Mr. Davidson has nailed it. We must from time to time step out of our space world and pay attention to what Mr. Davidson and others see. If we cannot adequately respond to what they are telling us, if we cannot find a way to demonstrate that space can answer their legitimate concerns, then-and I know this will be a tough one for us in the space community to accept-perhaps space should not be a national goal, at least not at this time.

Sometimes Facts Just Don't Matter

All of us have heard the arguments against spending public money on space programs. The arguments are based on assumptions that the public's money can be better used for something on Earth, usually a program that is important to the person engaged in the argument or discussion. After hearing this complaint again and again, I decided to research what does happen to public money spent on space compared to public money spent on other federal programs.

My research results support the creation and funding of high-quality publicly financed space programs. I compared a successful public infrastructure revenue-generating project, Hoover Dam; a successful entitlement program, the federally funded School Breakfast Program (SBP); and a successful public space program, Apollo. I wanted to see what impact such programs had and still have on the economy. I wanted to know if there was any validity to the many claims suggesting better uses of taxpayer money than spending it on space.

According to David Moore, P.E., for its first forty years of operation, from its opening in 1936 through 1976, Hoover Dam had an economic multiplier of 2.29:1. [1] For every $1 dollar spent by the government on the project, the government received $2.29 from power generation revenue. It is noteworthy that this number does not include any value represented by regional economic development, which is ongoing today, nor does it represent added value for the contribution of Hoover Dam to our winning World War II through the abundant production of power to Pacific Coast shipyards along with producing major supplies of magnesium which was used for war munitions. [2] Hoover Dam was originally financed by 4% bonds in the amount of $165 million. The bonds were paid in full and retired two years prior to their due date. Hoover Dam is now part of a larger regional power authority, the Western Area Power Administration (WAPA), which is part of the Boulder Canyon Project. Though WAPA financial losses have occurred, the overall benefits and contributions of Hoover Dam to the nation and the region remain undisputed except by a few who challenge large dam construction projects for environmental and other reasons.

The federal School Breakfast Program (SBP) is an entitlement program that provides breakfasts to certain children of primary school age. To qualify, children must come from low-income families; suffer particular hardships at home; or face long, difficult transportation to school early in the morning. Through SBP, more than 8.1 million children regularly received breakfast in 2002. [3] The public's investment in this program through congressional funding and allocation is now in excess of $1.63 billion. [4]

Determining the economic multiplier for SBP proved illusive because it is an entitlement program designed to meet social objectives, not generate revenue, create jobs, or support industrial development. Following the money flow of an entitlement program would not provide the same type of data as the other two projects. While the program's funding can be thought of as an investment for the nation, several very different issues would need to be evaluated over a long period of time to even attempt to determine a potential economic return for the country. Also, the choosing of a control group might lead to ethical questions. Nonetheless, I decided to use an entitlement program for comparison in my mini-study because of the common objection to public space funding, "The money would be better used for an entitlement program."

In selecting Apollo for the space program, I discovered that not all public space programs are created equally. Apollo was and still is the most successful of the public space programs. The economic multiplier for Apollo varies from 5-7:1, returning between $5 and $7 to the taxpayer for each $1 invested in the program. [5] Estimates reach 14:1, 20:1, and even 340:1 for Apollo's specialized technologies and spin-offs. Unfortunately, the higher multiplier returns cannot be substantiated by the Government Accounting Office (GAO) because of the excessive number of variables in the econometric equations used by the economists. [6]

Other facts regarding the Apollo program include the creation of more than 400,000 jobs at its peak in 1965. What is so interesting about Apollo is that it continues to have a positive impact on our economy today, long after the program was shut down, funding and all. I like to poll Space Show guests to determine what caused them to enter a space- related field, and the vast majority tell me about Apollo's impact on their education, their careers, and even their outlook in life. This phenomenon exists as well for the guests that I have interviewed living outside the United States. Were the SBP funding and were Hoover Dam's repairs, maintenance, and upgrades to cease, within a short time the economic and humanitarian benefits of the project would also come to an end. Conversely, although Apollo ended in 1974, the program still generates significant benefits. This continued return of benefits distinguishes Apollo from other federal programs whose benefits cease once funding, maintenance, and improvements are discontinued. The economic, educational, and inspirational results from Apollo were most likely unintentional. But I believe the continued role of Apollo as a wealth builder for individuals and the nation sets Apollo apart from the other programs in this study. Apollo stands alone in what it has accomplished and in what it is still accomplishing today, scientifically, economically, and inspirationally.

Ironically, as I began sharing my findings with Space Show listeners, attendees at space conferences, and the public, I noticed that while people were interested, the facts made little difference in their opinions about space. People didn't seem to care much about facts as they are driven far more by their agendas, beliefs, and ideology.

What Constitutes a Quality Public Space Program?

From my study of the Hoover Dam, SBP, and Apollo programs, I have been able to identify certain characteristics that make for a quality public program, especially a space program. Understanding these attributes can help us to reach more people with today's space programming, so long as we communicate effectively and encourage the American public to become involved. Though some of the initial success of Apollo was spurred by the zeitgeist-the program was jump-started by a charismatic President Kennedy when the United States was considered to be lagging behind the Soviet Union in space, science, and math-there were other characteristics of Apollo that contributed to its own success. Based on my analysis of the Apollo program, I've listed eleven essential characteristics that make for a quality public space program:

1. Vision. The program must be visionary, and the vision has to speak to everyone in the country. It also has to be obtainable. Individual citizens should be able to easily claim ownership of the vision.

2. Inspiration and motivation. The program must inspire and motivate all people, not just those working on the project or involved in the industry.

3. Leadership. The program must have strong, ethical leadership. The leadership must drive the program to reach its objectives. The leadership must be accountable for all results.

4. Clear goals. The program must have a strong purpose with clearly stated and obtainable goals.

5. Quantifiable results. Progress must be quantifiable and measurable so that successes and failures can be evaluated. This needs to be an open process, not a classified process. While leadership alone assumes the final responsibility for the project, everyone associated with the project has to have accountability for his distinct role.

6. A good investment. The project should be able to pay for itself over time by producing wealth for the country and its people. All cost data must be disclosed and variables clearly identified. All accounting must be transparent. The project cost should be viewed as an investment in the future of the nation, bringing many different payback possibilities from a variety of different sources.

7. A plausible challenge. The program has to be challenging, but also plausible.

8. A service for the people. The program needs to serve the American people, not the other way around.

9. Economic benefits. The economic benefits must be apparent and broad based. Special expertise or status as an insider need not be a requisite for understanding or experiencing benefits.

10. A nonpartisan orientation. The space program must be nonpartisan. It needs to appeal to conservatives, liberals, and independents, and it needs rise above the political fray. This is especially important today in our polarized political environment.

11. Ongoing benefits. The program should be designed and managed so that benefits continue to flow even when the program has been closed down. In this way, future generations will be motivated and inspired to continue the progress of those who came before them. The fact that the program should continue to produce economic benefits after funding has ceased is fundamental to bringing the people a high return on their investment and for creating important national and individual wealth.

Why Not Market the Space Program?

The marketing of space programs is a frequent topic on The Space Show(r). Some guests assert that NASA and the Jet Propulsion Laboratory (JPL) do a lousy job of promoting, marketing, selling, and educating people about the true value of space and space programming. NASA and JPL advocates, on the other hand, are usually quick to point out that the law forbids a federal agency from promoting or advertising itself. These agencies prefer to win supporters by coalition building rather than formal marketing.

Clearly, there is a preference to focus on science and mission facts when talking about and promoting space programs. Space Show guests and listeners alike are always making the case for both science and marketing because one tactic does not have to preclude the other. As for promoting or advertising a governmental program, listeners are quick to point out that the armed services always use slick, professionally created ads to promote themselves in print, radio, and TV. Surely a balance can be struck to excite Americans about the space program and still obey existing laws.

In earlier days, NASA employed a different style of communicating its purposes to the public. Werner Von Braun made himself available to the American people. For example, the July 1971 issue of Popular Science has Von Braun on the cover with the headline "Apollo 15: What We Will Learn This Time." His essay in that issue is a direct, simply stated presentation to the people about the Apollo mission. Another recent Space Show guest Jim McDade-a space education outreach professional, media pro, and space historian-suggests that the current NASA administration has forgotten how to talk to people because the organization has become arrogant, even in its educational materials. The promotion of space as a national goal is difficult enough without NASA creating a distorted image.

Another example of how space could be better promoted involves the Mars-themed Space Show program mentioned earlier in this chapter. During that particular program a listener in Nebraska asked about marketing the Mars mission and Vision for Space Exploration (VSE). He couldn't understand why NASA or JPL didn't focus on the controversial Cydonia region of Mars and the possibility of life on the red planet to stimulate the interest of the American people.

Millions of Americans would be fascinated by a mission to investigate the Cydonia region of Mars. For a variety of reasons, however, Cydonia causes problems for scientists probably stemming from the 1976 Viking pictures showing what appears to be a face on the Cydonia plateau and the "Face on Mars" subculture that has arisen as a result. The caller to the program was not emphasizing Cydonia for the sake of Cydonia; he was recognizing that Cydonia had marketing power for the millions of people who would lend support to an important public space program. Instead, these millions of people-who, by the way, pay taxes to help pay for the space program-are ignored because scientists at NASA, JPL, and elsewhere discount the validity of anything important at Cydonia. The guest for this show replied to the Nebraska listener that it would not be ethical to promote Mars by marketing Cydonia because nothing of interest was at Cydonia. Disregard of public interest, as in this example, is one of the problems that I often hear expressed on The Space Show(r). It goes something like this: "The scientists and mission planners are elitist, out of touch with society, and we are getting ripped off. We have no input into the system, so why should we care about it. They don't care about anything other than what they deem to be important." In the almost five years that I have been hosting The Space Show(r), the frequency of this complaint has not diminished!

I am not suggesting even for a minute that science bend to the public will, because that would severely impair scientific development and advancement. To pay attention, however, to the millions who are curious about Cydonia is not an ethical issue or a threat to science. Early Mars missions need to locate resources, places for settlement, and determine the crucial landing points for future Mars exploration. But it would not be a stretch to say to Americans: "We know that many people have an interest in the Cydonia region of Mars. While we doubt that the region contains any vital scientific information, because the space program is for all Americans we will visit Cydonia to determine what's there." To say and do something like this is not a breach of ethics. It is inclusive; it brings people into the program. Such an approach is not alienating; rather, it builds coalitions. It is smart, low-key marketing and public relations. But more than that, it shows that the space agencies do care what Americans want. It also shows that the space agency realizes that this segment of the population should be included in the process.

The Nebraska listener's second point was to use the probability of life being discovered on Mars to interest people. Again, this likelihood is downplayed by NASA and JPL, thus alienating a large part of the population. There is very strong evidence of water on Mars, and this may well suggest some sort of primitive biological life in the past or even now. Credible scientists have also studied Martian photographs and found amazing evidence of fossil-like rocks. Again, why not market a future Mars mission to investigate all this, to include the interests of millions of Americans who would like to feel as if the space program were their space program? Instead, millions of Americans must accept the directives of public agencies and they feel alienated by it all.

I am fully aware of the controversial nature of these suggestions. But I've also learned that Americans need to be sold and want to be sold. The public has an abiding interest in space. Let's find a way to reach people without sacrificing the long-term goals of the space program.

Summarizing and Moving Forward

To elevate space exploration and development to a national priority will be a challenge. The goal implies a return to the enthusiasm for space in the early 1960s, but 2006 is undeniably a different time with different conditions. Because so many compelling interests compete for scarce public resources, we must clearly demonstrate how the space program reflects real long-term value.

In this chapter, I have shared some of what I have learned as host of The Space Show(r). You may be curious to know what others in the space community have said in response to this information. I can sum it up quickly: The feedback has been both positive and defensive. All sorts of reasons have been given to discount what I have shared with you, among these are that The Space Show(r) audience is too small to be statistically significant, that as a host I am too negative, that I seek out these types of listeners, or that The Space Show(r) simply appeals to opinionated people. Yet another thing I've learned is that because the space community is not always open minded, and I can expect rationalizations to deflect unwanted information.

The Space Show(r) has changed me, not just in how I think and see space development, but in all sorts of ways I never imagined. I know how easy it is to be seduced by space-its promises, toys, and adventure, and learning to be grounded in reality is, I believe, an important quality for spacers to possess.

For me, the proof of what I say is simple. When will the United States be comfortable with its role as a space-faring nation? How much real support is there for the Vision for Space Exploration? Will it continue when President Bush leaves office? Will Congress continue to finance the space program, and at what level? What about the embryonic New Space Industry? Will it confirm the existence of markets and demand? If so, when? Will NASA open up its programs to be more inclusive for both the developing New Space Industry and the American people? Will space programs start connecting with more Americans? I believe that we can have positive answers and outcomes to many, if not all, of these questions, but I also believe that we need to change the way we do things, talk about space, reach out to people, run our education programs, and prioritize space as a national objective.

If we continue with business as usual, I suspect that we will see some advancement, mission successes, and positive headlines, but a relatively slow evolution in our space program. But if we learn to reexamine the benefits of space development and bring more people into our vision, I know that space will finally become the important national goal that it deserves to be.

References

(1) Moore, David, PE; "The Hoover Dam: A World Renowned Concrete Monument," 1999, www.romanconcrete.com/docs/hooverdam/hooverdam.htm, accessed 9 July 2004.
(2) "Hoover Dam," Eugene School District 4J, Eugene, Oregon; http://schools.4j.lane.edu/spencerbutte/StudentProjects/Rivers/hoover.html, accessed 9 July 2004.
(3) "The School Breakfast Program," Nutrition Program Facts from the Food and Nutrition Service, p. 2, http://www.fns.usda.gov/cnd/breakfast/AboutBFast/FactSheet.pdf, accessed 29 December 2004.
(4) Ibid, p.3.
(5) "Apollo Program Benefits," ftp://ftp.seds.org/pub/spacecraft/APOLLO/Apollo.benefits, pub. June 1989, accessed 29 December 2004.
(6) Kraemer, Sylvia K., *R&D Investment and Economic Growth in the 20th Century*, March 26-28, 1999, Haas Business School, University of California - Berkeley, "Policy and Practice: Patenting NASA Inventions," NASA office of Policy and Plans,

About the Author

Dr. David Livingston is the founder and host of The Space Show(r), the nation's only talk radio show focusing on increasing space commerce, developing space tourism, and facilitating our move to a space-faring culture. The Space Show is broadcast three times per week on radio and the internet. Past show archives and listening information can be found at www.thespaceshow.com. Dr. Livingston is also an adjunct professor at the University of North Dakota Department of Space Studies, both on campus and in their distant learning program. He earned his BA from the University of Arizona, his MBA in International Business Management from Golden Gate University in San Francisco, and his Doctorate in Business Administration (DBA) also at Golden Gate University. His doctoral dissertation was titled Outer Space Commerce: Its History and Prospects.

Livingston has spoken at or had his papers presented at various international space conferences, including Space and Robotics, the Mars Society, the Lunar Development Conference, the IAA 2000, the Cato Institute, the National Space Society Conference, the World Space Conference in Houston, and more.

His lecture topics include business ethics and corporate responsibility for off-Earth business ventures and New Space Industries, and he has written a Code of Ethics for Off-Earth Commerce.

Chapter 13

Harnessing Bacterial Intelligence:
A Prerequisite for Human Habitation of Space

By Eshel Ben-Jacob

"Man, being the servant and interpreter of Nature, can do and understand so much and so much only as he has observed in fact or in thought of the course of nature. Beyond this he neither knows anything nor can do anything,"
Francis Bacon

Our Best Friends

Eons before we came into existence, bacteria inhabited the then hostile planet Earth. Being the first form of life here, they had to devise ways to counter the spontaneous course of increasing entropy and convert high-entropy, inorganic substances into low-entropy, organic molecules. Acting jointly, these tiny organisms also paved the way for other forms of life by changing its harsh conditions into the life-sustaining environment we know. With their impressive engineering skills, bacteria changed the atmosphere above us to be oxygen rich, and the water and soil to be loaded with nutrients, resulting in the Biosphere that supports all life on Earth [1-5].

Four billion years have passed, and the existence of higher organisms still depends on the unique bacterial know-how that converts between inanimate and living matter. With all our scientific knowledge and technological advances, the ways that bacteria act as Maxwell demons against the second law of thermodynamics is still a mystery. This makes bacteria our best friends on Earth, indispensable friends we simply cannot do without. If we seek a future for the human race in space, we must take bacteria along for the ride, as none other can prepare the setting for us. They will quickly learn how to thrive in any new environment, and make use of whatever it offers to synthesize life-sustaining organic molecules and to recycle waste products for further use.

Our Worst Enemies

But, as we know, the same best friends are also our worst enemies. In our rush to free the human race from deadly bacterial diseases, we created a major health problem worldwide: bacteria are becoming increasingly resistant to antibiotics. Unaware of bacteria's cooperative behavior and social intelligence, which allow them to learn from experience to solve new problems and then share their newly acquired skills, we recklessly used, and still use, antibiotics to fight them. As a result, we are now witnessing the resurgence of strains of disease-causing bacteria believed to have been vanquished long ago; only now they come armed with multiple drug resistance, and we can't invent new drugs fast enough.

And it gets worse when we venture beyond Earth. The very same conditions that suppress humans' immune systems and general well-being seem only to awaken bacteria to the challenge: they become smarter. Since bacteria are inevitable companions in human space exploration, as it is impossible to sterilize the crew and spacecraft, space programs dedicate much research effort to the effect of space flight on microorganisms and on host-microbe interactions. It turns out that exposure to microgravity and to a high level of radiation causes bacteria to undergo faster genetic alteration while adapting to new conditions. New pathogenic strains may develop that way, some of which could possibly be very different from any we have ever encountered on Earth, and for which we have no ready response. Bacteria can also cause much damage by ingesting electronic components and communication lines, blocking pipelines with stable biofilms, and producing toxins.

The New Strategy

If we want to survive the challenges that bacteria pose to our space voyages, we need to learn how to outsmart them, how to harness their adaptation capabilities to our benefit, and at the same time to tame those capabilities to reduce the health risk.

For this to happen, we must first understand that bacteria are not the simple, solitary creatures of limited capabilities they were long believed to be. These most fundamental of all organisms are smart, cooperative beasts that use advanced communication to lead complex social lives in colonies of enormous populations [4,6-17]. They know how to glean information from the environment, talk with each other, distribute tasks, generate collective memory, and turn their colony into a massive "brain" that can process information, learn from past experience and might even create new genes to better cope with new challenges [18].

In this chapter, I will try to demonstrate how we can go about recruiting bacterial intelligence to our Space quest. I suggest sticking to a model that has already been proven to work once: let bacteria go there first to prepare the ground, and then go ourselves with them at our side. To prepare the pioneering troops, bacteria should go through a training program during which they will be exposed, here on Earth, to a variety of conditions they are likely to encounter in the new places we intend to inhabit. And if we gain understanding of how bacteria learn from experience to solve unfamiliar problems, we can even speed up the training process by boosting their intelligence in ways much like those we employ to improve our own intelligence.

Under harsh conditions, these versatile organisms work as a team and benefit from the power of cooperation. By acting jointly, they can live on any available source of energy and thermodynamic imbalances the environment offers, from deep inside the Earth's crust to nuclear reactors, and from freezing icebergs to sulfuric hot springs. Since we do not know to do the trick, we will need bacteria to operate as Maxwell demons at our service to sense the new conditions, perform their secret information processing to extract the biologically relevant information, and then to convert the local matter into breathable and edible substances for us.

The Engineering Skills Of Bacteria

The idea that bacteria act as unsophisticated, solitary creatures stems from years of laboratory experiments in which they are grown in Petri dishes in benign conditions. They can be tempted to reveal their tricks by, for example, growing them on nutrient-poor hard surfaces. The bacteria you see in fig. 1 coped with this situation by collectively producing a lubricating layer of fluid, which allowed them to swim on the hard surface. As they swim, the individual bacteria at the front push the layer forward so as to pave the way for the colony to expand. By carefully adjusting the lubricant viscosity, the bacteria stick together and keep the colony dense enough for protection [4,6,15-17].

Under conditions somewhat more favorable to motion, such as softer substrate, the bacteria engineer radically different classes of colony patterns. In this situation, the branches exhibit macroscopic chirality, always curling in the same direction (handedness). Accompanying the colonial structure is a designed genome change; the bacteria are now programmed to become much longer, which helps them to move in a coordinated motion within the branches [4,6,15-17].

To achieve even greater efficiency, bacteria invented the clever mechanism of chemotactic signaling, in which the individual bacteria send chemical messages to tell their peers in which directions to move. For example, when detecting a rich source of food they call their peers to join the meal by sending attractive chemotactic signals. On the other hand, bacteria that detect regions of low food or harmful chemical imbalances send out a repulsive chemical to signal the others to stay away [4,6,15-17].

Using these self-engineering strategies, the individual cells collectively manipulate the overall colony organization for the group benefit, as is reflected by the tantalizing colonial patterns shown in Fig.13.1. (Also see color section.)

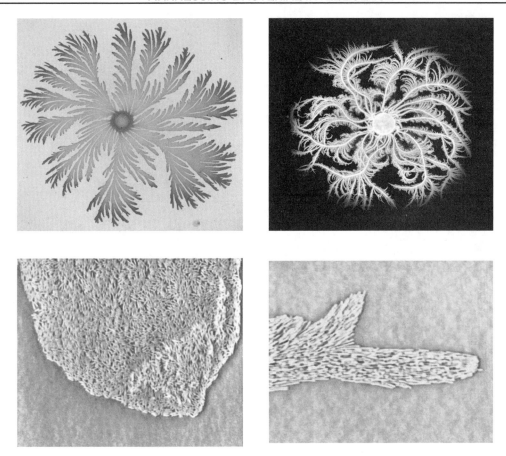

Fig. 13.1: Cell Colony Organization

Patterns of Paenibacillus dendritiformis bacteria form when grown on nutrient-poor, hard substrate. Far from being shapes of mere aesthetic beauty, these colonial structures reflect the self-engineering skills of bacteria. The spreading patterns help the colony access more of the scarce food in the most efficient way under the given conditions. Ordinary branching pattern is shown on the left (a), and the chiral one (with broken left-right symmetry) is shown on the right (b). The top pictures show the colony patterns. Each colony is a few inches in size and has more bacteria than the number of people on Earth. The bottom pictures (c) and (d) show the individual bacteria (the small bars) at the branch tips with x500 magnification for (a) and (b) respectivly.

Clearly, bacteria cannot contain in their genes the information for creating all the patterns they might need to survive in unexpected situations. Well, they don't need to; they only need to have coded genetic information to provide them with the strategic design principles and the tools for communication, for information processing, and for changing themselves accordingly. Using these tools, they can design new creative shapes [4,6,15-17].

Bacterial Communities

Bacterial engineering creativity is further manifested when forced to grow on very hard surfaces. The colony is now formed from new building blocks - the vortices shown in Fig. 13.2. It becomes much like multi-cellular organisms, with cell differentiation and distributed tasks.

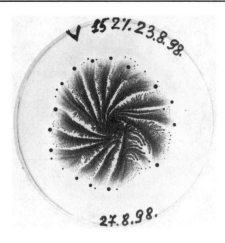
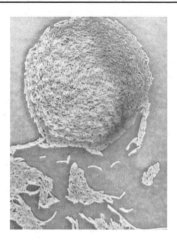

Fig. 13.2: Bacterial Engineering Creativity

Patterns of the Paenibacillus vortex are formed during growth on very hard surface. In these colonies (a), foraging vortices of rotating bacteria shoot out to conquer the hard agar, lubricating the way for their followers. The dynamics is fascinating: a vortex (b) grows and moves, producing a trail of bacteria and being pushed forward by the very same bacteria left behind. At some point, the process stalls and this is the signal for the generation of a new vortex behind the original one; the latter leaves home (the trail) as a new entity toward the colonization of new territory.

In fact, bacteria can go a step higher and form a community (biofilm) of many cooperating colonies [8,9]. Each colony in the community acts as an organism that communicates with the other colonies for coordination and distribution of tasks, for the benefit of the community as a whole. To have an idea of the complexity involved, let us look at our oral cavity, which hosts a biofilm composed of hundreds colonies of different bacteria species, each consisting of tens of billions of bacteria.

Yet bacteria of all those colonies communicate for tropism in shared tasks, coordinated activities, and even exchange of relevant genetic bacterial information. For that to happen, cells should be able to talk and make sense of chemical messages they receive within a chattering of a huge crowd that is about thousand times larger than the number of people on Earth. In linguistic terms, the cells have multi-lingual skills, and each cell should be able to identify messages from its peers to the colony but at the same time also understand some of the messages from other colonies.

For that, bacteria have developed intricate chemical signaling mechanisms using a broad repertoire of bio-chemical messages - from simple molecules to "cassettes of genetic materials" (plasmids). More recently, it was realized that to conduct social life, bacteria use the chemical messages much like a language, including the seman-tic (the assignment of meaning or interpretation of messages) and pragmatic (conduction of a dialogue) levels of linguistics [15].

Learning From Bacterial Intelligent Mating

Bacterial communication methods have endured through evolution and are now crucial for the successful survivable of almost all organisms. For example, chemotactic signaling is the mechanism used for the wiring of our brain during its embryonic development. Sperm cells navigate towards the egg also using the methods of chemotactic signaling invented by bacteria. Even the pheromones used to attract partners and for mating have evolved from the pre-pheromones bacteria made for courtship before conjugation.

Some bacteria assume the role of Information Keepers for the well-being of the community. The stored infor-mation, say resistance to antibiotics, is given to other bacteria by mating (conjugation, which is direct injection of genetic material from the giving to the receiving cell). Prior to mating, the information-carrying cell sends courtship messages to make potential partners aware of the valuable load it carries [8]. A bacterium in need of that information responds by sending pheromone-like peptides to declare its willingness to mate and what it can offer in return. Then, if the other cell accepts, it begins foreplay by emitting competence peptides to modify the membrane of the partner cell to form a pipe like segment. The latter is then used for direct injection of the genetic information. During this process, some valuable proteins and metabolic materials are given back in return for the information.

Bacterial Intelligence - A Metaphor Or Overlooked Reality

In fact, bacteria also invented the foundations of intelligence and sociality [5,15]: to generate individual and group identity, to recognize the identity of other colonies, to make individual decisions, perform information processing, to conduct a dialogue for collective decision-making and to generate common memory to learn from past experience..48 The use of the term bacterial intelligence reflects the recognition that these features are the fundamental (primitive) elements of cognition that any living being must possess [15,19].

However, based on Gödel's theorem, it was suggested that for an organism to be intelligent the genome must function as a cybernetic system -perform information processing to solve problems and then change itself and even create new genes according to the outcome of the computation [18,19]. Some special strains of ciliates provided direct proof that indeed the non-coding part of the genome (Junk DNA) can perform computer-like computations and also build new genes [20].

This daring idea about the cybernetic genome received strong support from the findings of the human genome project. I refer to the facts that our genome has only 30,000 genes while some bacteria strains have 8,000 genes, and that about 10% of our genes came almost unchanged from bacteria. These discoveries led the Celera team to exclaim in wonder:

"Taken together, the new findings show the human genome to be far more than a mere sequence of biological code written on a twisted strand of DNA. It is a dynamic and vibrant ecosystem of its own, reminiscent of the thriving world of tiny Whos that Dr. Seuss' elephant, Horton, discovered on a speck of dust . . . One of the bigger surprises to come out of the new analysis, some of the "junk" DNA scattered throughout the genome that scientists had written off as genetic detritus apparently plays an important role after all."

Bacteria can perform the same "tricks," and more. A colony as a whole, by using its communication capabilities, the exchange of genetic messages and the internal information processing in each of its hundreds of billions of cells, acts as a massive brain [18]. This colonial brain, with more-than-supercomputer capabilities, can generate new genes to allow the bacteria to cope with new environments.

Learning from Past Mistakes - Respect Your Enemy

Overlooking bacterial intelligence, we made a colossal mistake - the hasty use of antibiotics for people and especially in agriculture [16]. Now we understand that by doing so we led to a surprising evolutionary phenomenon in the microbial world [18,21]. New strains of more sophisticated bacteria are rapidly appearing. These new strains have multiple drug resistance and, moreover, can learn to develop resistance to new drugs at an alarming rate. In effect, what we did is to boost bacterial intelligence by forcing them to cooperate and challenging them with increasingly harder problems to solve (advanced drugs to cope with).

This cardinal mistake will inspire us to elevate bacterial cooperation and boost their intelligence when they are tricked into functioning for our benefit. At the same time, we can also learn from this mistake to tame bacteria and limit the risks posed by them during space exploration.

Taming the Bacteria

Studies show that bacteria fare well aboard spaceships. In microgravity, bacteria's growth rate increases and large, stable multi-colonial biofilms are formed. The combination of microgravity and strong radiation seems to elevate bacteria's intelligence and creativity.

Another expression of bacterial increased creativity is the heightened resistance to antibiotics [22,23]. The nature of this mechanism of resistance is unknown to us, but we can deduce that it is specially designed to fit spaceflight conditions, as resistance is quickly lost upon return to Earth. Hence it is reasonable to expect that the acquired resistance is not just a result of a higher rate of random mutations, but rather the outcome of adaptive self-improvement processes.

Bacterial increased resistance, together with human immune suppression, is one of the more severe risks faced in space. Only by understanding the foundations of bacterial intelligence we will be able to outsmart them and protect our health. For example, we should not wait for resistant strains to appear following the use of antibiotics and then try to develop new antibiotic drugs to replace the old ones; it may be too late! Instead, we should use the old war strategy of cutting the enemy communication lines, or even confuse it by sending wrong messages. Doing so, we will render the bacteria more vulnerable to the antibiotic drugs, which will allow us to use lower doses. At the same time, in the absence of communication, the bacteria will not be able to process information for the redevelopment of new resistant strains.

A simple and direct way to interfere with communication could be by blocking the receivers (membrane receptors) they use for reading the incoming chemical messages. Or we can block the transmitters (membrane channels) used by the bacteria for broadcasting the chemical messages.

Once we learn to talk bacteria language (identify the different molecules that convey different messages), we can turn to more efficient and elegant tricks. For example, we could paralyze the bacteria by spreading the message "let's sporulate." Or we could drive them to engage in purposeless group sex rather than go about their routine tasks and protect their colony. This would be done by exposure to a "love potion" concocted from both "male pheromones" secreted by bacteria on the look for partners to transfer genetic information to and "female pheromones" broadcast by bacteria seeking conjugation with a bearer of needed information.

These are just two of the many creative ways we can scheme to use bacteria's intelligence against them. However, for this to happen we must first learn to decode the chemical messages that bacteria are using.

Keeping Our Immune System in Shape

While bacteria flourish in space, human physiology and general well-being decline. Human immune suppression in space was first observed in the 1960s and 70s during the Apollo missions. Notably, 15 of 29 Apollo astronauts reported a bacterial or viral infection during a mission, immediately after, or within 1 week of landing back on Earth. In 1991, experiments on STS-40, the first US space shuttle mission dedicated to medical research, identified T-cells as the particular components of immune function that were compromised [23,24].

Most of us are aware of the important role of the immune system in protecting us against external invaders - microbes and viruses. Less appreciated is the immune system's role in maintaining the synchronization and harmony between our different cell types, each with its own self-identity and specific tasks. Coordination and enforcement of disciplined behavior of the cells are needed to ensure efficient functioning of the body as a whole [25].

To perform these roles, the immune system developed advanced modes of communication between the immune cells and the other cells of our body. Much like in bacterial colonies but in a far more advanced manner, the communication is used to glean information about the state of the body and assaults by intruders, to perform distributed information processing, learn from past experience to solve newly encountered problems, and to generate an inner common self-identity of our body.

Studies on human physiology in space show that normal cells behave differently than they do on earth. This can cause the immune system to be confused about the identities of our own cells during voyages to space. The result can be devastating; it will trigger the autoimmune system to attack various internal organs, giving rise to avalanches of known and new autoimmune disorder. At the same time, weakening of the immune system can severely reduced its ability to identify abnormal cells. The outcome of this malfunctioning will be an alarming increased rate of emergence of malignant cells that will be misidentified and missed by the immune system, leading to bursts of multiple tumors.

To prevent these frightening prospects we must rely on the bacteria to help keep our immune system in shape. More specifically, they are needed to maintain its self-awareness, the abilities to tell self from non-self and its intelligence from decaying. The radical idea is to keep training the immune system by a designed continuous exposure of crew members to a wide variety of different bacteria, as opposed to the current attempts to keep the conditions as sterile as possible.

This counter-intuitive idea stems from our cognitive perspective of the immune system and the lesson learned from bacteria: to boost a system's intelligence, it has to be challenged with problems to solve and learn from experience.

Tricking Bacteria to Work Harder

Having dealt with the destructive aspects of bacteria, we can now consider how to exploit their constructive capabilities. As an example, let us take Geobacter, microbes that like to dwell where there is plenty of iron and no oxygen. They also have the interesting ability to transfer electrons into metal, that is, to produce electricity while processing waste. These bacteria have even been shown to be able to generate electricity by decomposing body waste. Obviously, building bacterial power plants that can both convert wastes to usable materials and at the same time produce electric power is almost like a dream come true. However, letting the bacteria do the magic at a pace that suites their needs is far too slow for ours.

We cannot force the bacteria to work harder, but we can use their own intelligence in order to trick them into doing so. Suppose we'd like the bacteria to produce some particular enzymes at a higher rate (e.g. needed to decompose pop, process sugar into alcohol or any other need) or some human hormones (e.g. insulin) or any other useful material. We can prepare a plasmid (a cassette of genetic information that is used in genetic engineering) that includes both the genes for production of the material we'd like the bacteria to make for us, and genes for resistance to a specific antibiotic. Now, if we expose the bacteria to this antibiotic they will make many copies of the plasmid to protect themselves from the antibiotic and as a byproduct they will also make the substance we want. This illustrates how we can use bacterial skills to work for us once we understand the essence of bacterial intelligence.

Bacteria Training Program

As mentioned earlier, bacteria have a proven ability to make a hostile environment into a habitable one - planet Earth. Sine the way they did it is still a mystery to us, let us send bacteria first to the new habitats to prepare the ground, and then afterwards go ourselves with them at our side.

To prepare the pioneering bacteria troops for the voyage, they should go through a training program during which they will be exposed, here on Earth, to a variety of conditions that they are likely to encounter in the new places we intend to inhabit. Armed with the new understanding how bacteria learn from experience to solve unfamiliar problems, we can plan an efficient training program much like the methods we would employ to improve our own problem solving skills - exposure of the bacteria to a sequence of varying conditions at increasing levels of difficulties. Moreover, we can apply what we learned from the mistake we made in the past that caused the rapid evolution of antibiotic resistant strains of bacteria to speed up the training program. For example, we can expose the bacteria to non-lethal levels of antibiotics during the training program to elevate the cooperation, "motivation," and "team spirit" of those pioneers.

Sending Bacteria to Pave The Way

The idea is to train and send three different families of new bacterial strains, or strains that will function as specialized of task forces to achieve three missions:

1. Bacteria that will work with us inside the settlement bubble. These bacteria will be used for recycling, for the production of new materials from substances to be brought from the external environment, for supporting the life of other organisms, and to keep our immune system in shape and all other functions bacteria do to sustain life on Earth.

2. Bacteria that will be spread outside the bubble as our pioneering troops to start changing the environment and to learn how to use the available substances and energy sources for life inside the bubble.

3. Bacteria that will form a giant biofilm interfaced within the bubble wall. What I have in mind is a cyborg bubble built of several layers of materials and several different layers of bacteria. The bacteria will thus function as Maxwell demons, using their secret skills and knowledge for proper exchange of materials, energy, and bio-information between the bubble and the environment.

The parts and tools for building the bubble will be sent alongside the bacteria. It will be built on site, and then the bacteria and the bubble itself will co-evolve to form a new gentler biosphere for us and other organisms to move into.

I would like to conclude with another quote from Francis Bacon (could he have had Space travel in mind?)

"It would be an unsound fancy and self-contradictory to expect that things which have never yet been done can be done except by means which never have yet been tried."

References

Additional relevant publications, pictures of bacterial colonies and video clips of bacterial movements can be found at my Home page http://star.tau.ac.il/~eshel/

(1) Lovelock, J. (1995) Gaia: A new look at Life on Earth Oxford University Press.
(2) Margulies, L. and Dolan, M.F. (2002) Early life: Evolution on the Precambrian Earth Jones and Bartlett
(3) Liebes, S., Sahtouris, E., Swimme, and Liebes, S. (1998) A Walk Through Time: From Stardust to Us Wiley
(4) Ben-Jacob, E. (2003) Bacterial self-organization: co-enhancement of complexification and adaptability in a dynamic environment. Phil. Trans. R. Soc. Lond. A361, 1283-1312,
(5) Ben Jacob, E., Shapira, Y. and Tauber, A.I. (2006) Seeking the foundations of cognition in bacteria: From Schrödinger's negative entropy to latent information Physica A vol. 359 ; 495-524
(6) Ben-Jacob, E. et al. (1994) Generic modeling of cooperative growth patterns in bacterial colonies. Nature 368, 46-49
(7) Shapiro, J.A. and Dworkin, M. (1997) Bacteria as Multicellular Organisms, Oxford University Press
(8) Wirth, R. et al.. (1996) The Role of Pheromones in Bacterial Interactions. Trends Microbiol. 4, 96-103
(9) Rosenberg, E. (Ed.) (1999) Microbial ecology and infectious disease, ASM Press Washington DC
(10) Velicer, G.J. et al (2000) Developmental cheating in the social bacterium Myxococcus xanthus. Nature 404, 598-601
(11) Crespi, B.J. (2001) The evolution of social behavior in microorganisms. TrendsEcol. Evol. 16, 178-183
(12) Bassler, B.L. (2002) Small talk: cell-to-cell communication in bacteria. Cell 109, 421-424
(13) Kolenbrander, P.E. et al (2002) Communication among oral bacteria. Microbiol. Mol. Biol. Rev. 66, 486-505
(14) Velicer, G.J. (2003) Social strife in the microbial world. Trends Microbiol. 7, 330-337
(15) Ben Jacob, E. et al (2004) Bacterial Linguistic Communication and Social Intelligence Trends in Microbiology 12 (8) 366-372
(16) Ben Jacob, E., Aharonov, Y. and Shapira, Y. (2005) Bacteria harnessing complexity Biofilms
(17) Ben Jacob, E. and Levine, H. (2006) Self engineering capabilities of bacteria Interface.Published online. Rsif
(18) Ben-Jacob, E. (1998) Bacterial wisdom, Godel's theorem and creative genomic webs. Physica A 248, 57-76
(19) Ben-Jacob, E. and Shapira, Y. (2004) Meaning-Based Natural Intelligence vs. Information-Based Artificial Intelligence. The Cradle of Creativity Edited by H. Nen Nun, Saarei Tzedk Jerusalem
(20) Landweber, L.F. and Kari, L. (1999) The evolution of cellular computing: nature's solution to a computational problem, Biosystems 52, 3-13
(21) Shapiro, J.A. (1992) Natural genetic engineering in evolution. Genetica 86, 99-111
(22) Todd, P.(1989) Gravity-Dependent Phenomena at the Scale of a Single Cell. American Society for Gravitational and Space Biology. 2:95-113,
(23) Cioletti, L.A.; Pierson, D.L.; Mishra, S.K. (1991) Microbial Growth and Physiology in Space: A Review. SAE Technical Paper Series No. 911512.
(23) Nicogossian, A.E.; Huntoon, C.L.; Pool, S.L., eds. (1994) Space Physiology and Medicine, 3rd Edition. Philadelphia, PA: Lea & Ferbiger,
(24) Konstantinova, I.V.; Rykova, M.P.; Lesnyak, A.T.; Antropova, E.A.(1993) Immune Changes During Long-Duration Missions. J Leukocyte Biol 54: 189-201,
(25) Tauber,A. I. (1994) The immune Self: Theory or Metaphor? Cambridge University Press

About the Author

Eshel Ben-Jacob is a Professor of Physics and the Maguy-Glass Professor in Physics of Complex Systems. He served as the vice President (1998-2001) and as the President (2001-2004) of the Israel Physical Society. He is an expert in nonlinear dynamics, pattern formation and self-organization and bio-complexity. He pioneered the new studies of bacterial self-organization and discovered the foundations of intelligence and cognition in bacteria. During last decade, he also studies neural networks and the human brain activity. Recently, together with his student, he invented the functional holography analysis method and put forth the holographic principle for bio-computing. Professor Ben Jacob is a fellow of the American Physical Society and the World Institute of Physics.

Chapter 14

Biotech: A Near Future Revolution from Space

By Lynn Harper

"The aggregate of all our joys and sufferings, thousands of confident religions, ideologies and economic doctrines, every hunter and forager, every hero and coward, every creator and destroyer of civilizations, every king and peasant, every young couple in love, every hopeful child, every mother and father, every inventor and explorer, every teacher of morals, every corrupt politician, every superstar, every supreme leader, every saint and sinner in the history of our species, lived there on a mote of dust, suspended in a sunbeam."
Carl Sagan, Pale Blue Dot

But is that the only option for the future?

Despite light years of searching, the only life in the universe we know comes from Earth. In this profound and vibrant isolation, exaltations are whispered - and often unheard - that life is a treasure, rare and miraculous. In the comet-battered landscape of alien worlds, warnings are shouted - and often unheeded — that living worlds are fragile and few and vulnerable. Still, all of our eggs - literally - lay uneasily in this one beautiful precarious basket called Earth. Because, after all, what alternatives do we have?

Four billion years of evolution in the gravity field of Earth has shaped us, inside and out, from each tiny cell to all the organs that enable us to live and think. The space environment is profoundly different from any encountered during the evolution of life on Earth. It is characterized by significant variations in gravitational force, magnetic fields, and radiation. But of all the environmental variables that changed on Earth to shape life over the past 4 billion years, gravity alone did not change. Even in the close orbital environment of Earth's low-Earth-orbit front yard, life experiences a thousand-fold to million-fold reduction in gravity, itself one of the most fundamental organizing forces of nature. The last time an environmental change of this scope was encountered by Earth life was when the first organisms emerged from the sea to the land.

So April 12, 1961 was truly an extraordinary day. On this day, Cosmonaut Yuri Gagarin was the first human being to leave Earth - and — he did not die. Since his historic flight, a wide range of bacteria, plants, animals, fish, insects, primates and, of course, humans have journeyed in space; none of them have died as a result of the biological changes induced by leaving Earth. Only mechanical failures and design flaws have been responsible for the deaths of space travelers. This is important, because life succeeds in ecologies not as individual species, so in order for us to thrive in space over long periods of time a wide range of organisms in addition to humans must be able to live for generations beyond Earth. So far, there have been no biological causes of death in space, and with that fact, alternatives emerge.

However, there is a need for extreme caution as we plan long forays beyond Earth. Life exists in space by changing its biology in response to the truly strange environmental cues elicited from the unusual gravity environments, and some of these changes are worrisome. As of this writing, only microbes, a couple of plant species, and insects have lived for more than one generation beyond Earth. When pregnant animals have flown in space, their developing offspring exhibited unusual responses. In fact, from microbe to human, whether living in space for short times or long times, for parts of a lifetime or over generations, all life that has flown in space has experienced minor to profound biological changes. But in this fact, opportunities arise.

From a biological perspective, space can also be thought of as a new "extreme" environment - extremely low gravity. Because of this, space biosciences research is likely to reveal aspects of terran life that literally cannot be seen on Earth, because they simply cannot emerge while living within the constraints of Earth's gravity. Biotech industry practitioners seek out some of Earth's least habitable environments to harvest the buff life forms that reside there because such extremophiles (creatures who thrive in extreme environments) exhibit biological solutions that can be and have been exploited for medical and commercial benefit in non-extreme environments. The extreme hypogravity of space offers at least the same - and possibly even superior - discovery and wealth-generating potential.

It's not a bug, it's a feature!

From the dawn of manned space flight, space medical experts have treated the biological changes in astronauts and cosmonauts as pathologies to overcome rather than as phenomena to explore.

Developing solutions to space medical problems was significantly hampered because scientists lacked the tools necessary to determine the causes of the problems they saw, and to rapidly evaluate potential solutions. Life is complex, space is one of the most unusual environments life has encountered, and for the past forty years the tools available were inadequate for the job. Then, in the last decade of the last millennium all that changed.

The toolkits available to biologists can be divided into two great epochs - before the Human Genome Project, and after the Human Genome Project. Before the Human Genome Project hit its stride in 1995, scientists had developed a number of hypotheses about why life responded so oddly in space. They were often right, especially about gross effects, but equally often they were spectacularly wrong, especially about subtle effects. Space biologists just scratched their heads about some responses, unable to formulate a sensible explanation of the biological responses they were observing.

Before the Human Genome Project, trying to understand what was happening to life in space was like trying to figure out what was wrong with your car by staring at the hood. At some point, you have to get inside and look at the machinery, but the tools that could delve into cells and tissues and illuminate what's happening there simply did not exist before 1995.

But amazing new tool kits have been developed over the past ten years - and the biotech revolution is still accelerating with new devices offered daily. Three technical revolutions—the biotech revolution, the infotech revolution, and the revolution in superminiaturized machines—each of extraordinary significance, and each offering exceptional capabilities, converged to form the most powerful investigative toolkits in the history of biology.

For decades, the U.S. government has invested billions of dollars to reduce the cost of a pound of payload to space by a factor of ten. It has never succeeded. However, the purpose of a space mission is not to launch the rocket; the purpose is to deliver people and equipment to extraterrestrial locations for the purposes of discovery and development. And this is where the biotech/infotech/miniaturization revolution changes all the value equations for space. Use of these new tools amplifies the value of a space biosciences payload pound a million times or more over what was possible before 1995.

Discoveries and Dollars — A Life Saving, Wealth Generating Story

This convergence provides new meaning—and a wealth of new possibilities—to the already compelling story of life. Using the tools spawned by the Human Genome Project and its prolific offspring, we can tell a new traveler's tale: the biological story of the only life in the universe we know in its first generations beyond the planet of origin.

This story is told not only in words and pictures, but also in the dialect of life itself - the language of genes and proteins. This space saga can generate wealth. It's biological message can save lives. And it will tell us whether life from Earth is planetbound, or what biological costs and opportunities will be available to us if we live for generations beyond our planet of origin.

For just as scientists seek extreme environments on Earth in their search for novel biological solutions that can be applied to terrestrial problems, the space environment is an unexplored extreme environment, the extreme of very low gravity, which offers exceptional potential for life saving and wealth generating advances.

The opportunity to bring the biotech revolution to space is an historic first, and is likely to reveal features of terrestrial life that literally cannot be seen on Earth. It also allows us to determine whether life from Earth is biologically bound to this world, and to explore the biological costs and opportunities inherent in living beyond Earth. And that's just the beginning.

Recent results show that the space environment yields unique knowledge that is medically important and commercially interesting. In the near future, commercial space biolabs can be used to develop and test new intellectual property products for high yield markets, especially in infectivity research, tissue/organ cultures and products, and insights into combating some of the debilitating effects of aging. Together, these three areas tap into a market valued at more than $100 billion annually. This is serious money to be made from "doing well by doing good".

Small space "cottage" industries have produced space habitats for the model organisms that pioneered the Human Genome Project as well as the analytical laboratories for studying them.

Amazingly, the biotech revolution has not yet been used in space - except for a few cases. But these very cases caused jaws to drop and history to be made, as we will see below.

The New Case for Space Biotech.

Early concepts for commercial involvement in orbiting laboratories such as the International Space Station envisioned that companies would pay to use manufacturing facilities in space to exploit the unusual features of the microgravity environment to create new products in situ. This vision has never been realized for a variety of reasons and it is not proposed here—at least not for the near future.

In the last five years new technologies have emerged from the biotech revolution that could be applied to research in space. A new entrepreneurial paradigm is emerging and will provide a valuable new product in the biotech portfolio - patentable intellectual property based on space research using commercially available biotechnologies adapted to space. Four case studies are described here to profile the major opportunities foreseen: Cell and tissue cultures, infectivity, aging, and agriculture.

Cell and tissue cultures in space. Good cell and tissue cultures can accelerate by years the discovery of the causes of and cures for diseases. This can potentially save millions, if not billions, of dollars that would otherwise be spent on unproductive research. It can also generate millions or billions of dollars of revenue. Most importantly to most of us, it accelerates the development of life saving pharmaceuticals. But the value of a cell/tissue culture depends on how well that culture mimics what really happens in the body. For many diseases on Earth, there are no good cell/tissue culture models - yet. Surprisingly, space appears to hold one of the keys to this problem.

More than a decade ago, transplant surgeon Timothy Hammond, M.D. of Tulane University was looking for a tissue culture model for kidney disease. Kidney disease is one of the most expensive diseases to treat because there are only two treatment options for the advanced stages of the disease—dialysis or transplant. Kidney disease incidence increases with age, and so is rising as a national health cost as baby boomers age. 100,000 people per year in the US are diagnosed with kidney failure, and this country spends $20 billion/year treating this disease. There was no good cell/tissue culture for kidney disease until Hammond connected with what is currently known as the Biological Systems Office (BSO) at NASA Johnson Space Center. Hammond collaborated with Dr. Neil Pellis, subsequently the Co-Director of the Cell Biology program, and was introduced to the technology called the Rotating Wall Vessel (RWV).

The RWV, invented by David Wolf, Ray Schwarz, and Tinh Trinh, of the NASA Johnson Space Center, mimics the effects of microgravity on cells. When Hammond grew kidney cells in the Rotating Wall Vessel, the results were dramatically better than any others that had been grown. Tissues grown in the RWV began to re-acquire their three dimensional structure and biochemistry, which had been missing in standard terrestrial cultures, and electron microscopy of the cells showed that the microvilli, an important characteristic of kidney cells in the body but absent in standard cultures, had returned in abundance. The team reasoned that if the RWV was good, then space was the gold standard, and three additional pioneering investigations in space confirmed and extended their early findings in the RWV and showed even greater promise. Their results were published in prestigious peer reviewed journals (REF) and Stelsys, the entrepreneurial arm of Johnson and Johnson, became the first paying customer on the ISS as a result of both RWV and space flight data.

The reason that the space environment and RWV yield such improvements in cell cultures is because they more closely approximate the "cues" actually given cells as they grow in the body, which cells grown in Petri dishes on Earth cannot. To grow a tissue culture on Earth, cells are removed from the body and placed in a Petri dish where they grow flat. Cells are not smart, but they are very adaptable, and they obtain their information from the top, the sides and the bottom. In flat Petri dishes, the bottom information is missing and consequently the cells grow very differently than they do in the body. Scientists know this, and so in an attempt to overcome that problem, they suspended the cells in fluid. To prevent them from settling in suspension, they placed the cells in a mechanical mixer.

Unfortunately, the mechanical force needed to maintain the cells in suspension was so great that cell aggregates fragmented, and again, the cells did not grow the way they do in the body.

Space—and to a lesser extent the RWV—allow the three-dimensional structure of the tissues to emerge in cell cultures. In space, this occurs because nutrients and wastes do not separate on the basis of density differences, the cells do not settle to the bottom of the culture system, and nutrients and waste removal can be circulated with very gentle mixing that allows much larger cell aggregates—tissues—to form as a result. The RWV tumbles gently around a cylinder packed with nutrients, so it also reduces shear and turbulence in the mixing process, thus providing a gentler growth environment, also in three dimensions. (However, gravity does limit the size of the 3-dimensional tissues formed in the RWV when it's used on Earth.)

The biotech revolution allows researchers to read genomic instructions encoded in cells grown in the space environment, and to correlate these instructions with their physiological meaning, which is to say, to understand cause and effect to a much greater degree of precision that has been previously been possible.

This depth of understanding is important because most diseases—and treatments—are not single-element events. Rather, they are a symphony of multiple signals, transduction of environmental data, genomic instructions, protein responses, and feedback. Fortunately, the biotech revolution has produced superminiaturized analytical laboratories that can be fielded on spacecraft for on-orbit analysis, following which samples can then be fixed or flashfrozen and returned to Earth for more comprehensive examination. Companies can then take this information and, using contemporary biotech tools, engineer the organisms and systems needed to replicate the results on Earth.

Infectivity. Because of similar features in the cell-culture environment, bacterial cultures grown in the RWV exhibited over a 20-fold increase in infectivity compared with Petri-dish grown cultures. The surprised research team, led by Dr. Cheryl Nickerson, hunted for the biological explanation for this greater infectivity, and genomic and proteomic analysis revealed that it was caused by mechanisms that had not been predicted by current infectivity theory. Prior theory, in other words, was incomplete, a discovery made possible by the RWV.

The data were immediately published, of course, because the results revealed that research in this deadly and expensive medical area was not correctly informed. The terrestrial paradigm for infectivity was at best incomplete and at worst wrong—all because of limitations in standard culture conditions. The next step in this investigation is in queue, now waiting for a flight opportunity. (REF) Similarly, cancer, liver, brain, colon, bone, muscle, and other tissues have shown superior results when grown in the RWV, and are awaiting further investigation in space.

Aging. One of the most intriguing research areas is the role Earth-orbiting laboratories can play in providing insights to help combat aging. From mouse to man, when mammals live in space for a long period of time they lose bone and muscle mass, experience cardiovascular deconditioning, vestibular disturbances, hormonal imbalances, immune suppression, brain repatterning, and balance problems. The only other instance when all of these factors change simultaneously is during the aging process.

However, mice, men and women get better after they return to normal gravity. It is now possible to compare the processes of aging with the processes of space deconditioning by examining the molecules that begin the process, the cells changed by it, and the resulting impact on tissues, organs, systems, and whole organism. This particular class of work is important to every person on the planet.

Agriculture. Millions of dollars are spent each year to rid paper mills of lignin, a key structural component that occurs naturally in plants that interferes with paper making. The ecological cost to remove lignin is so significant on the paper-making process that paper companies developed expensive "knock-out" versions of the major plants used in paper production in an attempt to eliminate lignin.

A knock-out is a strain of the plant in which a single gene has been eliminated, knocked out. Doing so enables scientists to determine, in theory and sometimes in practice, the role of that particular gene.

However, life is not that simple; lignin, they found, is produced by a complex ensemble of genes that the knockouts didn't solve. In space, however, lignin production is significantly reduced, and by learning the genomic choreography by which this occurs new strains could be engineered on Earth, saving millions in environmental remediation costs.

Realizing the Potential

Given those interesting findings, why aren't the biotech companies and biosciences communities clamoring for more time on the ISS or other orbiting laboratories? There are several reasons.

One of the key factors that enabled the biotech industry on Earth to progress so quickly over the last decade is that the new tools provided the ability to perform very quick learning. Learn fast, learn often is their paradigm. Thus, in order for biotech to realize the potential that zero gravity—and for that matter lunar or Mars gravities offers—they need more frequent flights to do iterative studies. In biotech, a learning cycle is defined as the time it takes to define an experiment, develop the necessary hardware and protocols, perform the experiment, analyze the resulting data, and prepare for the next cycle. Typical learning cycles in biotech are measured in days and weeks.

Unfortunately, the learning cycles for space experiments are usually measured in years, a barrier that has slowed biotech utilization of the ISS to a trickle.

On the other hand, it's an opportunity, too, because meeting the emerging customer need for rapid learning cycles is what could make the emerging commercial space industries a success.

Make Space Research More Like Terrestrial Research

No one does biological research on Earth the way it must be done in space. For example, experiments in terrestrial labs are not begun by shaking them violently for several minutes, as occurs during lift-off and ascent to orbit. In fact, all parts of the space experience are significantly different from any type of terrestrial biological research. Expert guides are needed to translate commercial investigation procedures that work successfully on the ground to investigations that work successfully in the unusual environment, and under the unusual accommodations and constraints, of space.

But if we use the current NASA system, the time and effort that any company must invest to get an investigation to space is great, the paperwork burden crippling, and the space environment itself fraught with opportunities for experimental error that can invalidate any investigation's potential. Unless there are guides make this process much easier, companies will pursue other terrestrial avenues, even if they are not as effective, and all of us will suffer as a result. And while there are expert guides at the NASA centers and at NASA Centers of Excellence throughout the country, the entire NASA system is not organized to support the types of research—or the high-frequency learning cycles—that the new biotech revolutions call for.

From Earth Orbit to Mars, A Journey of Generations and Discoveries

When we go to another world, one of two equally profound events will occur. Either we will encounter alien life and know, finally, that we are not alone. Or, we ourselves will be the origin of life on that world.

This makes the study of Earth life in its first generations on other planetary bodies uniquely important for determining a fundamental aspect of life in the universe: can life evolve—not just live—beyond its planet of origin? Is life from Earth biologically bound to this one world? What are the implications for expanding life beyond Earth?

Soon we will have the opportunity to study Earth life in its first generations on other worlds, on the Moon and Mars. Each of these environments is unique, with features found nowhere else in the solar system. They differ significantly from our home planet in gravitational force, magnetic fields, day/night cycles, and radiation stresses.

Other than research conducted on the astronauts themselves, initial research in Earth orbit, on the Moon, and on Mars will start with the smallest simplest organisms, the best understood organisms on Earth, both because they are easiest to handle and because we need to build our knowledge from simplest to most complex. This research will identify specific biochemical mechanisms by which a wide range of organisms adapt to each new environment, from conception through maturity, reproduction, and death over multiple generations.

All life, from microbe to human, has shown fascinating and unpredicted changes as a result of spaceflight. Space biologists concur that the effect of different gravity levels, especially those below 1 G, are unknown and unpredictable from current paradigms, and that their study will reveal new knowledge about life on Earth, and empirically, about life in the universe.

Past Successes Herald Future Progress. Many people are alive today because we studied life in space, but few people know that procedures used in all intensive care wards worldwide are based on the technologies pioneered by Apollo. This is because intensive care wards rely on telemetry - the transmission of medical data from the patient to nurses and doctors in a way that alerts caregivers to life-threatening changes in patient condition. Perfection of these technologies was achieved because NASA doctors needed to monitor astronaut health remotely.

Even fewer know that the Micro-Electro-Mechanical Systems industry (MEMS) owes a significant fraction of its wealth and its products, currently yielding more than $5 billion per year in revenue, to breakthroughs achieved by Dr. Lynn Roylance's in developing a device to measure blood flow in the hearts of rats in space. Pacemakers, airbag crash sensors, and fetal surgery monitors are among the fruits of this research.

Implantable insulin pumps, shock trousers, telemedicine, remote surgery, 3-D observations for diagnostic and reconstructive medicine, implantable medical devices, and many, many more advances have resulted from the study of life in space.

Some argue that all of these advances would have occurred without the space program, and perhaps this is so, but theirs is a hypothesis without data because that's not how these advances were achieved. All who are alive today because space biosciences made these technologies available sooner rather later may feel that their penny-per-year investment in space biosciences was money well spent.

Based on actual history, then, we can expect that the knowledge and technology products developed as a result of our future efforts in space will offer important insights to help prevent death, enable life, ensure that a significant number of people attain a substantially higher quality life for much longer periods of time, and in the process, generate exceptional wealth and numerous jobs.

Clearly, then, it's time to bring the biotech revolution to space.

Oases Beyond other worlds, and the vast spaces between, them are biological laboratories on many levels. They hold unique clues about the events that enable life, change evolutionary trajectories, and ultimately create the two most mysterious forces of nature, life and intelligence. Space laboratories in orbit, on the Moon, on Mars, and beyond, will enable the first empirical tests of whether it is possible for life from one world to thrive on others, and will reveal features about life on Earth that cannot be seen from any other vantage point.

And while arguments may rage on many subjects about space exploration and life in the universe, this we know: Very different futures are available to a species that can thrive beyond their planet of origin as compared with those whose destinies are constrained to a single world.

Which future will you choose?

About the Author

Lynn D. Harper is Lead of Integrative Studies for the NASA Ames Research Center Space Portal, a consortium to promote commercial space development for public benefit. Harper was awarded NASA's Outstanding Leadership Medal for her role as one of the founders of the science of Astrobiology and her service as the first Lead for Astrobiology Advanced Concepts and Technologies. For several years, she was the Acting Chief and Deputy Chief of the Advanced Life Support Division at NASA Ames and oversaw the development of air and water regeneration systems that are now leading candidates for Space Station upgrades and bioregenerative life support projects that, among other things, produced world record wheat yields using hydroponic techniques. Harper served as Study Team Leader for Life Sciences, Life Support, and EVA on NASA's 90-Day Study for Human Exploration of the Moon and Mars; was a member of NASA's Decadal Planning Team for Exploration as well as the NASA Space Architect Team that provided the technical foundation for the President's Vision for Space Exploration. Harper was the Program Manager for Advanced Missions and Special Projects in the Space Life Sciences Division at NASA Headquarters between 1986 and 1989. During her tenure she initiated, established and managed the Controlled Ecological Life Support System Flight Program and the Exobiology Flight Program. She Served as Program Manager (at NASA HQ) and Deputy Project Manager (at NASA Ames) for the Search for Extraterrestrial Intelligence Project (SETI). Harper has managed science instrument development programs for Space Shuttle, Mir, Space Station and unmanned planetary exploration spacecraft as well as computer and radio telescope development and applications programs for SETI. She was the first to support the development of aerogel instruments to collect intact fragments of cosmic dust, an investment that enabled the Stardust Mission and the first recovery of pristine samples from a comet. Harper has a deep commitment to education and has initiated and managed several pioneering educational projects for teachers and students using space to inspire interest in science, technology, engineering and math. Recently, she helped establish the Silicon Valley Space Club, a volunteer think tank that has produced several pioneering space development concepts. Harper is the recipient of numerous NASA awards and has been recognized in Who's Who in America and Who's Who in the World for years.

Chapter 15

Space Exploration and a New Paradigm for Education and Human Capital Development

By Michael J. Wiskerchen

Introduction

Beyond Earth - Future of Humans in Space formulates a vision that is rooted in the following precepts:

* Humans are genetically disposed to explore and to find new ways to survive;
* Space exploration, habitation and utilization by and for Earth's humanity will be an increasing phenomenon as the future unfolds;
* Human outward migration into space will present an opportunity to develop a rewarding and exciting future of global collaboration to capitalize on the lessons learned from human history on Earth;
* The successful migration of humans into space will be highly dependent on the development of a responsive and evolving educational system that focuses on the increasing influence of space on Earth and its humanity.

Can we study history and learn how to create a responsive and evolving educational system? Is it possible to use the lessons-learned of the space age to create a learning environment that could evolve as humans migrate into interstellar space? It is obvious that educational capabilities of the Apollo era have been transformed today into global information access and knowledge sharing systems. This rapidly advancing educational technology is outpacing the ability of people and organizations to use it productively. This Chapter will attempt to look at education and human capital development through a past and present "space environment" window to derive lessons-learned that will allow us to develop a responsive and evolving "interstellar" educational system

A Historical Perspective on the High Technology Human Capital Crisis in America

A Personal Perspective - The future of exploring and living in space will greatly depend on our ability to appropriately educate and train the next generation of explorers. During the past ten years, many experts have questioned the will and/or ability of the United States to lead this effort and as a long time space scientist and educator I have broached this issue from many directions. I have asked the question: Why does a young person select a science, technology or engineering career path? What are their motivations and/or influences? In an attempt to address these questions I have examined what influenced my own career choices and that of my children.

I was raised on a small Midwest dairy farm and educated in a small town of fewer than 300 citizens, an extremely different childhood than my children have experienced. In 1960 I was starting my first undergraduate year with a double major in physics and math and my choice of majors was based on a persistent childhood curiosity and lots of hands-on experience with nature and mechanical systems. I was also strongly influenced by my parents and the community we lived in which instilled a strong work ethic and acceptance of daily responsibility. My high school was very small and offered only a basic science and math curriculum from teachers that may or may not have had a degree with a minor or major in the courses they were teaching. But one thing was certain: everyone in the community knew you and your parents and they were there to mentor you.

The space program didn't exist for me until I entered college. I viewed it as an exciting venture but didn't think that a "farm boy" from the Midwest would ever be able to participate. Throughout my undergraduate years that viewpoint did not change but at the start of my graduate work the opportunity presented itself to actually get engaged in the space program. Looking back, my career was built on serendipity, grasping opportunities as they appeared: during the 60's many space-related opportunities appeared.

My children's formative years were considerably different as they were continuously bombarded with space-related people and events. Although one would have thought that this environment would stimulate their interest in a space-related career, it didn't. What was missing? They had a similar curiosity about nature and an interest in observing processes in nature, but they had very little day-to-day hands-on experience with nature and mechanical systems. The community connectivity and mentoring was hard to develop in the large urban environments where we lived, and establishing a strong work ethic or acceptance of daily responsibilities was not automatic in this environment either. The quality of K-12 science and math education, although on a much broader scale than what I had experienced, was, from my perspective, less stimulating or effective. The number of qualified science and math teachers (those with majors in the courses they taught) was about the same as I had. In summary then, the only apparent difference between my early science and technical learning environment and that of my children seems to be hands-on experiences, and the prevalence of community mentoring.

America's Human Capital Crisis

Congress, the Executive Branch, and a number of Federal and State Agencies have recognized the critical need for a skilled aerospace workforce supported by many K-12 science, technology, engineering and math (STEM) initiatives. This critical need is clearly stated in numerous reports and studies, including: "Report to the President: The Crisis in Human Capital" (Voinovich, 2000); the "Final Report of the Commission on the Future of the United States Aerospace Industry" (Walker, 2004); the "Roadmap for National Security" (Hart-Rudman, 2001); and "NASA's Human Capital Strategic Plan" (NASA, 2004). It reminds me of the old country saying, "Everybody talks about the weather but no one does anything about it". More than twenty years of reports and statements have lamented the crisis in our nation's educational, scientific, and technological infrastructure. In 2001, the Hart-Rudman commission on national security clearly stated the problem as follows: "The inadequacies of our system of research and education pose a greater threat to U.S. national security over the next quarter century than any potential conventional war that we might imagine." The weakening of science and technology in the United States will inevitably degrade its social and economic conditions and in particular erode the ability of its citizens to compete for high-quality jobs.

The time for bold and decisive action is long overdue so one might reasonably ask why corrective action has not been taken by some combination of the government, industry and academia. What's missing from the last two decades of human capital development efforts relating to science, technology, engineering, and mathematics careers? What's it going to take to turn this crisis around? What cultural and organizational change must occur in the government, industry, and academic sectors to make progress on this critical human capital issue? As the future unfolds, can the quest for space exploration, habitation and utilization by and for Earth's humanity be a driving motivation for today's youth to embrace science and technical learning?

Learning from History

It's instructive and relatively easy for those of us that started our scientific, technical and/or management careers in the 1960's to compare what's happening now to those Apollo era years. In 1958, this nation was stimulated by the Soviets' launch of Sputnik One into collective national action to upgrade its educational and technological infrastructure, as it was clear that the Soviet emphasis on math and science would put our Cold War adversary ahead of the United States in ten years.

Even though the United States had demanding fiscal constraints at that pre-Apollo time, Congress appropriated a billion-dollar National Defense Education Act ($6B in today's dollars) that emphasized the study of math, science, and foreign languages, and it led directly to the creation of NASA. NASA then developed unique and extensive partnerships with U.S. colleges and universities by establishing scholarship and fellowship programs and campus research facilities. Through procurements, NASA encouraged partnerships between industry and universities for R&D as well as workforce development. The number of science and engineering Ph.D. degrees awarded annually by U.S. colleges and universities rose from 8,600 in 1957 to almost 4 times that in 1973. That type of integrated public - private partnerships dissipated after the mid 1970s.

Historical Impediments to STEM Human Capital Development in the Government, Academic, and Industrial Sectors

Why did the political system so readily take action then yet find it so difficult to do the same thing today? One answer is the high level of public awareness of the Cold War impacts while today's declining technological prowess and economic security does not seem to generate the same public awareness or response. Without this pressing public awareness, many federal and state office holders do not perceive this as an important re-election issue.

Thus, it's important to examine the political or social environment that persists and hinders sustained state and federal government responses to this decline in our nation's high technology workforce that will be required for humanity to venture out to the space frontier. First, no long-term vision has been implanted into the general public psyche that requires this sustained effort. Also it's difficult to attack the STEM-related human capital problem since you must address the entire STEM workforce pipeline (K-12 through adult continuing education) to produce long-term results. This adds extreme bureaucratic complexity since it crosses multiple local, state, and federal agencies and institutions. The sustained general public support is necessary since most STEM career education and training programs will not produce quantifiable impacts in less than one to two decades. This is a much longer period of time than the re-election time scale for most federal and state elected officials. The collection of problems reveals a disturbing picture—a definite pattern of short-term thinking and insufficient long-term investment.

Another historical factor that is complicating the development of a competitive scientific and technical workforce is the United States' strong dependence on immigrants and foreign nationals to populate the high technology industry workforce as well as most major university science and engineering graduate programs. A decade ago this worked in our favor with most foreign students and H1-B Visa workers staying in the U.S. and contributing to its technical success. Given that 55% of the engineering doctoral students in the United States are foreign nationals and that many of these students are leaving the United States, the effect could be to drastically reduce the U.S. talent pool.

As a result of all these trends, human capital development is one of the major challenges facing the U.S. aerospace industry today. Historically the U.S. aerospace industry allocated minimal resources for developing and maintaining a trained workforce. Instead the industry has followed the practice of laying off workers with the wrong skills and hiring the new workers with the skill mix required by specific government contracts. This human resources practice was a natural outcome of the industry's dependence on government contracts for revenues.

A 21st Century Solution to the Aerospace Human Capital Problem

What does this tell us about the high technology and aerospace human capital problem? What are the implications to the "Future of Humans in Space"? My views are as follows:

* The political, social, and organizational forces of today are far different than during Apollo, and consequently most Apollo-era human capital approaches will not work today;
* The education and learning must be driven by a common long-term vision of the future for humanity;
* Solutions must address the entire human capital "pipeline" (K-12 to life-long learning);
* Solutions must be based a "Systems Approach" to human capital problems crossing government, industry, and academia;
* Any sustainable solution must develop and provide long-term budgetary stability;
* All education and human capital development programs must be directly linked to the economic development strategic plan of the local, regional, and national communities.

My more than thirty years in the space program, in both the private and public sectors, has taught me to focus on the long-term economic issue first. Vision and funding issues are similar to the old chicken and egg problem. If you don't have a believable and accepted vision no one will provide the necessary resources to implement the vision. Visionary ideas without funding are worse than useless—they also waste good people's time and energy. For the Apollo-era vision, the federal government was the primary funding arm for most education and workforce training programs. Today, where we lack a common vision, depending only on government funding for education and human capital development is impractical.

A recent Report of the Government Accounting Office (GAO) to the U.S. House Committee on Rules entitled, "Higher Education—Federal Science, Technology, Engineering, and Mathematics Programs and Related Trends Report" (GAO, 2005), reports that 13 civilian federal agencies allocated a total of $2.8B in 2004 for promoting STEM-related education and training programs. Compared to the National Defense Education Act of 1958 ($1B or $6B in 2004 dollars), the Federal Government is spending far less ($3.2B less) on science and engineering education. With the seriousness of the STEM human capital problem, this is very disturbing.

So the question becomes, *"How do you create and sustain an environment where the government (local, state, federal) and private investment jointly provide stable long-term funding for human capital development?"* What is the role of government in this new "business model"?

First, the government will still have to provide at least 50% of the funding through a proposal - peer review process that teams the private sector with educational institutions. Second, the government should include a mandatory human capital development requirement in every Request for Proposal (RFP) issued, which should reserve (fence off) between 5 and 10% of the total proposed budget for human capital development.

But beyond this, an integrated approach must be adopted.

An Integrated Approach to Human Capital Development

In 1988 Congress established the National Space Grant College and Fellowship Program (Space Grant) in the NASA Authorization Act. The NASA managed Space Grant Program, through the network of 52 Space Grant consortia in all 50 states, Puerto Rico, and the District of Columbia, was designed to play an important support role in the nation's science and technology mission particularly in areas that pertain to aerospace-related research, education, human capital development, and public outreach. In carrying out its mission, Space Grant partners with academe, industry, and government agencies (Federal, state, local) to provide educators, researchers, students, and the general public with the experiences that capitalize on the excitement of United States' unique aerospace research, exploration, and discovery environment.

Each year, the Space Grant programs involve over 500 Colleges/Universities, provide over 2000 student awards, performs over 500 student-mentor programs and engages over 40 State/Local government partners and over 100 aerospace industry partners. The extensive Space Grant network of affiliates has become the de facto national facilitator/coordinator in the quest to develop an integrated approach to the aerospace human capital crisis.

The Space Grant human capital development effort involves two integrated elements relating to education and training. The first engages students (K-12, higher education and life-long learning) in formal and accredited science and engineering education programs. This includes incentives for many of the higher education students to become K-12 teachers of math and science, inspiring even more students to pursue these fields.

A second critical element in a human capital development strategy is the implementation of an experiential learning environment. I have named this type of training effort, "student - mentor". Student - mentor programs, as developed by Space Grant, demonstrate baseline characteristics that industry, NASA, and the government agencies have realized are critical for solving the high technology workforce problems. Those baseline characteristics are as follows:

* Team participation of students from K-12, undergraduate, and graduate levels (pipeline);
* Experiential learning through "hands-on" aerospace-related projects;
* Highest priority given to workforce skill development of students & mentors;
* Emphasizes that students should experience the full mission life cycle including mission definition, design, build, fly, and analyze in less than two years;
* Community-based private - public partnerships involving industry, government, and academia for mentors, facilities, and investment;
* Addresses "Human Capital" issues of the local, State, and Federal Government.

Space Grant and the Human Capital Development Model

Over the past 15 years, Space Grant consortia across the nation have implemented aerospace human capital development efforts that involve science, engineering, and management student teams at all levels in hands-on aerospace projects. These programs were created to provide students with practical experience and scholarships/training grants while under the guidance of mentors from the industrial, academic, and government sectors. Many of these projects involve unique aspects of public - private partnerships, government procurement processes, intellectual property agreements, and human resources efforts across the government, industry, and academe. It is instructive to fully examine several of these efforts that were implemented by the California Space Grant organization.

First, a few words about the structure of the NASA sponsored Space Grant organization in California. There is the congressionally designated and NASA managed part, the California Space Grant Consortium (CaSGC) that consists of more than 25 California universities and colleges with the management lead directed by UC San Diego. From a procurement point of view, the only legal aspect of the Consortium is the NASA grant administered by UC San Diego. The Consortium itself has no legal organizational identity so that the enactment of any government procurements with the entire Consortium requires a legally recognized organization to represent all of the CaSGC affiliates. A separate California non-profit 501(c) 3 organization called the California Space Grant Foundation (or the Foundation) was formed.

The Foundation is structured to assist Space Grant in enabling students of all ethic and financial backgrounds to attain high skill technical and professional careers through education and exciting programs. It creates, facilitates, manages and integrates K-12, college and university and life long learning opportunities built around real world space, land and sea projects.

On a national level the same legal situation affected the 52 Space Grant Consortia, which led to the formation of the National Space Grant Foundation (NSGF). The NSGF is a tax exempt 501(c)(3) organization whose purpose is to support and enhance the Space Grant Consortia in every state to carry out education, research, and public outreach activities in science, mathematics, engineering, and technology and additional fields related to space, aeronautics, aviation, and Earth system science. The partnership between the California and National Space Grant Foundations provides an effective coordination/facilitation mechanism to address local, regional and national human capital development efforts.

Applying the Human Capital Development Model — The AERO Institute

With this organizational structure in place, Space Grant was prepared to address the aerospace human capital issues in a pilot effort in California. Through the opportune conjunction of a number people and organizations, the pilot location was picked to be the "High Desert" of California. The High Desert, an hour north of Los Angeles, is the home of Edwards Air Force Base, NASA Dryden Flight Research Center, the Department of Defense Plant 42, Mojave Test area, China Lake, and the fast growing cities of Palmdale and Lancaster. The one community college is Antelope Valley Community College and there are no 4-year institutions. Over the last two years, a Space Grant led team of dedicated people and organizations has created the Aerospace Education, Research and Operations Institute (AERO Institute - non-profit educational partnership) which has proven to be an excellent environment to test and evaluate what works or fails in the Human Capital development model.

The vision of the AERO Institute was to enable competitive U.S. utilization of industry, university and government assets by:

* Leveraging the assets of NASA Dryden, local, state, & Federal government agencies, Space Grant Universities and local industry;
* Creating strategic private/public human capital and education partnerships;
* Leveraging the intellectual capital of the academic community.

The AERO Institute's long-term goals are to act as innovator, facilitator and integrator for joint government (local, state, federal), university and industry projects. It will focus on:

* Human Capital development;
* Educational outreach;
* Applied Research and;
* Operations Improvement.

To that end AERO will:

* Provide comprehensive technical, undergraduate and graduate education;
* Conduct leading edge aerospace research and provide an experiential learning environment for students;
* Incubate, stimulate, and commercialize new intellectual property;
* Promote aerospace science and engineering.

With all of these lofty visions and goals, what became reality and why? The reality part was amazing but the why is more instructive to enable propagation of the model to other parts of the nation. The following is a partial list of accomplishments over the past two years:

* Completed a Space Act Agreement signed between NASA Dryden and the California Space Grant Foundation creating the AERO Institute;
* Created a partnership between the City of Palmdale and the AERO Institute - a 20,000 sq. ft. building was provided by Palmdale to the AERO Institute for education and human capital development of the High Desert;
* Relocated the NASA Dryden Office of Academic Investments to the AERO Institute. This Office serves as an education center providing K-12 students with NASA Education materials and resources, serving as a Higher Education center providing on-site and distance-learning access to universities, industry, government, and serving as a focal point of general public aerospace-related education;
* Established NASA Dryden - AERO Institute Space Act Agreement research partnerships with Space Grant universities (across the nation) exceeding $3,000,000, including a faculty and graduate/ undergraduate student intern program;
* Implemented an effective Intergovernmental Personnel Act (IPA) program to allow NASA and Air Force personnel to transfer and work into the AERO Institute;
* Initiated a distance education program offering Engineering Masters Degrees through Purdue University, University of Southern California, and University of California, San Diego, and MBA Degrees through CalPoly Pomona and University of Southern California to government and aerospace industry personnel;
* Established regional partnerships with Workforce Investment organizations at the state and Federal level;
* Initiated a partnership with the Air Force at Vandenberg AFB to provide engineering education and hands-on training for payload integration and launch operations.

Thus, the AERO Institute has been more successful than I could have ever imagined. Since I have been exercising parts of the model for more than 20 years with varying degrees of success, I asked myself why was it working so well this time. The simple answer is that the right people at the right time are working together in a government, industry, and academic collaborative environment.

A more complete answer is given below that explains what I mean by the "right people" and a "collaborative environment."

Who are the "right people" who made this vision a reality? Some were visionaries, others were educators/researchers and a number were experienced business and organizational people. A common thread that linked all of the key team members was that they had "history" together and they all could be identified as "passionate advocates" for the AERO vision. They had already developed a respect and trust relationship that made the collaboration process efficient and productive.

The "collaborative environment" is also an intricate combination of elements that are necessary and sufficient to produce a successful outcome. A few key necessary and sufficient elements are listed below:

* A "fair broker" facilitator/coordinator organization must be established to operate and maintain the enterprise. For AERO, the provision of management services was assigned to a non-profit facilitator organization - California Space Grant Foundation (low overhead business office and "passionate advocate" for aerospace-related research, education, human capital development, and public outreach);
* It is important to have a "passionate advocate" team member representing each key organization (government, academia, industry). This team member must be able to influence the upper administration of their organization;
* The utilization of the Space Act Agreement procurement vehicle was critical in establishing the public - private partnership in a timely way. This procurement mechanism created a shared resources environment (facilities, funding, and personnel);
* The AERO Institute was, both in reality and perception, an "inclusive" partnership organization. It can never become a competitor to its partners and its principal role is to serve as the team facilitator/coordinator between industry, government, and academia;
* Intergovernmental Personnel Act (IPA) was employed as an excellent mechanism for placing key civil servants within the AERO Institute. They retain all of their government status but work day-to-day in the Institute;

* Industry and academic personnel can also be temporarily assigned to the Institute for varying periods of time (several months to two years) similar to IPAs;
* The Institute budgeted 5% of its gross revenues for human capital development (scholarships, fellowships, and training grants);
* The Institute plays the role of regional facilitator/coordinator for K-12 outreach, university level education, distance learning, and life-long learning for STEM related careers;
* Institute budget was derived from multiple private and public sources including local, state, and Federal government agencies, industry, academia, and philanthropic organizations.

ACES - Reapplying the Human Capital Development Model

The amazing part of the AERO Institute story is that the concept and initial discussions took place in late October 2003 - that was just two years ago. One would therefore ask if this model could be implemented elsewhere with different initial conditions. That opportunity appeared in the summer of 2005 relating to an effort by NASA Ames Research Center to address commercial enterprises in space. The California Space Grant organization has responded to this opportunity by taking the lessons learned from the AERO Institute to promote the creation of the Alliance for Commercial Enterprises in Space (ACES - a 501 (3) C performance based private - public partnership for commercial space enterprise).

The ACES was formed to advance the development of the low Earth orbit environment for all users - scientific, technological, and commercial, in order to more efficiently advance scientific knowledge, technological capability, and commerce on Earth as a gateway to 21st Century exploration and development of space.

It's purpose is thus to engage a new business model for achieving research and development (R & D) in low-Earth orbit in a way that is consistent with the present and future goals of the U.S. Space Program. The goal is to create an Alliance that will aggressively pursue science, technology, human capital development, and commercial development programs with clearly defined roles for government, industry, and academic partners.

The ultimate success of ACES will depend equally on the efficient operation of existing and emerging space, transportation, and ground assets (laboratories, launch & retrieval vehicles, spacecraft and space station among others), and on the optimal utilization of those assets by the R & D, business, and private investment communities. As with any technical or business venture, the level of success will depend upon the long-term education and training of the workforce involved.

Although ACES could eventually be several orders of magnitude larger than the AERO Institute, the organizational and management structure, the human capital emphasis, and the budgetary breakout would remain similar. Successful implementation of the model will also depend strongly on having the right people at the right time working together in a government, industry, and academic collaborative environment. However, there are a number of "very" different initial conditions. These include:

* A strong historical government and industrial culture involving marginally successful commercial space ventures;

* Historically strong scientific, technical, and educational community (Silicon Valley);

* NASA and associated contractor support personnel experiencing human capital downsizing;

* Extremely constrained budgetary environment particularly in human capital development.

At this time the most that can be said is that the core group of "passionate advocates" is very experienced, talented, and determined: overcoming the existing and inhibiting organizational cultures is the greatest challenge they currently face.

Summary

In the AERO and ACES environments, I see the 85-12-3 Rule to be fully in play. The Rule refers to the different classes of people who exist in any organization that is undergoing cultural change (i.e. implementing the model). 85% of the people in the organization will prefer to keep the status quo. They generally do little to help or hinder your progress. 12% of the people in the organization will be adamantly opposed to any change, since change is generally perceived as a negative impact to their personal reward and/or risk systems. This group often attempts to impede the progress of the change. The third group (3%) is made up of the natural "change-makers" in every organization. These are the people who form the core planning and implementation group and become the "right people" that help create successful programs.

The above discussion was given to characterize the environment one faces when implementing the Space Grant Human Capital Development Model in a community of government, industry, and academic organizations. The "fair broker" facilitator/coordinator is the key element to success. It is the "glue" that that holds the private - public partnerships together. For a facilitator/coordinator to be successful one must identify the people in each organization that fit into the 85-12-3 classes. Strategies must be planned and exercised to involve each group at the appropriate time and manner and to anticipate and mitigate problems in a timely way. This is not as easy as it sounds.

The real question to ask at this point is whether the above pilot activities (AERO Institute & ACES) demonstrate any applicable answers that could be applied to the future space exploration, habitation and utilization by and for Earth's humanity. Does the model provide a framework that will address the "need to develop new ways to collaborate within competitive boundaries" as stated by Dr. Ken Cox? Also does the model provide a fundamental organizational structure that could be applied to the ATWG proposed "Interstellar Space University" (The Interstellar Space University mission is to create and archive space knowledge and discoveries; provide the networking focus for space academics, practitioners and students; and supervise research that will peacefully and productively weave the future of space exploration, habitation and utilization for the benefit of Earth's humanity)?

I have arrived at the following conclusions about the past and the future potential for human capital development as humans explore, inhabit, and utilize space:

PAST and PRESENT
The U.S. aerospace human capital problems are at a crisis level and no national action plan is presently being implemented;

* Apollo era human capital approaches are difficult or even impossible to implement in the political, social, and budgetary climate that exist today;

* Nationalism permeates human capital development programs while knowledge creation and sharing is networked globally;

* Government - industry, university partnerships for R&D and human capital development are severely constrained by historical procurement regulations and processes and budgets that are limited by insufficient political will;

* Existing STEM-related civilian agency programs are under-funded (by more than 50%) and many of those that are funded are the result of ill-conceived Congressional "Ear Marks". Most don't align with any unified national human capital development strategy and many have never been peer-reviewed for quality and/or STEM workforce impacts.

FUTURE
The Space Grant Human Capital Development Model possesses the following elements that are needed to foster and maintain a successful future human capital development program:

* Grass roots, community-based human capital development goals and objectives based on aerospace-related projects;

* Private - public shared resources environment; and

* Emphasis on hands-on student - mentor learning.

* The migration of humans into space will require a global educational and human capital development infrastructure that emphasizes shared knowledge and resources;
* Although the future education and Human Capital Development Model will be practiced and shared globally, all implementations will be driven by local considerations (organizations, resources, personnel, and outcomes).

My major personal conclusion is that every local community must stop looking to their government for the answer. Government responds to the strongest demands of its citizens, and if the local citizens, whether from the private or public sectors, speak with a unified voice concerning STEM education and training, resources will be available and significant progress will be made. The Future of Humans in Space will depend on each one of us taking action. As members of the human race, the future is ours to win or lose.

References

GAO Report, (October 2005), Report to the Chairman, Committee on Rules, House of Representatives, Higher Education - Federal Science, Technology, Engineering, and Mathematics Programs and Related Trends

Hart-Rudman Commission, (February 2001) Road Map for National Security: Imperative for Change, The Phase III Report of the U.S. Commission on National Security/21st Century

NASA, (2004), Human Capital Strategic Implementation Plan http://nasapeople.nasa.gov/HCM/

Voinovich, George V., (December 2000) Report to the President: The Crisis in Human Capital - Subcommittee on Oversight of Government Management, Restructuring, and the District of Columbia Committee on Governmental Affairs

Walker, Robert S., (November 2002) Final Report of the Commission on the Future of the United States Aerospace Industry, www.aerospacecommission.gov

About the Author

Dr. Michael Wiskerchen is a faculty member in the Department of Mechanical & Aerospace Engineering at UC San Diego and Director of the California Space Grant Consortium (NASA sponsored K-12, undergraduate, and graduate educational program). Over the past 35 years, Dr. Wiskerchen has had a diverse academic and research career in aerospace-related science and engineering while in government, industry, and academic organizations. His recent efforts have been focused on the development, application, and operations of aerospace-related projects involving an alliance between industry, university, and government partners.

He has published or presented over 100 articles on space-related research, human capital development, and organization modeling. He is recognized as a national leader in hands-on career training programs at the high school, university, and industry levels. These programs emphasize the effective use of state-of-the-art distance learning (Internet based multimedia curriculum) techniques in a collaborative environment involving students and academic, industry and government mentors.

E-mail: mwiskerchen@ucsd.edu

Chapter 16

Music and Arts for Humans in Space

By Bob Krone

*"A discovery, a work of art, or a noble act
enrich the mind of all humanity."*

Michael Polyani, The Lindsay Memorial Lectures
University College of North Staffordshire in
The Study of Man, University of Chicago Press, 1959.

Throughout human history, artistic expression has enabled humans to clarify and deal with important experiences, as well as a means to examine our values and aspirations, and express them for others. The stars, the heavens, and outer space have inspired artists since the earliest days and nights of human consciousness, they still do so today, and they will certainly continue to do so in the centuries and millennia to come. So as humans venture into space in ever larger numbers, the arts will come with them, and will have as significant a role to play in space settlements as they do in Earth-bound civilizations.

My own relationship with the arts is a deeply personal one, as my father, Dr. Max T. Krone, earned his Ph.D. in Music at Northwestern University in Chicago in 1940 when I was a young boy. He had married music educator Beatrice Perham in 1936, and their collaboration over the next thirty-five years produced a music and arts legacy that lives on. Max Krone was appointed the Dean of the University of Southern California's School of Music in 1940, and during World War II he planned the creation of the Idyllwild School of Music and the Arts in the San Jacinto Mountains near Palm Springs. In 1949 Krone was appointed Dean of the USC Institute of the Arts, where music, dance, theater, visual arts, ceramics, and photography-cinematography have been taught to many generations of students. I was one of the forty students who attended the opening Summer Session in 1950.

By the end of 2005 the Music and Arts School that Max and Beatrice Krone created had served 6,561 alumni from thirty nations of the world, and forty-six of the United States.50 Now celebrating its 57th year, Idyllwild, as it is now called, is one of the world's most successful schools for music and the arts. (See Idyllwildarts.org). [1]

The values and principles Max and Bee Krone brought to their life's work was the fundamental reason for their personal successes, and for the survival and growth of the Idyllwild Campus through the early difficult years from 1950 - 1967. They also developed a library of music publications that eventually grew to over 300 compositions, including a series of folk music arrangements from around the globe entitled "A World in Tune." Each booklet began with the following:

*"Building a world in tune upon a background of hatreds, jealousies, suspicions,
misunderstanding and bitterness, will be the task of us who sing these songs
today. Barriers of language, creed, color and race must be broken down.
Somewhere, somehow, we must find a common ground of understanding, of mutual
respect, of enjoyment and enthusiasm in doing something together.*

*"Singing together is such a natural way to begin; and nowhere do we prove that
we are more alike than in our songs. We all - everywhere - sing our babies to
sleep, we sing as we play and work, we sing when we feel the need of strength
and help from the great Source, we sing to and for the ones we love, we sing
of the joy and sorrow of life, and a song follows us to our final resting place. All
over the world the same songs, different tongues, different tunes, but the same
Songs"*

The Krones: Beatrice and Max, A World in Tune: Folk Songs of Our
Inter-Americana Southern Neighbors, Book III, Kjos Music Company, 1945.

These thoughts express universal truths, and reflect an underlying hypothesis of this book, *Beyond Earth*: *"We will find a common ground of understanding of mutual respect, of enjoyment and enthusiasm in doing something together."* The something for the next human adventure of exploring, living and working in space is a revolutionary and grand undertaking that will certainly require A World in Tune. Repeating historical scenarios of the space race, or making the science fiction of Star Wars battlegrounds into a reality are dystopias to be prevented. With this in mind, let's take a brief survey to see how themes of space have been expressed in the arts.

✳✳✳

Let's start with quantity. A search of the internet via Google in December 2005 yielded a mere 149 million hits for a query on "Space Art," and 143 million hits for "Space Music." Amazon.com showed 981 "Space Music" books and 181 "Space Art" books (using a broad definition of "space"). So the instances are massive. A more narrowly focused source, the NASA/European Space Agency Hubble Space Telescope web site (www.spacetelescope.org/), returned 633 images in response to a search simply for "Art."

And what about the history of art and music? There are countless compositions of classical music that make reference to space, such as Holst's *"The Planets,"* Mozart's *"Jupiter Symphony"* (#41) and, Hindemith's *"Die Harmonie der Welt."* Popular music makes constant reference to the stars, space, and the cosmos, as exemplified by an early hit song called *"Telstar"* by The Tornados named after one of the first communication satellites; the song reached no 1 in 1962.

"Starry Night" is one of Van Gogh's most famous paintings, a vivid depiction of the night sky; he painted a *"Moonrise"* in 1889. Vermeer's *"The Astronomer,"* painted, in1668, depicts the dawn of the new Copernican world view.

Science fiction is a staple of literature, from the stories of Jules Verne to more contemporary writers such as Isaac Asimov, Robert Heinlein, Ursula Leguin, and Arthur C. Clarke. Space figures prominently in the popular imagination, as HG Wells demonstrated with his 1938 radio broadcast of The War of the Worlds, which captivated the nation and gave voice to the latent fears of millions in the gathering storm of World War II, a powerful example of how the arts can capture the popular imagination.

Non-fiction literature also links the arts with space, as exemplified by Arthur I. Miller's Einstein, Picasso: Space, Time, and the Beauty That Causes Havoc, [2] about which Martin R. Kalfatovic writing in Library Journal had this to say:

"During the span of a few years shortly after the start of the 20th century, roughly from 1904 to 1908, two quiet revolutions in how we perceive the world were underway. In Switzerland, Einstein was working on the nature of time and space. In Paris, Picasso tackled a similar problem in the creation of the seminal Cubist work "Les Demoiselles d'Avignon." Miller examines the two men and the revolutions they initiated, pulling together the lives of the physicist and the painter, as well as the band of friends, colleagues, influences, and lovers that surrounded them at that time. Miller creates a compelling argument for the confluence of aesthetics and science."

And of course there are the movies - from the B movies of the 1950s depicting aliens in flying saucers, to Stanley Kubrick's eloquent film version of Clarke's *"2001: A Space Odyssey,"* the big screen brings the Earthbound a bit closer to the experience of space. Soon after Kubrick's landmark, George Lucas introduced the world to Star Wars, which soon became one of the most lucrative film franchises of all time. Today you can even buy Star Wars M&Ms candies! Steven Spielberg then gave us a deeper look at the unexpected confrontation with other life forms in *"ET: The Extra-terrestrial"* and *"Close Encounters of the Third Kind."*

Meanwhile, Star Trek developed a loyal following on TV and then in the movies, expressing Gene Roddenbury's humanist vision of the human future in space. The large-film IMAX format, with its massive screen and surround sound, brings the experience even closer to real with numerous movies about the space, the Space Shuttle, and the Space Station.

Photography is also a means by which artists have expressed our connection with space. Some of the most famous photographs of all time are images of Earth as recorded on film by astronauts in flight. Recently, astronaut Don Petit lived about the International Space Station for six months in 2002 - 2003, and made a systematic photo survey of the world's cities at night, producing a series of hauntingly beautiful images.

Mythology, an art form that explores the depths of the human psyche, is of course intimately linked with the stars, and Raymond and Cheryl Garbos do an excellent job of recounting human relationships with the imagined heavens in Chapter 20, *The Meaning of the Heavens to Humankind through History.*" Mythology, of course, is a deeply metaphorical expression, which historian and mythologist Joseph Campbell evokes in his comment that, "Getting into harmony and tune with the universe and staying there is the primary function of myth." And then Campbell gives us a metaphor of his own, *Mythology was the song of the universe—the music of the spheres.*"

Space inspires artists, and space organizations such as NASA sometimes commission artists to express space-related themes. Since 1963 NASA has had a Space Art Program, a public relations scheme through which artists are commissioned to portray space missions that are being planned and executed. Image 16.1 shows painting by Rick Guidice that depicts the work of the NASA/ASEE Summer Study on the Feasibility of Using Machine Intelligence in Space Applications, University of Santa Clara, 1980. The study participants were tasked to define "Advanced Machine Intelligence," and then apply the definition to four original missions:

* Intelligent Earth-Sensing Information System
* Space Exploration to Titan
* Automated Space Manufacturing
* A Self-Replicating Lunar Factory.

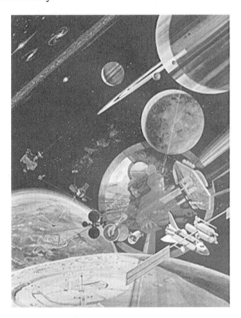

Figure 16.1 Rick Guidice 1980 Painting (See this painting in the color section)

Two Beyond Earth authors, Joel Isaacson and myself, were participants in that summer research.

Among the earliest forms of artistic expression that remain with us today are human creations in architecture such as the great circles of Stonehenge, which apparently functioned as an observatory. The Greeks and Romans built countless temples to their gods, the Gothic cathedrals reached skyward, and even today our architects design church spires that reach toward the heavens.

The purpose here has been simply to convey a sense of the enormous scope of artistic expression that has been inspired by visions - real and imaginary - of space, and of course countless other examples could be cited.

Space also presents new opportunities for artistic experience, an obvious example being dance in microgravity, which has already exploited in parabolic aircraft flight by choreographers in both Europe and Japan. Space dwellers will have unprecedented opportunities to experiment with all forms of expression as well as all forms of experience. (3)

We have countless painting of space, but how will painting in space bring us images never seen before? If Picasso lived on a space station, what would he see? Once space architecture moves beyond the purely utilitarian, what new forms of beauty will space architects create? Will films actually made in space differ from the images of space that are produced today using special effects and computers? Would a young Mozart, born to a space-faring family, compose music differently than his European counterpart? What would space-hip-hop sound like? If Ursula Le Guin lived on an asteroid, what would she write about? How would Baryshnikov perform Giselle or The Nutcracker on Mars?

If the many authors of Beyond Earth are correct in our conviction that humanity is destined to live in space, then the coming centuries will give us answers to all these questions. And even now on the Space Shuttle flights, astronauts asleep in orbit are awakened with music - from rock to classical, their choice.

Throughout Max Krone's tenure at USC (1940 - 1967) and as Founder/Director of the Idyllwild School of Music and the Arts (1950 - 1967) the following phrase was printed on the bottom of his stationery:

"The greatest use of a life is to spend it for something that outlasts it."

What this meant to him, of course, was that great art far outlives it creators, and that each of us can all aspire to express the universal truths of our humanity in the works we leave behind, and also in the way that we live our lives.

The authors and publisher of *Beyond Earth: The Future of Humans in Space* believe that we are also contributing to what H.G. Wells described in 1902 as "The greatness of human destiny." (4) And although we are as yet far removed from a world in tune, we continue to search for and work toward harmony, and we believe that the journey to space can help to bring us closer to the harmony that we all seek. The arts, in all their wonderful manifestations, will help us to understand and communicate the hopes and aspirations that lie in the profound depths of human longing and joy. (5)

References

(1) Thanks to Theresa McCaughey, Idyllwild Arts Director of Parent & Alumni Communications, and Lissa Claussen, Director of Annual Giving, for the research to find these numbers.
(2) Arthur I. Miller. *Einstein, Picasso: Space, Time, and the Beauty That Causes Havoc*. Basic Books, 2002.
(3) Thanks to Professor Jim Burke, CalTech University for the information in this paragraph.
(4) H.G. Wells, Lecture at Royal Institution of London, 1902. www.sylviaengdahl.com.
(5) I appreciate the editing and additional content contributed by Langdon Morris.

About the Author

Dr. Bob Krone is Editor of *Beyond Earth: The Future of Humans in Space*. He is in his third career as a global Air Force pilot, educator, author, and consultant in Advanced Management theory and practice; an Emeritus Professor of the University of Southern California in Los Angeles, U.S.A. (1975-1993); a Distinguished Visiting Professor in the School of Business at La Sierra University in Riverside, California, U.S.A.(1992-present); and an Adjunct Professor for Doctoral Programs (Ph.D. and DBA) for the International Graduate School of Business at the University of South Australia (1995-present). In this latter capacity he has supervised the PhD and DBA studies of forty globally based doctoral candidates as of 2006. His Ph.D. from UCLA in 1972 was in Political Science and Policy Sciences. He is a Fellow Member of the American Society for Quality (ASQ). His book and journal publications are extensive over the past three decades.

Bob's music and arts exposure began in the Max & Beatrice Krone home. His BA Degree was in Cinematography from USC in 1952. His senior project was a 22-minute color sound film, The Ballad of Idyllwild, showing the 1951 Summer classes and programs on the Idyllwild School of Music and the Arts Campus. That film was used to market the school until 1964. Bob and Sue Krone have been Board of Directors members for Idyllwild Arts (the name changed in 1985) since 1967.

Bob's first career was as a Fighter Pilot, F-105 Jet Squadron Commander flying combat in Vietnam, Headquarters USAF Personnel Officer and NATO International Staff Officer in the United States Air Force (1952-1975). He joined the Aerospace Technology Working Group (ATWG) in 1990 to keep in touch with the space community. Bob and his wife, Sue, have been continuous world travelers. They have lived and worked in thirteen countries. BobKrone@aol.com.

Chapter 17

A Code of Ethics for Humans in Space

By K.T. Connor, Lawrence Downing, & Bob Krone

Science fiction writers who use space travel as their literary vehicle frequently depict those in the far-off regions of the cosmos as villains, misfits, and weirdoes. The age-old conflict between right and wrong, the good guys and the bad guys, is the basis of Star Wars, Star Trek, Flash Gordon and countless other space fantasies. However, the authors of this chapter do not believe evil is an inevitable product of human habitation. We hold to the belief, perhaps an over optimistic one, that it's possible for human beings to occupy space and live together in a spirit of cooperation, respect, and peace.

It is not our intent to establish a Utopian society, and we understand that the humans who journey into space share the same traits as those they leave behind. They will be subject to the same inclinations, foibles, and weaknesses as everyone else. But given the vulnerabilities and challenges that those living in space will be subjected to, it becomes even more essential that an accepted Code of Ethics provide a strong foundation for a society in which an exceptional degree of mutual dependence will be the norm rather than the exception.

Such a Code of Ethics, if it is to be effective, must be accepted by people of all faiths and cultures, by who are Jewish, Buddhist, Christian, Muslim, Atheist, and any number of other religious and philosophical affiliations. And despite their differences, the one constant that connects each to the other is their humanity, so the Code of Ethics must therefore build on innate human values and respond to the human way of thinking. Describing such a Code is the intent of this chapter.

The Challenge

The exploration and habitation of space will succeed only with the collaboration of many nations, cultures, and minds. Gone are the days when one country or one church can set out alone to conquer distant frontiers. The global village has not only arrived, it has engendered the vision of universe-village, the habitation and enrichment of which will require that wealth of diversity of which our planet is bountiful. It requires a mosaic of the cultural wealth of many different minds. As these minds will be making decisions for and with one another, achieving decision-making integration will require nothing less than an ethic that is boundary-less, yet firm and engaging.

This challenge to launch an intensely collaborative venture beyond our planet is a humbling one. It calls upon us to reflect on our common humanity and to summon the best of the goodness that is within us to infuse our actions and decisions. We can squander this opportunity if we allow ourselves to be limited by the biases of a single culture, class, or country, or if we settle for less than that goodness.

Only if we fashion our efforts within an ethical system that is universally valid can we succeed in this, but accomplishing this is not as self-evident as it sounds. We will be tempted to search for "the perfect values," or "the perfect philosophical system," or "the perfect summary of what the majority of humans believe is important," but none of these will suffice. At worst they will lead us to never-ending conflict over which of the many options is "the perfect" option; at best they will provide us with an incomplete picture, even a shifting one.

What is required is an ethical system that is universally valid, one that reflects a multitude of cultural, national, religious, philosophical, and even corporate principles through a single, universal system, not because it's a statistical model built inductively out of collected values data, but because it's derived from the logic of the human psyche.

In order for it to survive into perpetuity rather than be influenced by changing mores and world views, this ethical system must have internal fitness and rigor. Otherwise it will fail to withstand the inevitable pressure for modification by external cultural, religious, and political pressures. It requires, in fact, the same rigor that is characteristic of the aeronautical and technological thinking that is fashioning the means for space exploration.

However, ethics is often considered to be a "soft" discipline, one more philosophical or religion-based than scientific. If our ethical system is to share the same rigor as other scientific contributions to the space strategy, it is important that this view of ethics as "soft" change. It can do so, we anticipate, through careful conception and design.

Does such a scientifically ethical system exist? As reasonable as such requirements sound, can they be realistically met? We say they can. Using Value Science as a base, we describe here an ethical system which is:

Universal-defining unifying ethical principles that cut across national, religious, personal, and class boundaries;
Objective-rooted in a mathematical and logical system, and independent of any one particular religious, cultural, or philosophical system of ethics;
Measurable-expressed in terminology that is reduced to logical expression;
Unifying-expressed in terms whose truth is intuitively obvious and independent of any one ethical system or expression;
Applicable-producing rules for making ethical decisions;
Binding-eliciting general commitment and engagement.

Rather than starting with maxims, this approach reaches maxims by applying universally derived principles having rigorous internal logic and precision.

The Major Source

Robert S. Hartman was a judge in Hitler's Germany, and in dismay he watched his former classmates rise to power and effect, as he put it, "*the organization of evil*." Hartman decided then to devote himself to identifying how to organize good. He lamented the fact that the technological miracles that the human mind had brought about were not matched in the moral and ethical realm, and his devotion gave birth to a dimension of value science that opens up for us an understanding of ethics which is rich and transcending, yet eminently practical. He was able to fill the gap he saw: the need for a rigorously scientific way to organize ethical behavior.

The basis for it is simple.

Human beings understand the value of things and ideas according to a hierarchy: some things are more valuable than others. As there is a deeply embedded logic to what this hierarchy is, this logic is amenable even to mathematical formulation.

Humans tend to agree that people are more important than things, and things are more important than rules or concepts. Most people would save a person from a burning building before they would save a burning chair. There is an internal logic, mathematically describable, that accounts for these judgments. This is not to say that we are all unanimous or uniform in our expression of this logic all of the time, for variation is part of humanness. What we cannot deny, however, is a logic of valuing that exists, and that in our clearest and best moments it is recognized universally and implicitly.

The hierarchy exists not because a majority of the human race were polled and agreed to the values, but because of a simple axiom that is self-evident. We do recognize that there are people who do not hold these beliefs and who do not place the high value on people, but society, in general, considers those who value things above people to be pathological.

The fact that as individuals we vary on their acceptance of these principles at one time or another does not change the logic. That thing which most fulfills its idealized concept is more valuable than that which fulfills it to a lesser degree; that which possesses most richly the qualities proper to it is that which is the most worthy and the most valued.

Hartman reasoned that there were three kinds of value: Intrinsic, Extrinsic, and Systemic. **Intrinsic values** are infinite and can not be exhausted. Nor can they be subtracted from. John is always John, yet you will never exhaust the dimensions of John. Moreover you can never make John less John. Because intrinsic realities are unique, only one of each exists, and that one has all the characteristics of its definition.

Extrinsic values are also infinite but can be added to and subtracted from. A chair by definition has a limited number of characteristics but there are an infinite number of chairs that could fit that definition: tall chairs, short chairs, wood chairs, metal chairs, and so on. Thus, chairs can be more or less good in being chairs. Some could only have soft cushions, others more durable seats, or have rockers. Such a chair might be deemed a better chair than another.

Systemic values are finite. Only one thing passes for a circle, and if those conditions defining a circle are not met, it is not a circle. It is perhaps a good arc, but not a circle. Systemic values either are or are not. It's not possible to be more or less a circle in the systemic sense. Of course a child's drawing of a circle can be a bad circle. But that is an extrinsic reality, not a conceptual or systemic reality. Conceptually, we know that a drawn circle that looks like a watermelon is not a true circle.

Applied to Ethics

All three value dimensions are important to an ethical environment.

Systemic thinking provides necessary definition and consistency, and clarifies moral code. It determines whether or not something is right or perfect.

Extrinsic thinking allows us to compare one value with another and choose the better option. It determines whether or not something fits a need.

Intrinsic thinking establishes the value of the object (often a person) in itself, and creates in us an ownership and identification with the choice. It determines whether or not something is unique.

Moral choice, then, depends on all three kinds of value. In the ideal ethical situation, moral code determines the boundaries, differing actions are weighed and choices made, and all the while each person, true to him or herself, owns their potential for good and bad, and commits to their principles. And those principles add value and protect the relative value of the logical hierarchy: first people, then things, then concepts.

Based on this foundation, we summarize as follows:

Our hypothesis: An acceptable Code of Ethics can be designed for those who migrate to space.
We believe: A code of ethics is needed for humans in space.
Our conclusion: Without an accepted code of ethics, humanity will export its history of competitive survival-of-the-strongest into space, and some will continue to exploit and denigrate other human beings.

Ethics for Space

In space and on the way to space we will be faced with decisions requiring us to sort through difficult options. Upon what can these decisions rest in order to make these decisions ethically, that is, with integrity and responsibility? When controversies arise, how does one discriminate among the choices to make the most ethical decision? In space this is a particularly critical question, because of the probability of meeting situations for which there is no earthly precedent.

We propose, as a response to this question, a system of decision-making that takes advantage of the logical thoroughness of the axiological process (from "axios" meaning worth, worthy, valuable; thus, a science pertaining to or studying values) and yields guidelines for ethical decision-making. Given the complexity of "reality," we need a decision-making framework that reflects such complexity while enabling us to see the issues at hand with more clarity.

Our model reflects intrinsic dimensions, extrinsic dimensions, and systemic dimensions, as noted in the vertical columns. The horizontal rows are labeled "Uniqueness," "Function," and "Structure." Uniqueness, because for any concept it is possible to identify the results when the uniqueness of that concept is valued intrinsically, extrinsically, and systemically. The same applies to function, which includes action, parts and steps; and structure, which includes principles, order, and meaning. In looking at the concept of "ethics for space" the following systems will result from a matrix view of these factors, charted accordingly:

	Valued Intrinsically	Valued Extrinsically	Valued Systemically
Uniqueness	Life System	Individual Ethical System	Growth System
Function	Social Ethical System	Operations System	Logistics System
Structure	Personal Principle System	Mission System	Codes and Standards System

Source: ©Wayne Carpenter, Axiometrics International, Inc.

Figure 17.1

The components of these systems will be guidelines for actions, and will follow the logical hierarchy inherent in Formal Axiology, or Value Science. According to this reasoning, intrinsic is more valuable than extrinsic, which is more valuable than systemic.

A clear set of prioritized guidelines will result, looking like this:

1. Life system
Guidelines for identifying the value of life, for respecting and valuing life.

2. Individual Ethical System
Guidelines for building moral and ethical decisions.

3. Growth System
Guidelines for managing innovation, change, and risk

4. Social Ethical System
Guidelines for maintaining social cohesion

5. Operational System
Guidelines for prioritizing day-to-day tasks

6. Logistical System
Guidelines for implementing strategies and principles

7. Personal Principle System
Guidelines for integrating mission requirements and personal commitments

8. Mission System
Guidelines for prioritizing actions according to the vision and purpose of the mission

9. Codes and Standards System
Guidelines for defining and storing principles, rules, and codes.

Each of these systems is then divided into 9 guidelines, three intrinsic, three extrinsic, and three systemic.

These guidelines in turn generate decision applications. Because of the structured basis of the systems and guidelines, it would be possible to create a computer generated reservoir of the logical/ethical metasystem that could produce guidelines for decision-making, thereby providing a neutral "third-party" counsel. In this way, the desired universality of the ethical system can be retained and decisions removed from the arena of particular cultural and ethnic perspectives.

To illustrate the process on ever more specific guidelines being produced, consider the first system, the Life System.

	Valued Intrinsically	Valued Extrinsically	Valued Systemically
Uniqueness	Life is unique and valuable in itself; it is irreplaceable and infinite.	Life is communal; it exists in groups which share and duplicate its uniqueness, function, and purpose.	Life is evolving and creative; it exists in a time sequence flow which grows and changes.
Function	Life is autonomous and independent, bounded by a uniqueness, function and purpose which defines and sustains its existence.	Life has concrete existence; it has physical presence.	Life is purposeful; it evolves according to a blueprint and has a destiny.
Structure	Life is conscious and interconnected through a unity of conscious experience.	Life is bounded by space and time and responds in the context of space and time.	Life is structured; it reflects a blueprint, strategy and model which defines and sustains its specific forms and changes.

Source: ©Wayne Carpenter, Axiometrics International, Inc.

Figure 17.2 The Life System

Corollaries flowing from these would be as follows:

Human Life

PRINCIPLE: Life is unique and valuable in itself; it is irreplaceable and infinite.

COROLLARIES:
Life as found in space deserves respect and reverence until it proves itself harmful to other life.

No society, nation, universe, or other special grouping of life is more valuable than each individual expression of life.

No principle, code, dogma, rule, principle, or collection of them is more valuable than each individual expression of life or more valuable than a society, nation, universe, or other social grouping of life.

PRINCIPLE: Life is autonomous and independent, bounded by a uniqueness, function and purpose which defines and sustains its existence.

COROLLARIES:
All living things, including the universe of all things, have unique and irreplaceable value.

All actions will be based on the responsibility to preserve the uniqueness and worth of all living things.

No action that destroys or disrespects life is acceptable.

No experiment to study and understand foreign or alien objects is acceptable if such study destroys the value of the living thing.

No action taken in the name of progress or no advancement in exploration will be acceptable if such progress and advancement violates the unique worth of that which is studied.

No idealistic dream, dogma or principle such as "manifest destiny" will be employed to justify the destruction, disvaluation, or disrespect of living things unless those living things harm others.

PRINCIPLE: Life is conscious and interconnected through a unity of conscious experience.

COROLLARIES:

All forms of thinking, all ideas, all principles and theories will be respected, as long as they do not violate the unique, infinite worth of life or do not violate the unique, irreplaceable worth of living things.

No prejudice or bias against any organization or definition of society will be permitted.
Communication of beliefs, values and principles will reflect the interconnectedness of the common conscious experience.

Divisive forms of communication will not be acceptable.

A common language will be available to all.

Rules governing belief systems, principles and theories will take into account the unity within the diversity of thinking and believing.

Punishment for diversity of belief is not acceptable.

Personal Ethical System

When the Personal Ethical system is analyzed according to this logic, it will yield a framework that is something like this:

	Valued Intrinsically	Valued Extrinsically	Valued Systemically
Uniqueness	Love self	Own your actions.	Embrace growth.
Function	Love others.	Produce good results and harm no-one.	Act efficiently.
Structure	Love life.	Let principle guide your action.	Be clear about your purpose.

Source: ©Wayne Carpenter and K.T. Connor

Figure 17.3 Personal Ethical System

The full development of a code of ethics would follow this structure to arrive at a complete expression that would account for all nine boxes on the matrix of systems.

And as this is being done, the question may arise as to what role God and religion have in all this. In fact, we see these as two different questions. Religion is a culture-bound way to make sense out of the Source of life. God is understood to be that source, however defined. Since our minds are wired to apply the same logic to God, we see a nurturing parent, a provider, and a rule-giver, these being Intrinsic, Extrinsic, and Systemic realities. We respond in similar ways: we love God, we ask God for favors, we obey God, in line with whatever we consider God to be, Intrinsically, Extrinsically, and Systemically. Through all this, God is the source of life, and life is the keystone of ethics. The various religions, responding to this hard wiring, tend to emphasize one or more of these concepts and responses at various times. That is why a logical framework is so critical for a universal ethic: it takes it from the realm of religions so that it is applicable to all, regardless of their religion, even as it is consistent with the best in each one of them.

Consequences of Failure

We can also look at this from the opposite perspective: what if we fail to achieve consensus on a system of morality and ethics for space? Then, clearly, the pathologies that have plagued humanity on Earth for millennia will propagate to space.

Yehezkel Dror, a key thinker in the field of Global Policy Sciences, and author of Chapter 5, *"Governance for a Human Future in Space,"* has described the relationship between improvement in the human condition and our process of governance on Earth this way:

"If we want to achieve changes in basic social realities, such as reducing human suffering, eliminating warfare, and increasing global equity, and if we want to do so democratically and in accordance with the will of citizens, governance must influence accepted notions of a 'good life'."

Yehezkel Dror. The Capacity to Govern: A Report to the Club of Rome,
Frank Cass, London, Portland, Oregon, p. 19. 1994

We must define the 'good life' in a way that is consistent with the long term well being of humanity and of all life, and such a definition requires by its very nature an effective, understood, and well applied ethical system.

Conclusions and Future Visions

Our early 21st century era is characterized by uncertainty, complexity, novelty, and adversity. There is much dysfunctional about the family of humanity, and we are daily witness to unethical and immoral acts that result in destruction. Humanity's greatest fear is that there are inadequate constraints against the immoral unleashing of weapons of mass killing, and history gives us reason for serious concern about the feasibility or universal acceptance of a Code of Ethics-even for the few who may choose to live and work in space. But our very project, the expansion of human presence outward, itself offers a new hope:

"For we now have the extraordinary, encouraging circumstance of being able in effect to restart civilization under the best circumstances possible, learning and demonstrating how the evolving New World could become a peaceful one for all of its inhabitants."

Thomas F. Rogers, Chapter 7, *"Creating the First City on the Moon"*

Our challenge in planning for future space civilizations is thus to envision space settlements as positive role models for Earth-bound societies to emulate; defining ethics and morality needs to be a key to planning, governance, law, commerce and living in space.

We believe that *"When in doubt, choose optimism, then manage wisely to achieve a self-fulfilling prophecy."* This book is filled with optimistic views for the future of humans in space leading to revolutionary changes for humanity's betterment on Earth, but planning for the future of people in space cannot be left to unguided, natural evolution. It must be purposefully directed.

As people migrate into space, sociology and politics will go with them, and ethics must as well. The Governance Model of Yehezkel Dror, presented in Chapter 5, will be an influential driver for the ethical code, and the Space Law of George S. Robinson in Chapter 6 will facilitate its implementation. Humans will modify their politics, laws and codes as life in space progresses; but as we wish to avoid exporting conflict to space, a carefully thought out and broadly accepted ethical code is needed from the outset, and is well within our reach.

About the Authors

K.T. Connor, PhD, founder of the Center for Applied AxioMetrics, is an OD (organizational development) specialist with a focus on Organizational Climate and Performance. For the past 15 years she has been a practicing axiologist, applying the metrics of axiology to the analysis of decision making and thinking processes. She has been involved in extensive research in Value Science through Wayne Carpenter and Axiometrics International, Inc. Her client list includes Merrill Lynch, IBM, GTE, PSEG Power and others. Her PhD is from the University of Southern California and her Masters is from Case Western University. She did her undergraduate work at D'Youville College. She is author of Rethinking work, and numerous articles on decision making and thinking. She resides on St. Simons Island, GA from which she travels to serve clients around the country. http://cfaam.org

Lawrence Downing, D.Min. is senior pastor of the White Memorial Church in Los Angeles and serves as an adjunct instructor at La Sierra University School of Business and School of Religion. He has written numerous articles for various Adventist journals and has served in churches on both Coasts. His B. A. degree is from Pacific Union College. He has a M. A. and B. D. from Andrews University and a D. Min. from Lancaster Theological Seminary.

Bob Krone, Ph.D. See Chapter 16.

Chapter 18

Children's Visions of Our Future in Space

By Lonnie Jones Schorer

"My vision for space travel is that NASA and other space flight companies will make commercial launches to outer space and it will become as common as airplane flight. People will go on flights for business, pictures, reports, interviews, and just for fun. If all this happens, NASA will get more money, and we will be able to add to the space station or explore deeper into space. We will learn more about our Universe. It all benefits from this project. That is my vision for space travel."
Graham Haydon, age 10, Pennsylvania

I. Children's Visions

The purity and strength of thought expressed by children is born of dreams and wonder. Unfettered by political goals, economic constraints and technical feasibility, children's visions are often idealistic projections of our own. Free to travel to the outer reaches of imagination, to life in parallel galaxies and to worlds beyond our Universe, children's visions project unconditional hope and optimism for the future. They are not concerned about what has never been done before or about what others say can't be done. Each time we become bogged down by bureaucratic obstacles, were we to retain a child's sense of wonder, we might more easily return to the basics to focus on the essence of our ideas. To do something complicated, we have to go back and remember how to do something simple. By working with children, we may be able to travel to the future with a more insightful understanding of humankind's future presence in space.

An essential part of opening new frontiers is getting people interested and involved in the incremental steps of the planning process. Although there is historic precedent, we can't just round people up and ship them off to new worlds. A period of education, preparation and planning, as well as the desire to go, is an essential prerequisite. But a harsh evaluation, expressed by students ages 5-18, is that they do not feel they have a part in the planning process for the opening of the space frontier. They assume that exploring and accessing space is the exclusive prerogative of an elite few. To lay the groundwork and make the space frontier a tangible reality for future generations of space travelers, those who may be tomorrow's space pioneers should already be involved in contributing their visions so that they will understand and believe in pursuing the goals. If they become aware that there will be opportunities and that not everyone involved will be a rocket scientist, when the time comes to establish space communities, a broad spectrum of the populace, representing all trades and professions, will already be educated and prepared to participate.

In a *Kids to Space* effort to involve children in our decisions for tomorrow, between January and June 2005, six thousand students (ages 5-18) in schools throughout the US, responded to the question: "What is your vision for America's, or the world's, future in space?" Parts II, III and IV of this chapter summarize many of the children's responses, and Part V is the author's overview.

II. Visions of Going to Space

Younger children's imaginative and futuristic visions of space are more inspired by Star Trek, Star Wars, Space Jam, the Jetsons, Phil of the Future, Daul's *Charlie and the Great Glass Elevator*, Scholastic's *Magic School Bus*, Battlestar Galactica, and a wide selection of other-worldly video games than by the daring accomplishments of the Gemini and Apollo missions that took place long before they were born. Space, for elementary school students, is an innovative, sci-fi, warp speed fantasy, filled with aliens, galactic wars, laser beams and interstellar, intergalactic travel via teleportation and faster-than-speed-of-light starships.

Those who are five, six and seven, wonder if stars are sharp when you touch them, and they want to know if there will be in-flight movies and snacks. They pretend to be Buzz Lightyear, Luke Skywalker and Jedi knights. For them, space is bright light and lots of colors, with lines painted to connect the constellationsso we will not get lost. It's an interesting place with new things, mostly to read about and draw. They are not ready to think about leaving family members or being far from home.

When they reach the age of ten and eleven years old, their imaginations tumble with fearless feats of daring. They want to rocket to space with a bunch of kids to take a walk, float in weightlessness, do triple flips and play hide and seek on the Moon. They imagine being on spaceships surrounded by threatening aliens and then making a successful escape to Mars. They envision exploring solar cities and other worlds in hovercraft or with personal jetpacks. They wonder what there will be to do when they get there and whether their pets can come along.

At ages twelve and thirteen, they envision extreme sports such as skateboarding in lunar craters and rocket racing around cities in distant solar systems. Anxious to know that their familiar foods and 'amenities' will all be there, they envision malls and food courts, soccer fields, roller coasters, and hotdog stands. They would live in rocket-propelled houses. They want to know who else is going, how long it will take to get there, where they will stay when they get to space and how long they will be gone. They want to help with experiments and suggest that scientists should live in space so that they can be closer to their work.

As they mature, many of their visions become more socially oriented. At ages fourteen and fifteen, students are starting to think about the less privileged and the homeless and they wonder whether the US should be spending money on space efforts instead of addressing society's needs. Others of the same age recognize that overcrowding, pollution, global warming and depletion of resources on Earth may be urgent reasons to accelerate the exploration of space and that, although humans will still live primarily on Earth, it's just a matter of time before people will start to move off of Earth and onto other planets. They know that those born in space will someday exercise the option of spending their entire lives off-Earth. As they begin to outline the needs of the first family neighborhoods in space, these students envision that within thirty years, people will refer to hometowns on the Moon.

Most technically-oriented high school students are interested in the survival aspects of life in space, such as how to breathe, and protection from cosmic radiation. Those who would themselves choose to be space pioneers give practical suggestions for active settlements on the Moon, with cameras on planets and TV transmissions from the Moon so that the rest of us can get used to the idea and see what's going on. They see these first communities as necessary prerequisites to exploring greater portions of the Universe. This age group is also quite occupied with what they will wear, not only in terms of safety, but also in terms of appearance. How will they look to others?

Starting to focus on their interests and plans for future studies and pursuits, some sixteen to eighteen year old students hope for fuller utilization of facilities such as the International Space Station and would like to have space preserved as a purely scientific frontier, with laboratories funded by private companies for the purpose of scientific research and development of resources. Some feel that the only people permitted to travel to space should be those who are trained to pursue and enhance scientific knowledge.

Many older students hope that the opportunity to learn about spacewhile in spacewill one day become part of their own school's standard curriculum. Eager to participate in discovering the secrets of space, these students question the possibility of life in space. They seek ways to understand intangible concepts and are trying to grasp their own role and place in the Universe.

III. Money Making Visions

Some high school students believe that space will be a purely commercial endeavor - a weekend, spring-break destination, a major business enterprise, a space Las Vegas of the future, and the world's #1 tourist attraction. With casinos, basketball courts, golf courses, rock climbing walls, gravity parks, swimming facilities, hotels vacation resorts, restaurants, amusement parks, rovers, hovercraft, rocket cars and robots, it will be another great place to visit. They predict that space will be a Mecca for the entertainment and extreme sports worlds, as well as a corporate convention center for major businesses. One student expressed the idea thatas long as survival and commerce in space don't spark another reality TV show, she's all for it!

Figure 18.1 Pam Cameron, age 17, Arkansas, *Kids to Space*

As a regular tourist attraction, in spite of the initial high cost and probable discomfort of space travel in their lifetimes, the optimistic majority of high school students foresees that people will, nevertheless, pay for the experience just to find out for themselves what it's actually like to see the planet Earth, an orbital sunrise or the first footprint on the Moon. They will want to go for the memories and they fully expect to be influenced by the space experience for the rest of their lives.

Expressing impatience for the time when business initiatives provide backing to the commercial space effort, giving people the opportunity to travel on interplanetary cruiselines or to vacation at all inclusive orbiting hotel resorts, many high school students predict that as the space effort becomes more technical, access to space will become more affordable. As the airline industry switches to fuel efficient space planes and provides faster, safer ways to get to space and back, space travel will become safer and more efficient and, eventually, as commonplace as airline flights. They foresee space companies operating out of spaceports and privatized space launches taking people to space as often as they choose. These students hope that an average person will be able to visit space and experience being off Earth at least once in a lifetime.

As middle school and high school students imagine trips on space tour buses, with stop-offs at colonies along the asteroid belt, sightseeing tours of a huge space museum on the Moon, visits to bubble-domed colonies on Mars, time at intermediary planet-viewing stations, field trips with science classes to take part in space research projects in space station laboratories, and semesters abroad attending colleges and universities in space, some mention the endless possibilities for the advertising business. They also look forward to a high-tech products boom, and would-be space travelers admit to very high expectations as they project completion of complex transportation systems and infrastructures for space in the near future. The following comment is representative.

"Within the year 2048, the US should have at least established a moonbase. A spacecraft that could be reused should be achieved, and normal transport of ordinary citizens should be happening or about to begin. A craft able to go to Mars should be developed, and a Mars landing should be achieved by 2052. By 2070, a lunar colony with permanent citizens should be established, and by 2090 normal life on the Moon would be a normal happening. Beyond the year 2090, space is the limit with a permanent colony on Mars and beyond."
Taylor Blunt, age 16, Virginia

IV. Problems With Going

Figure 18.2 Makenzie Aafedt, age 14, North Dakota, *Kids to Space*

On the other side, however, insufficient resources, lack of cooperation and an America not ready for commercial spaceflight and exploration of the cosmos are some of the reasons that middle and high school students list for not going to space. Some don't think we currently have the technology to harvest space resources, and they judge our space efforts premature by at least two hundred years. They believe that if people were to visit space now we could cause problems and damage to ourselves as well as to the Moon and the planets.

Those few students who neither support nor believe in our future in space don't think it will ever happen. Problems will occur. It simply won't work. It seems expensive and not likely for the debt we have. Or, they say, it's a great thought but not likely for regular people. The end of the world will come before we ever live in space. Those who say 'no' to our future in space are joined by others who have no vision for our future in space. Statements by some thirteen to fifteen year old students convey a general indifference (probably not reserved for space) such as, *I don't really have a vision because it doesn't matter to me; it is not important to our everyday lives; I don't think about this subject very much; this is probably the first time I have thought of it, so I don't know; we were put on Earth for a reason, and I think that's where we should stay; what's there to envision, anyway? We know almost all there is to know about space.,*

Other students express that although it is great to be learning more about space as an interesting hobby or project, it's unethical to concentrate on what's happening outside this world when there is so much that still needs to be fixed inside this world. With space money chalked up as being a frivolous waste of time expenditure, they suggest we abandon or delay space programs and use the money for such things as helping the unemployed, the elderly and cancer research. They want to solve problems on our own planet and pursue goals that will make a difference on Earth before we go running off to explore the stars or move to Mars.

A pessimistic minority of high school students sees interest in space exploration waning, predicting that space will be empty because no one will want to go there. In the "been there, done that" mentality, someone has *already* landed on the Moon. These students feel that space has become a political issue, and they count on underfunding to lead to the failure of space programs. They do not hear a call to explore.

In this pessimistic minority, it is interesting to note that some of the more articulate students find themselves caught in a mental conundrum. While they clearly communicate the problems with going, they also allow for the possibility that evacuation to space could follow a natural disaster or if destruction of Earth's ecosystem forces us to look for a back-up planet. They foresee that when we deplete Earth's resourceswe will have to pack into space colonies, and they also foresee that in the future, as medicines become more advanced and people live longer, over-population on Earth may drive people to live in space. They forecast that people will eventually destroy Earth with chemicals, wars, or drugs, and will have to make a civilization on the Moon and, when that is destroyed, start all over again on Earth. They are not able to reconcile their beliefs about not allocating money for space with the idea of having technology and destinations already in place in order to move off Earth. For now, they just want to leave space alone, with no human intervention or change.

Students of every ageapproach the topic of space with fear—fear of the unknown, fear of heights, fear induced by media memory, fear of claustrophobia, fear of motion sickness, fear of leaving home, fear of being lonely, fear of dying, fear of taking over space and turning it into an orbital junkyard, fear that we'll just make it another tourist resort and ruin it. . . Their outlook for humankind in space is grim and they only think of space as dark, black, scary, big, cold, and without end—a huge infinity jar that goes on forever. They know about the Challenger and they witnessed the Columbia tragedy on TV and they associate space travel with being blown to smithereens on the way, 'like that teacher.'Their vision is that many people will be killed, and space operations will be cancelled. Afraid of the life-threatening risks, they are not against space exploration, but would send robots rather than people.

And there is another group, primarily high school boys, that predicts territorial warfare in space will be an inevitable consequence of nations pursuing their own interests, programs and 'greed-based schemes.' They envision every space-going nation having its own space station and space colony, with remote-controlled satellite destroyers in place to prevent orbital warfare. Some predict that in 2161 the world will go into a war that will completely eliminate the human race. Trying to monitor and prevent space warfare will be a deterrent to progress, an effort that will consume tremendous resources and time, so perhaps we shouldn't go at all. In hopes of getting there and co-existing peacefully, students want to know that there will be no missiles allowed in space; others express concern that America, dedicating funds to policing the world, will fall behind and leave space development to the Chinese.

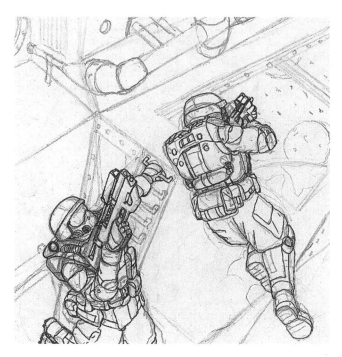

Figure 18.3 Christopher Pfluger, age 17, Virginia, *Kids to Space*.

V. Author's Overview

Because some elementary students are as informed about space and as able to express their visions as some high school students, it's difficult to categorize their visions by age group. In fact, students of all ages may be represented in each of the following paragraphs..

Most of the nearly 6,000 students who responded to the *Kids to Space* survey revealed that they view space exploration as a risky but inevitable adventure into the unknown for all humankind. With so much out there still to discover, students seek assurance that space will be not be the final frontier, but will be a limitlessly expanding frontier, as we take all we think we know, test it and continue to make new discoveries to add to our existing knowledge of the Universe. The idea of a final frontier implies to many that we are approaching the end of our journey of discoverybut they feel we are at the beginning of the journey, and that one day new galaxies will be the next frontiers. One day we will even reach another universe and be amazed by what we find. As we travel to the future, they want space knowledge to be easily accessible and continually utilized, for the benefit of all on Earth.

Many older students feel that our successful future in space will provide us with a better life and enhanced comfort here on Earth. In looking for ways to become more knowledgeable about space, they hope to learn as much as they can about Earth's past, present and future. They feel we could be a better, more advanced species by taking our studies to a higher level, and their hopes for eventual expansion and peaceful colonization of the entire Universe are often coupled with ideological demands for smoke-free, alcohol-free, drug-free, multi-national environments, devoid of national boundaries, political controversies and Earth's problems. They envision places where people from all over the world can work and live together in harmony, exploring space peacefully; places where geologists mine valuable minerals from different planets; scientists cooperate to find new cures for diseases; translators mingle purposefully; and all efforts are for beneficial purposes and the betterment of humankind. Their mantra is peace in space, with affordable space travel for all. According to their visions, people will stop fighting each other to go and explore the solar system.

Many students who feel strongly about our presence in space feel it is a mandate for us, as space explorers, to use probes to keep looking for possible life forms on other planets and beyond our solar system. They want to participate in more complete exploration of planets and moons to find destinations which may be inhabitable. They want to make unknown regions of space known to all mankind and compare the upcoming adventure to Columbus' voyages in search of new worlds. They believe in the vision that, for some of us, space will soon be a home where, in individual homes and facilities in cities just as on Earth, we will lead daily lives and have daily routines. And, although they would not want us to get carried away putting things in space, they foresee bringing soil, trees and other vegetation to the Moon in order to create some semblance of a familiar surrounding.

With America as a catalyst for technological expansion, high school students are in favor of utilizing our potential and our resources to expand perceived limits and to go for it! Irrepressible ingenuity, coupled with more advanced space technology, is a key ingredient in their expectations for the future. Students of all ages think of space as being cool and 'sciencey' as they foresee weird looking architecture inside oxygenated bubbles, domed craters with interconnecting airlock doors, greenhouses with gardens attached to houses that look like igloos, floating cylindrical space condos with lots of windows, self-propelled computerized houses, high tech shielded environments, oxygen tubes connecting Earth to planets and the Moon, gigantic residential space stations - miles in length and orbiting not far from Earth, space elevators, artificial planets, induced gravities, robotic assistants, magnetized shoes and scores of new inventions, tools and ways of helping to sustain human life. As futurists, many foresee fleets of bigger rockets that can hold more fuel, can go farther and can carry 100 people at a time; people buying and operating their own shuttles to engage in private exploration of space; mini computerized rocket cars; people flying with artificial wings. As the level of technology and knowledge increases, they see spaceships that become obsolete in favor of travel at enormous speeds to planets and to the Moon via time tunnels, or by using some sort of jump node device. They foresee us tripping into new dimensions and patiently project that if our technology continues to increase at the current rate, in millions of years we will find an alternate universe. Although this future will be very high tech, students predict that as space becomes part of our normal, known world, what initially seems like high tech stuff willeventuallyalways seem just normal.

"I believe that the future of America's space activity is bright. With continued innovations by the Government

and civilians alike, I think the frontiers of space are a lot closer to us than we think. To make these dreams come true I think scientists need to study faster and more efficient propulsion systems, and they need to experiment with creating atmospheres on inhospitable planets."
Chad Harker, age 17, Utah

Whatever the divergent ideas and visions for humankind's future in space, most students express a protective interest in Mother Earth. In terms of viewing Earth from space or in evaluating the effects of space travel on our environment, they are both curious and concerned about our own planet.

With respect to the future, many students eagerly anticipate things we cannot yet imagine, saying that in the past the idea of airplane flight wasn't even an idea—so in the future, anything is possible. They are ready for quantum leaps in technology and feel that as space efforts continue and we become more advanced, we will figure a lot more things out and will know a lot more than we do now. Expressing the belief that space travel will encourage young people to make great accomplishments and to do great things, the majority seems mentally ready for the adventure and the quest. Equating our future in space to the pioneers' westward expansion in the 1800s, the most adventuresome students express eagerness to experience new surroundings and new ways of life.

Most students anticipate that space travel will all be an amazingly exciting 'far out!' experience. They want to be able to go as children, and then they want to go again, taking their own children to the Moon for vacation. They hope that many more people can travel to space but at the same time, they don't want the experience to become so common that it's taken for granted. They think America in space will be very different than Earthbound America is now. As technology and space research improve and as we advance with further studies of the Universe, they believe that people involved with space will be very successful and will do a lot for us. Most of all, with the belief that there is no limit to what we can do together, students would like to be informed, integrated, contributing participants on our journey to the future.

Figure 18.4 Kids in Space. Compliments of MacDonald Douglas Space Systems

*For further insight into what students are currently thinking about space, including statistics on how many of those surveyed would personally choose to go, please refer to *Kids to Space: A Space Traveler's Guide* by Lonnie Jones Schorer, CG Publishing, Apogee Press, March 2006.

About the Author

A native of Connecticut and a resident of Virginia, Lonnie earned a BA in Russian from Connecticut College and a Masters of Architecture degree from Virginia Polytechnic Institute and State University. Living overseas for more than 20 years in Thailand, Turkey, Italy, Norway, and the former USSR in support of her husband's State Department career, she raised a family and was professionally engaged in UNESCO World Heritage programs, as well as in land planning, residential, commercial, and architectural preservation projects. In pursuit of visionary goals, as former head of design for a new-concept 43,000-ton community at sea, the World of Residensea, and currently as director of Global Space Travelers, Lonnie believes that the future of space travel, with spacecraft ferrying passengers and supplies to destination spaceports, could be successfully modeled on the cruise ship industry.

With her Apogee Press book, *Kids to Space: A Space Traveler's Guide*, April 2006, she hopes to inspire and involve today's youth in our space plans for tomorrow.

Chapter 19

Sowing Inspiration for Generations of Space Adventurers

By Becky Cross

The quest for the stars and our ability to make our dreams manifest is highly contingent upon our unwillingness to defer those dreams. We need to be inspired, determined, bold, and open-minded to make it happen. We also need to share what we know with those who don't including students, parents, and educators who are preparing for the future every day so that they will come to understand the opportunities that space may soon provide. Humans around the world have repeatedly shown a strong desire to help one another in times of adversity. Space offers unlimited potential benefits for all of humanity and the opportunity to do so without needing a catastrophe to make it happen!

Our efforts to explore space have already had an enormous influence on the technology we use and how we use it. Likewise current and new space technology has great potential for the development of new commercial space ventures that will lead to future jobs. Historically, the inspiration and push for space has also driven our education systems through its natural intriguing, innovative nature. Conquering the unknown is not only highly creative it can also be highly motivational for our educational environments and for spawning new commercial space ventures (future jobs). By sowing the seeds of inspiration and opening the doors of opportunity, we will see the growth of a new generation of space adventurers. We simply need to put our minds to it and make it happen!

Stimulating and Pushing Educational Boundaries

Over the past eight years I have interacted with more than 15,000 students from kindergarten through high school. I have presented them with images and ideas about the human adventure in space; discussed the possibilities for the future; and made surprising discoveries about their own beliefs, attitudes, ideas and aspirations concerning space.

For example, I understand that children have a natural fascination with space at a very early age (Pre-K-3rd grade). There is a dreamy quality to their perceptions and the nature of the vast unknown is not as daunting to the majority of them as it is to many adults. They simply take it in. Young children can readily picture themselves in a space environment and often ask questions about day to day activities like, "How do you go to the bathroom in space?" To them, day to day living in space is as easy to imagine as their day to day living here on Earth. The obstacles that often concern adults about future space are merely "things to figure out" to young children, and they pass them off with a shrug while absorbing new information at an incredible rate. They are open-minded and creative. They don't experience a sense of futility at the magnitude of the challenge, nor do they get fixated on budgetary limitations that prevent them from considering what it means to reach beyond our planet. Simply put, with this age group if they are interested, there are no obstacles in their way, and they simply want to go!

Many older school-aged children (4-6th grades) still look at space as something really cool, but they have the attitude that it is something that only a few lucky people get to do, and that it costs a lot! They, like many others, do not understand the extended benefits that space development offers our society. Many would love to go and they certainly think about what it would be like, but by this age most have already determined that the odds are stacked against them. It might surprise you to learn, however, that children in this particular age group are fascinated with money! They like to know what things cost (and why) and are excellent candidates for teaching good business skills to promote future space ventures because of their interest in the topic. Unfortunately, when these students are surrounded by parents, teachers, and mentors who have a lack of knowledge or even the slightest degree of negative attitude about current space, they easily become jaded. This in turn becomes a great deterrent to their creativity, imagination, and desire to get involved. They don't think they can.

There is a wealth of multimedia material on the topic of space that is well suited for this particular age group, but many don't connect with it. This group can be highly motivated, however, when given the opportunity, so it is essential that we bridge this gap by giving them the tools they need to get involved. Hands-on after-school programs that engage these kids in the topic is one way to do so.

Very few 7th-12th grade students see space as a career opportunity unless they are already intrinsically motivated or mentored by someone else who understands the potential. Most are not aware of the opportunities that are soon likely to emerge through organizations like NASA, nor do they realize the larger potential that's soon to be available through commercial space ventures around the world. A quick look at a Federal Aviation Administration (FAA) map of current and pending spaceports in the US has helped me open their eyes to how close our reach into space is and how different the world can be in only a matter of a few years. An overview of worldwide space efforts expands their view making the potential for future space activities seem closer to their own experiences. Efforts to share this knowledge more broadly can affect their perspective.

7th and 8th graders are a highly tumultuous and unpredictable age group, and most are not yet focused on their own futures. However, they can readily be encouraged to "take another look at space" under the right circumstances. Offering students of this age group new opportunities that they perceive to have personal value to them is one way to get them involved. When they get involved, they can be very dedicated. Many space education organizations including NASA, European Space Agency (ESA), and others, are currently seeking to gain their interest through fresh approaches that tap into their existing interests. These approaches include games and programs that meld space with their most popular past times including: sports, gaming, music, Web-surfing, and Internet chat rooms.

Once students enter high school most have already decided if they are going to enter a science (space-related) career. Based on current space planning by NASA, the FAA Office of Commercial Space, other international space organizations, and privately held space companies, within as little as 10-15 years a significant number of people could be living in space, and space travel (tourism) may become more commonplace. Based on the same planning, within 30 years, it is estimated that entire communities will exist in space, and organizations like NASA anticipate that humans will be living on the Moon and potentially, Mars. Space communities will require citizens with education and skills in an enormous variety of fields, and anticipating these needs in the workforce while disseminating information through the education systems of the world will itself influence the future of space and the pace of its development. Unfortunately most student mentors are unaware of current space development and that there are many fields not directly related to science that could still bring our future workforce (today's students) into intimate contact with space. It is essential that we reduce this gap in understanding by examining career options for the near-future space workforce and provide it to those who mentor students in career opportunities.

Considering the "What's in it for Me?" Perspective

Perhaps the greatest obstacle we face in promoting space is that on a broad scale the general public, including many in our education system, are simply unaware of how close we are to fulfilling our desire to reach space, or how close they may be to participating in it.

We live in a society that wants and needs to be convinced why they should take time to understand and get involved in something new. Many times these questions circle the issue of money: spending it and making it. We live in a world of vast opportunity and immediate gratification through technology and high speed Internet connections. Life has so many opportunities for the individual living in a technological society, that it is up to the space community to convince them that they should get involved. I have found that K-12 children are the same way and need the same kind of convincing.

Even young children ask the seemingly belligerent questions, "Why should I do it?" and "What's in it for me?" Adults all around them consider these same questions every day as part of their daily routines. As a result, kids appear to be apathetic concerning their own education as well as the kind of future lives they will have. Nevertheless, if we can demonstrate to them why they should look at space as a future that they may be a part of, not only with they rise to the challenges set before them, but they will do so with amazing elegance and ease. They simply need to be informed and invited.

Sowing Public Support through New Business Ventures

Technologies developed in our quest to go to space have helped society in many ways through technology transfer (spin-off). To date there have been more than 1500 successful spin-offs from NASA-developed technology being used in the private sector. Some common spin-offs include smoke detectors, ear thermometers, pacemakers, and strong lightweight materials used in bike helmets. [See also chapter 14.] Many of these products have improved our daily lives, and others have the potential to help solve larger issues facing the worldwide community today. Spin-offs also offer great potential for the entrepreneur interested in getting involved in future space initiatives. This is particularly important when considering how to inspire generations of future space workers. For students interested in business, understanding how to use space technology for potential business ventures by learning business planning and the associated space law could open a whole new realm for existing space programs and add breadth to the job pool for those wanting to be a part of this exciting future.

The space community needs the support of a large portion of the public to stir the reach out into the solar system; not just because of the need of our education system as potential future workforce, but also because of the economic stimulus space provides. Understanding the extended (spin-off and economic) benefits to society will help in that way. Simply put, dollars spent on the reach for space equals jobs! Jobs help individuals pay their bills. The push to go to space also stimulates the development of new technology, and in turn that technology improves the quality of our lives here on Earth.

Another key element of the space movement is the fact that the average age of those working in aerospace community today is 50 years old. It is essential to develop a future workforce to replace those who will be reaching retirement age soon to support NASA's mission to "Moon, Mars, and Beyond," as well as for ventures in the commercial space community. [See also Chapter 15, Space Exploration and a New Paradigm for Education and Human Capital Development.]

Thus, if people are apathetic about the future, distracted by the many opportunities available in the modern technological world, or resisting because they don't see what's in it for them, it is up to space enthusiasts to convince them that they should get involved. Spin-off technology is a great tool to help do so!

Going Places

To accelerate our progress in the development of space exploration and in space industry we need to expose children, teachers, guidance counselors, parents, and businesses to the technology that is available now, to invite them to use their own creativity and imaginations to reach beyond where we have gone already, and to help them bridge the gap between being a dreamer and making it happen.

As we give them the tools they need we will all work together to get there ... because we should and because we can!

About the Author

Becky Cross has written 2 children's books, including I Am A Space Shuttle. I LOVE TO FLY!. She is also founder of The IGNITE Foundation (www.ignitefoundation.org), an educational enrichment organization operating in New England. Through IGNITE programs she has interacted with more than 15,000 students from kindergarten through high school on the topic of space. Ms. Cross has spoken at large on the topics of future space and the benefits of inspiring future generations of space adventurers. She presented at the 2005 International Space Development Conference on the topic of the Benefits of Space.

Chapter 20

The Meaning of the Heavens to Humankind through History

By Cheryl and Raymond Garbos

Introduction

Today words like "outer space," "the heavens," and "the cosmos" conjure up images such as Sputnik; astronauts Yuri Gagarin, Alan Shepard and Gus Grissom; huge Saturn rockets thrusting the Apollo astronauts toward the moon; the blazing exhaust of Shuttle launches; 77 year-old John Glenn making his second trip into outer space; and humans living on space stations. These events are an integral part of life in our century, but they are only a handful of the instances of humankind's experience of outer space. In fact, humans in all times and places have been looking to the heavens for millennia, so before we look to the future we take a few moments here to shine a light upon our past.

We are aware that writing about human history requires us to select just a few examples from a vast number of possibilities, and therefore reflects our personal point of view. Two authors had a major impact on our decisions: John Michell's definitive book, *A Little History of Astro-Archaeology* [10] takes an intriguing look at humanity and its relationship to the heavens, examining the link between astronomy and archaeology. In addition, the works of Joseph Campbell add a mythological and spiritual dimension to our exploration. [5] [6]

Most important, it is not our intention to present history as a recitation of facts, but rather as the story of an exciting and often perplexing adventure. Surprisingly, humans all around the world have adopted similar practices with respect to the wonders of the sky, even when they were thousands of miles and hundreds or even millennia of years apart. Hence, the dates we have included are a means to compare the timing of these unusual facts. We have used the modern dating convention of B.C.E. (Before the Common Era) and C.E. (Common Era).

Mythology and Evolution

Michell tells us of two historical world views which illuminate our understanding of humanity's experience of the heavens. *"The modern view, informed by the theory of evolutionary progress, is of civilization as a recent and unique phenomenon [with humans becoming ever more developed over time]. Against this is the older orthodoxy of ancient philosophers who believed that civilization proceeds in cycles, from primitive settlement, through the development of agriculture and technology, to empire, decadence and oblivion."* [10] He believes that the view of continual human evolution does not exclude the idea that human civilizations advance and decline, with human capabilities often lost for a time only to be found again under new circumstances.

Our lives are integrally woven with myths and spiritual concepts, as Joseph Campbell makes very clear: *"Getting into harmony and tune with the universe and staying there is the primary function of myth."* His teachings hold that myths are not fiction, but rather the stories that we live by. *"In them we are not talking about a search for the meaning of life, rather we are searching for the experience of being alive."* In an evolutionary view, Campbell also argued that *"truth comes from the ground of my being, the unconscious [or, as Jung put it, the "collective unconscious"] that I have inherited from all that has come before me."* [5] [6]

Another underlying thread of human evolution is the balance between spirituality/mythology/ religion, and science. Religion codified the spiritual into specific attitudes and practices, but as science developed the balance between what is not seen (faith) and what is seen (science) became a struggle. The relationship between religion and science was sometimes harmonious, and at others filled with tensions and even total separation among belief systems. We have found that our understanding of the heavens is at the center of this debate.

Early Humanity

We begin our story with early humankind. There is a great deal of debate as to when our ancestors can really be considered "human." The definition of humanity centers around attributes such as the ability to stand erect (roughly 3.2 million years ago), the making of tools (500,000 years ago), the ability to speak, and/or the degree of cognitive ability.

Various strains of hominids spread throughout Africa, Europe and the Near East, as early as 3 million years ago, all by 1.8 million years ago. While Western education's focus on Europe and recent American has often ignored developments in the Far East and in earliest America, human fossils have been found in China that are over 3 million years old, and there is evidence that the earliest Americans crossed a land bridge connecting Siberia with Alaska between 70,000 and 11,000 years ago, long before the "discovery" of the Americas by Europeans.

Campbell suggested that the criteria by which early hominids could be called "human" is *the birth, you might say, of the spiritual life such as no animal would ever have invented.* The evidence shows that hominids with brains roughly the same size as our own were making tools around 500,000 BCE, and at these early sites humans buried members of their clans and kept caches of cave bear skulls which suggest the worship of animals. Campbell believed that these were the first signs of not just intelligence, but also mythological thinking.

Picture a cave dweller seeing the frightening power of a lightning strike, yet picking up a resultant burning branch and realizing that he had found a wonderful tool, fire, which could be harnessed for warmth and cooking. We might call this intelligence. Or imagine a cave woman walking out under the stars and seeing a shooting star. In an experience like our own, she might look with awe at the expanse of the heavens above and come to the profound recognition that there must be a power or powers greater than ourselves, an experience we might call spirituality. These were huge cognitive steps for humankind.

"Animistic," polytheistic religions were among the earliest belief systems. Early on, Earth, air, fire and water were considered major powers, and these were soon joined by deities for any power which the people wished to influence or appease: gods and goddesses of rain, thunder, the stars, the sun and the moon, and many more. As early as 15,000 BCE naturalistic religion was especially strong among hundreds of Native American tribes in both North and South America. One of the most stunning aspects of the Native American experience is their star stories, 900 years before similar myths were created in the valley of the Euphrates, and more than 14,000 years before the Greek and Roman constellations. Hundreds of tribes gave us thousands of these stories explaining the mysteries of the heavens such as this one, which explains not only the existence, but also the human relationship with the star-filled heavens.

There was a man who went around each night carefully placing the stars in the sky. One night he fell asleep during his rounds, and his constant companion, a wolf, took the man's star bag and shook it out, scattering stars haphazardly throughout all the heavens. From then on the man spent his evenings lighting individual stars in the places where they had landed.

Many different civilizations that existed over a wide expanse of time created the same deities, but gave them different names. In Egypt major gods and goddesses included Re (god of the sun), Amon (or Amun, the creation deity; sometimes combined Ra as Amon-Ra, creator of all the gods), Horus (god of the sky), Khensu (god of the moon), Osiris (god of Earth and vegetation). Nut (goddess of the sky and heavens) is usually pictured arched over her husband Ge (god of Earth). The Sun is a good example of the variety of names given a common deity in different cultures: the Sun is Amun-Re in Egypt, Helios in Greece, Apollo in Rome, Tezcatlipotca among the Aztecs, Ansa to the Hindus, and Shen Yi to the Chinese.

Burial Practices

Burial sites, artifacts, and rituals are crucial in virtually every culture throughout history, and many show the connection between life, death, and the heavens. Forms of burial changed over time and from culture to culture, but all expressed one theme: the desire to approach the gods in an appropriate way so that a dead person could attain life after death.

The mound was a common burial form. The earliest ones appeared in the Americas around 7900 BCE, and vary in function from graves for the common people to burial sites for clan leaders, and later, even higher officials such as kings. These mounds were the preferred place of burial in the Americas at some places until 100 CE. They were certainly abundant—around 20,000 Native American mounds have been identified in Wisconsin alone.

The first mounds do not appear in Europe and Asia until 3100 BCE, exemplified by the well-preserved mound at Newgrange in Ireland. As is typical of burial mounds, there is long tunnel with graves along each side, and a primary grave at the far end. Especially noteworthy is that Newgrange is among the earliest examples of a connection between graves and the heavens. A "roof box"—an opening over the mound's entrance—is aligned so that the sun shines through the box on the winter solstice all the way to the primary grave at the other end of the mound.

A different approach to mound burials is found at Stonehenge in England. It is not a burial site, but the center point of more than 350 graves of different types spanning more than 1000 years. A henge, unique to the British Isles, is generally a circular pre-historic enclosure with a surrounding ditch. Some of the standing stones point to spring and fall equinoxes, but most are aligned with the summer and/or winter solstices. Exactly how sites such as Stonehenge were used is still a mystery, but clearly they addressed humankind's connection with the heavens, and may also have served as agricultural calendars. Figure 20.1 is a photo we took recently when we visited Stonehenge, a painting on the entrance wall to the park which shows what Stonehenge may have look like with all the stones in place.

Figure 20.1 Stonehenge

Of course, Egyptian burial practices are the most familiar to us. The Great Pyramids at Giza were built between 2551 and 1472 BCE as the burial sites of some of the most powerful Pharaohs. A priesthood regularly brought food and sacrifices to the tombs for the Pharoahs' use in the afterlife. The sides of the Great Pyramids were aligned perfectly with the cardinal points: north/south and east/west. Recently in one of the large pyramids a small opening was found with a shaft which may have pointed to a specific star where the pharaoh's spirit might travel. Evidence like this has convinced us that the stars were considered an important part of the Pharaohs' life in the hereafter.

However, pyramids are by no means unique to Egypt. The oldest mummies in the world, and some of the most sophisticated ones anywhere, come not from the Nile Valley but from Chile. Emerald Mound, the second largest ceremonial earthwork in the United States, was built more than two centuries before Columbus sailed into the Caribbean. There and elsewhere in the Americas, terraced temple-pyramids, often with shrines to their gods and goddesses on top, were common and were built over a wide span of history. They served as tombs for the people's leaders and places for religious ritual.

Around 300 BCE pyramids to the sun and moon were built at Teotihuacan, Mexico's first great city. The base of the Temple of the Sun was only slightly smaller than Khufu's Great Pyramid in Giza. Mound burial sites built by many different Native American tribes were located in dozens of places spanning the period from 2390 to 1425 BCE. Later civilizations such as the Aztecs, Mayans, and Incans built the same sort of terraced pyramids. There is evidence that these plazas were the site of human sacrifice meant to appease the god. Prime examples are the Mayan Tomb of Pakal in Mexico (built around 675 CE), the Monk's Mound, largest of hundreds of pyramids in the complex at Cahokia IL, USA (950 - 1050 CE), and the Great Temple of the Aztecs at Tenochtitlan (now Mexico City (c. 1325 CE).

18,000 years ago the Upper Cave Men of China were also burying the deceased members of their clans with artifacts such as food and clothing for use in the afterlife. . One of the most fascinating and grandest burial sites in the world is at Mount Li, built in 210 B.C.E as the tomb of China's first emperor Qin Shihuangdi, who lived in such fear of death that he employed 300 astrologers. The emperor's concern for the afterlife and his incarnation drove him to build a tomb which is, in fact, an entire mountain. The writings of China's first historian, Sima Qian, tell us that the tomb was designed to reproduce the universe below ground. The great waterways of China and the ocean were represented by mercury, which was made to flow mechanically. Heavenly constellations were depicted on the ceilings. Trees and grass were planted to represent the Earth, and a full city was created with temples, halls, and administration buildings. In one pit of the tomb, 6000 terracotta soldiers made up the emperor's army, including chariots and horses. Though not as grandiose as the Tomb at Mount Li, the Chinese continued to build tombs in hills for their emperors all the way to the end of the Ming dynasty in 1644 CE.

Evidence from many of the monuments built by ancient peoples all over the world, such as the alignment of burial sites with the stars and pyramids reaching far into the sky, indicate the peoples' strong connection with the heavens and their desire to reach beyond themselves and into the future.

The Development of Civilizations

The next major step in human development was the emergence of communities. Between 10,000 and 6,000 BCE a dramatic shift from nomadic Hunter-Gatherer groups to agrarian societies began. The shift happened in the Tigris and Euphrates valleys of Mesopotamia (now Iraq), what we know as the "cradle of civilization," and in Egypt along the Nile. Recent archeology has shown that the agrarian shift also occurred in India along the Indus River and in China on the banks of the Yellow River. Grain crops were cultivated, including wheat, rice, rye, oats and others. By 5,000 BCE maize (corn) was being cultivated in Mexico along the Rio Grande.

One of the most fascinating Egyptian myths is the story of Isis and Osiris. Osiris, a god of the underworld and earthly god of vegetation, was so hated by his brother Set (or Seth, god of chaos and evil) that Set cut Osiris into pieces and scattered him across all of Egypt. Isis, wife of Osiris, goddess of fertility and motherhood, was so disturbed by this that she gathered all the parts of Osiris' body and breathed new life into him. The power of this myth is in its relationship to the life of the Egyptian people. Isis, in her fertility role, discovered the grains of oat and barley and Osiris was the god who taught the people how to plant their seeds. Osiris was killed at the end of the planting season, the time of dormancy of crops, and resurrected after the annual inundation of the Nile when new crops were planted. So, Isis and Osiris explained a new wonder, agriculture, and were worshipped by the people of Egypt with the hope of gaining divine support for their crops.

Although we recognize that the shift to an agrarian society marks the beginning of a great deal more science to come, this is a time when spirituality, much of it based on observation of the heavens, was much more prominent than science. Put another way, the ancients would not have recognized anything like our idea of "science." They believed that things like rain or cultivation of the soil were controlled by and were gifts from the gods. So the "science" of the time, the beginnings of agriculture, was fully subsumed within worship.

However, more detailed observations of the heavens began creating another shift in human understanding. Through experimentation, humankind developed the science of astronomy. Observations of the heavens and aspects of life on Earth led to specific spiritual ideas. Goddess-centered worship was among the earliest, one that continues to modern times. In addressing the mystery of the origins of human life, the ancients recognized the importance of childbearing and saw that the rhythm of life was tied to the cycles of women's menstruation and the moon. So, it is thus no surprise that the earliest calendars were lunar.

Very early, in 5000 BCE, the Egyptians developed a calendar with a year that was regulated by the moon and the sun, consisting of 12 cycles of 30 days each (months), a year of 360 days. The development of such calendars was another great leap forward, enabling them to anticipate with some certainty when rivers flooded each year, depositing rich soil on their banks. Both the Egyptians and the Mesopotamians quickly learned to take advantage of the life force of these rivers by using irrigation. Calendars could help them prepare to plant, and the ability to predict annual events became critical as people around the world established civilizations along the banks of major rivers.

Sometime between 3200 and 2800 BCE the calendar was developed in Mesopotamia, and then copied by the Greeks. Their year was divided into 12 months, each beginning when a new crescent moon was first sighted low on the western horizon at sunset. These were among early "observational calendars" determined by priests who watched the sky so that they could declare the first day of the new month.

Around 2500 BCE the equinoxes and solstices were discovered, leading to calendars that were based on both the moon and the sun, or on the sun alone. The problem with these early calendars was that the total number of days they marked fell short of an actual year, so various schemes were invented to compensate for this discrepancy. Around 2000 BCE the Babylonian's calendar added an extra month every four years, while others attempted to correct the discrepancy by adjusting the duration of some months. In 46 BCE in Rome, the Julian calendar of 365.25 days was introduced with an extra day in February every four years, and it is basically the calendar we have today. Other cultures took longer to accept this model because it meant moving from an observational calendar which was under the control of state religions, to a scientific one.

In most places, astronomers joined priests as the keepers of the heavens. Science and religion lived in harmony, though support by the state opened the field for more discoveries and a better understanding of the heavens.

Tombs, Palaces, Temples and Colossi

Throughout antiquity there was major investment of human capital to build not just pyramids, but also palaces, temples, and colossal figures that had both astrological and religious significance. Notable monuments such as the Great Sphinx at Giza (c. 2500 BCE), the four colossal statues of Rameses II in the Great Temple of Abu Simbel, Egypt (c. 1285 - 1265 BCE), the Colossal Stone Heads of the Olmec on the Gulf Coast of Mexico (1200 - 600 BCE), the Statue of Zeus at Olympia, Greece (one of the seven ancient wonders of the world, c. 430 CE), the Colossal Buddha of Bamiyan in Afghanistan (7th century CE), and the huge Easter Island (Rapa Nui) statues in eastern Polynesia (1000 - 1500 CE) are just a few examples. These monuments usually required that huge blocks of stone be moved, usually over great distances. Quarrying, moving, and shaping the huge stones, and then putting them in place, required large crews of workers. How these amazing structures were built, and how they have survived for thousands of years are, to this day, great mysteries of human engineering.

As modern architects and engineers analyze the engineering processes and logistics required to build such structures they have become more and more impressed with the commitment and organizational skills their building required. Despite the development in scientific understanding behind these engineering practices, the major focus was still on their mythical and religious significance, which continually pointed to the heavens. So many of these monuments were aligned to the cardinal points, particularly at the equinoxes and solstices, that a new discipline has been created, "Astro-Archeology."

The Cycle of Civilizations

Over thousands of years, civilizations in Mesopotamia and elsewhere rose and fell. They were controlled by powerful leaders who not only raised conquering armies, but also codified laws, supported the development of writing, and established trade routes for their surplus goods. Stone and hardened clay tablets tell us of developments in mathematics and astronomy that helped these societies to become strong, sophisticated civilizations.

As one civilization reached its peak and began to disappear, another took its place and began its own rise to power. The people of earlier civilizations died off, were killed or became a part of the new civilization. . Sometimes the monuments of older cultures were destroyed, sometimes they were left to become the ruins of the march of time, and sometimes new civilizations took them over and made them fit the circumstances of the new civilization. So, we look at the development of humankind we see a continual ebb and flow into new circumstances. Yet everywhere there was continual improvement of human skills and understandings We particularly see evidence of this in the edifices that have been left behind, but how is it that nearly identical monuments were built centuries apart in places that were oceans away from one another? Perhaps the fact that all of humankind lives beneath the same sun, the same moon, and the same stars explains this to a certain degree, but it does not explain how such strong similarities in types of monuments and worship came to be. Could it be that human need for a connection to the great beyond is so strong that it evoked the same ideas from the shared mythos of intelligent human life?

The Quest for Science and Religion

Joseph Campbell notes that, *"Myths are the 'masks of God' through which [humans] everywhere have sought to relate themselves to the wonders of existence."* [6] In his view, this is a "Hero's Quest." In ancient times, the quest for meaning centered on spirituality, which was often codified as religion. Also, as we have seen, many monuments were associated with aspects of the heavens, both for worship and as practical ways to determine the seasons of the agricultural year. Our desire to develop currencies, enumerate trade goods, and, of course, to understand the heavens next drove a continual effort to improve the sciences. Notably, the Babylonians collected detailed planetary records, established a base 60 mathematical system that was used in navigation and time (degrees, minutes, seconds), and predicted the first recorded solar eclipse (763 BCE). The Indians and Chinese recorded the first known comet. In Africa, the Ishango people of Zaire used lunar phases for seasonal measurements, and the Mayans made astronomical inscriptions for agricultural purposes.

The coexistence of spirituality/religion and science in mid-antiquity was possible because the state supported both. Science, usually overseen by priests, was considered a part of the religious world view. However, with the development of the Greek and Roman empires from about 500 to 300 BCE, we see a major change evolving. Although they still had huge pantheons of gods at the center of their lives, government became more secular and individuals other than religious leaders began making significant advances in science and mathematics. Religion and science were still in alignment with the state, but advances in mathematics and astronomy began to chafe at the constraints of established religion. Pythagoras (circa 580-500 BCE), known for the Pythagorean theorem, actually led a group, that today we would consider a cult, complete with secret symbols and handshakes. They assigned significance to integers as a way to connect mathematics to a higher spirituality: #1 - reason, #2 - man, #3 - woman, #4 - justice, #5 - marriage (#2 man + #3 woman), #6 - creation, etc. Pythagoras also believed that the circle was perfect, and that God made the seven known planets (including the moon and the sun) that moved in circular, concentric orbits around the circular Earth. Meton of Athens (circa 440 BCE) calculated that the sun and moon moved in a 19 year period, which lead to the ability to predict eclipses.

Aristotle (384-322 BCE) was also one of the first to propose that the universe was eternal, that the Earth was spherical, and that it was the center of the universe. Aristotelian theory also provided "scientific" proof that there is a god. It was believed that since heavenly objects could not move by themselves, and since the planets move around Earth, they must be moved by God, who must therefore exist. The Earth, being at the center and not rotating, was considered critical to the unique relationship between the gods and humans: the universe was created for humans, who therefore must be at its center. This geocentric view became the dominant view of the universe for almost 2000 years. Figure 20.2 shows this early Earth-centered view of the universe, in which the stars were thought to be located on a sphere beyond the outer-most planet.

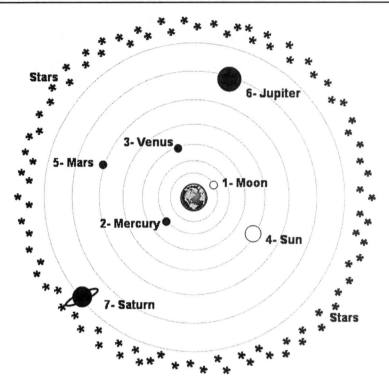

Figure 20.2 Earth-centered Universe

Scientific discovery continued around the world. In 352 BCE, the Chinese reported the solar event that is now know as a supernova and created star maps that were used for several hundred years. As more and more observations of planetary motions were made, it was discovered that they were not in Earth-centered concentric orbits. Aristarchus of Samos (c. 280 BCE, almost 2000 years before Copernicus and Galileo) was one of the first to propose a heliocentric or sun-centered universe. Another Greek, Eratosthenes (276-194 BCE), is believed to have calculated the circumference of the Earth as 250,000 stadia by using shadows from the sun taken at two different locations on the same longitude. Although his calculation was within a few percent of the actual distance, it was generally rejected. At this point, religion and science began to be in significant tension with one another. Sadly, human understanding was often codified as religion, causing a rift between the spiritual and the observed universe.

Astronomer, mathematician, and geographer Claudius Ptolemy (100-170 CE) attempted to validate the religious view that God made the earth-centered and, more important, human-centered universe, and to reject the idea of a heliocentric (sun-centered) universe by proposing the idea of an eccentric solar motion, by which the planets moved in small circles on larger circular orbits. Ptolemy rejected the calculations of Eratosthenes, and drew a world map with Earth's circumference at around 180,000 stadia, about 25% smaller than the actual distance. Many believe that Columbus had access to Ptolemy's map, which may have influenced his incorrect expectations about the length of his voyage.

In the early centuries of the Common Era, religion itself made radical shifts. Although Greece and Rome maintained huge pantheons of deities, the newly predominant religions, Judaism, Christianity, and Islam, were monotheistic. In 300 CE, the Emperor Constantine converted to Christianity, which then became the state religion of the Roman Empire. God was thought to reside just beyond the stars where humans could not communicate with him directly. To communicate with God, people had to pray through an angelic hierarchy representing the seven known planets, starting with the nearer cherubim and seraphim and ending with the farthest, the archangels. Ptolemy's Earth-centered universe became imbedded not just in the Christian faith, but also as the major world view which dominated European thinking until the sixteenth century.

Meanwhile, the Chinese developed a mechanical clock in 725 CE, block printing in 868, and around 800 CE, the "House of Wisdom" was established in Baghdad, a learning center that included a functioning astronomical observatory.

The Black Plague decimated Europe in what is now known as the Dark Ages, extending from the mid thirteenth century to the end of the fourteenth century, leaving the Middle East and Asia as the world's centers of education and science. Buddhist philosophy posited that the universe is periodically destroyed and recreated, while Islamic and Chinese cultures also continued to pursue science. As science became more precise, the gap between religious and scientific world views widened.

The end of the Black Death in Europe brought the beginning of the Renaissance, and soon a scientific revolution emerged. Nicolaus Copernicus (1473-1543) again proposed a heliocentric theory of the universe, but Protestants and Catholics alike strongly rejected this idea, claiming that the Bible's many references to the Earth-centered nature of the universe proved their views.

Galileo Galileo (1564-1642) used the newly invented Dutch telescope to observe the moons of Jupiter and concluded that the Copernician heliocentric theory was correct. Persecuted by the Church, he claimed that the Bible should not be taken literally and that religious leaders should accept scientific proof. His challenge to both the authority of Scripture and of the leadership of the Church led the Church to the denounce science as an evil profession. In 1616, the Church ordered Galileo to abstain from promoting the heliocentric theory, and eventually he was forced to confess his ideas as errors. In 1633 he was put under house arrest for the rest of his life.

There was now a clear break between scientific and religious views of creation and the universe, between science through which the physical world was studied, and faith through which the "heavens," the spiritual world, were revealed. The heliocentric theory was eventually accepted by the Christian Church, but it was not until 1992 that Pope John Paul formally acknowledged that Galileo had been treated improperly by Catholicism.

Around the same time as Galileo, Johannes Kepler (1571-1630 CE) determined that planetary orbits were not circular, as Ptolemy had established, but elliptical, and that the duration of planetary orbits was related to their distance from the sun, a far more accurate model of the solar system. The Church reconciled its acceptance of such views by stating that science was a method for describing the physical world, but that the tenets of religion and faith about God's creation were appropriate to describe the spiritual world.

Since then, many people have tried to bridge the growing gap between the science and religion. For example, in 1714 William Derham published "Astro-theology," which argued that the existence of God was in keeping with astronomy.

But major gaps between religion and science have continued since then, and today some still reject basic scientific views. Many Americans believe, for example, that the Earth is thousands of years old, not billions, while others do not see a schism between religion and science, believing that everything in our universe is a product of God's creation. So the dialogue continues, perhaps best exemplified by the debate among those who believe in the "Big Bang," evolution, creationism, or Intelligent Design. Modern secularism has added a new voice to the debate that is now affecting, for example, what is taught in schools about such issues. It seems that we are far from resolving these arguments.

In the future, new scientific discoveries will likely present challenges to our present understanding of both science and religion. For instance, we may find significant life on another planet, or we might detect things beyond "our" universe. Surely such discoveries would fuel new arguments about what is "true." Still, it is clear that the heavens which surround us, and the ideas and images which they continue to evoke in us, will continue to influence our views of the universe.

Campbell noted that, *"Myths are the 'masks of God' through which [humans] everywhere have sought to relate themselves to the wonders of existence."* (5) So we are still on a Hero's Quest, seeking to understand these wonders, to go where no one has gone before, each trying through our individual experiences of the spiritual and the finite to understand our own truths, and to tell our own stories.

References

(1) *"A Brief History of Astronomy,"* http://library.thinkquest.org/J002040F/history_of_ astronomy.htm, 2000

(2) *"A Brief History of China,"* Retrieved November 2005 from: http://www.travelchinaguide.com/intro/china.htm

(3) *"Babylonian Astronomy,"* Retrieved, December 2005 from: http://library.thinkquest.org/J002040F/history_of_astronomy.htm

(4) Burton, David M., *The History of Mathematics: An Introduction* (Fourth Edition. Boston, McGraw-Hill, 1999)

(5) Campbell, Joseph, *The Hero's Journey: Joseph Campbell on His Life and* Work (San Francisco, Harper & Row Publishers, 1990)

(6) Campbell, Joseph, *Transformations of Myth Through* Time (New York, Harper & Row Publishers, 1990)

(7) Cooke, Jean; Kramer, Ann; and Rowland-Entwistle, *History's Timeline: A 40,000 Year Chronology of Civilization* (New York, Crescent Books, 1981)

(8) Grun, Bernard, The *Timetables of History: A Horizontal Linkage of People and Events* (New York, Simon And Schuster, 1982).

(9) *"History of the Americas,"* http://en.wikipedia.org/wiki/ History_of_the_Americas#Migration_into_the_continents, 2005

(10) Michell, John, *A Little History of Astro-Archeology* (Updated and Enlarged Edition, (New York, Thames And Hudson, 1989)

(11) Scarre, Chris, Editor, *The Seventy Wonders of the Ancient World: Great Monuments and How They Were Built* (London, Thames and Hudson, 1999)

(11) Schroeder, Gerald L., The *Science of God: The Convergence of Scientific and Biblical,* Wisdom New York, Broadway Books, 1997)

(12) *"World History Timeline,"* http://www.wwnorton.com/ college/history/worldciv/reference/wrldtime.htm (New York, W. W, Norton & Company, 1997)

About the Authors

Cheryl and Raymond Garbos are partners in Math and Science education. Cheryl has a masters degree in theology with emphases on education and ancient history. Ray is an Affiliate Associate Professor at the University of New Hampshire and an Engineering Fellow.

Chapter 21

Space, Ideology and the Soul:
A Personal Journey

By Paul J. Werbos

1. Introduction

In Chapter 23, I lay out a basic vision of space development as a totally rational enterprise, aimed at achieving self-sustaining growth. The goal is simply to maximize the probability that humans do eventually achieve a kind of self-sustaining growth in our activities beyond the Earth. Given more time, I would have added pages and pages on the needs for new efforts to develop many critical enabling technologies, important to the viability of life in space.

But there was a major gap in that discussion: what about the inner human urges that make us want to achieve that goal in the first place? Why should we feel driven to achieve this? ? Why should we place a heavy value on settling space, as a basic goal in its own right?

In the end, objective reality and objective truth can never tell us what goals we "should" give priority to. It has to come from inside ourselves. (In my personal files, I have a letter from the logician/philosopher Bertrand Russell, sent to me when I was in high school, reminding me that it is simply impossible to deduce a sentence with the word "should" in it unless one starts with axioms or fixed assumptions which already contain that word. Thus anyone who claims to do so is cheating, engaging in false logic.) It is a personal matter for each of us. Objective understanding can be very important in understanding ourselves better, so that we can be truer to ourselves and more sane. But still, it is a personal matter. Thus in this chapter I will discuss my own personal reasons—and the objective views which support them—for supporting the space movement.

In actuality, the space movement is properly very diverse. If the quest for sustainable exploration and development of space is to become the "property" of all humanity, the motivations and the feelings about it must become as diverse as what exists on this planet. That is why some of us DO need to develop plans which are rational and objective in a way which is consistent with different personal viewpoints. I do hope that you, the reader, will still be able to respect the rational plans of another chapter even if you disagree with my personal motivations.

For me, the motivations have evolved and changed a lot as I grew and learned more about humanity and where it is going.

2. Journey Through Non-Spiritual, Rational Motivations

Like most of us in the space movement, I began with childhood wonder at some exciting things in books and in the news. There was a little book "All About the Stars" and then Hoyle's "Nature of the Universe" which enthralled me when I was eight - and undermined the Catholicism I had been devoted to at that time. As I read the first part of Hoyle's book, I remember saying to my druidic Irish Catholic mother "Ah, the glories of God's universe."

At that time, I knew that some of the nuns in the local Catholic school were not such glorious or spiritual people. it was easy for any child to see pettiness and anger and narrow blindness and robotic rigidity for what they truly are. But I did not see any reason why I should be bothered by their warped representation of Catholicism when I could see my mother and some of the better nuns as a model for "real" Catholicism. There was an overwhelming projection of love, love as the true expression of Jesus. There were the words of Jesus read in the church, which clearly said the kinds of things my mother was saying, not the nasty things of the nuns who warned us we would turn into a pillar of salt if we went to communion without going to confession the night before. There was a prominent picture somewhere of Jesus with a kind of open heart, stressing "the sacred heart." And there was a kind of natural openness and sensitivity all around. There were hours and hours of conversations in our home reinforcing all this (and accidentally supporting some intellectual growth as a byproduct). There was a consistent devotion to the application of love as a way to solve all problems—and an underlying distaste for anything not good and loving, handled gently but strongly with clarity and love.

And so - when I got to the last chapter of Hoyle's book, it was a real shocker for me. He basically said: "And, oh yes, there is another little planet I forgot to say anything about so far. This tiny planet Earth, way out at the far fringes of a mediocre galaxy, did not deserve attention yet, but it is amusing enough to warrant a few words at the end. It is populated by a bunch of odd-shaped two legged creatures which look a bit like robots, who suffer from the delusion that this entire universe was made only for them and that the very fabric of space and time was itself constructed by a gigantic member of their own species. They subscribe to weird and transparently inconsistent folk beliefs, like the worship of sacred totem poles, sacred cows, dictatorship of the proletariat and the holy self-sacrificing bankers—plus magical numerology, trading rules and ways to flip I Ching coins to concoct numerical scores to tell you where to put research money." (Well... Since I don't remember the details, I have elaborated a bit here, in the spirit of the book.)

I was totally shocked. Could the great universe truly be such a dark and impersonal place as all of that, without the kind of all-defining great love written into its very fabric? I certainly did not want to accept such a dark idea as that, and I firmly rejected it. Yet I did not forget it. I remember one night, as I walked down the hall into the bathroom, and turned on the light, asking myself: "How do I know? What is the basis for the knowledge?" (In hindsight, I now feel that perhaps the "I" which put that question into my mind was more like the larger "I" I often feel in touch with in such quiet times. Of course, I needed to ask the question, regardless of how little I knew at the time.)

As I thought about that question more and more - I began to think "My basis for believing this is that the nuns have told it to me." And realizing that was a kind of wake-up call. I was aware not only of my mother, but also of my hard-working German father, whose business and knowledge of engineering and science I had great respect for. When it came to sources of actual knowledge, of objective truth - I realized that my father and the world of science had it all over the fuzzy and unreliable stuff that came from the nuns. And as I followed the logic to its utter conclusion, with a very intense and sincere honesty - at eight, I became an extreme atheist, and utterly uncomfortable about going to a Catholic school where they now made me violate my moral principles of honesty.

And so, by various means, I arrived at age nine at Chestnut Hill Academy in the then most elite suburbs of Philadelphia, where I renounced that old Catholic culture, and totally embraced the world of Anglo-Saxon high culture (with a heavy extracurricular dose of my German heritage). By age 12, I embraced a kind of existentialist version of the philosophy of Utilitarianism, as formulated by John Stuart Mill. Bear in mind, Philadelphia was where they wrote the Constitution and the Declaration of Independence, and the old Philadelphia families still remembered all that very clearly. (Even my father said he had copies of the bar bills for Madeira wine signed by folks like Jefferson and Franklin in a bar owned by a member of my mother's family, where they wrote much of the Constitution. My two best friends, Richard Colgate Dale and Ben Robert, really exemplified the best of that culture. And so on.)We remembered the United States as something we set up, to embody the enlightened philosophy expressed by John Stuart Mill, a combination of enlightenment and freedom which is a true watershed in human history. And our glorification of this was so strong that maybe I fell into overestimating our uniqueness as a nation; it is a bit harder to live up to those ideals in everyday life in Washington than I understood back then.

From the viewpoint of utilitarianism - the supreme goal of life and politics is the greatest good of the greatest number, the sum of all human happiness. And so, to "strike it big" by that score, the biggest opportunity before us is to settle the entire galaxy (or as much of it as we can reach) with as many happy human beings as possible. (Those of you who study philosophy will immediately note that Mill's emphasis on human happiness is really an outgrowth of things that Aristotle said.) And a vast galaxy could be a great place to develop a vastly diverse open society. Indeed, as Hoyle pointed out, this planet really is a very tiny place in the larger scheme of things, and it is very limiting to be only here. It would be a tragic waste of potential to leave most of the galaxy dead. (If it IS dead, or devoid of intelligent life. If not, our survival might depend on our growth for other reasons.) And of course, we were all aware that the idealistic spirit of going out to settle the frontier had been a key factor in establishing and continuing the spirit of freedom. Whatever the imperfections of those old days, and the challenges before us now, we should never forget these basic positive facts which underlie our best hopes for the future.

But - not everyone at Chestnut Hill Academy was so optimistic back then. Though I was an outspoken atheist, the dominant culture was of formerly Quaker families that had turned Episcopalian. They were my friends. All of us were strongly committed to moral values of some kind. And the old families had a sad feeling that a lot of the old moral values and principles and philosophy which created the United States and its freedom were being eroded by grossly immoral, unprincipled, showy, crude, nasty, narcissistic nouveau riche people, particularly some of the folks in New York. My father was more or less nouveau himself (though highly principled in his own way), and my mother sometimes laughed when one of my classmate friends said melancholy things which she called "full of beans." But still, I did listen a bit when classmates said they were worried that the United States might be going the way of the Roman Republic. They got me to read Toynbee's famous classic of the rise and fall of great civilizations. Being proud of my German heritage, I of course had to read Spengler's Decline of the West to balance that out. And it did worry me a bit, just as Hoyle had worried me earlier.

I wondered: "Could it be that our grand efforts to design the ideal society are a bit unreal, like castles in the sand? Could it be that there are forces at work in human society and human evolution which simply make it impossible to create the world we choose, and make it a challenge even to survive at all?"

As I started to worry about this - I became a stronger supporter of the space movement. (And I remember Newt Gingrich's background as an historian! But I never read his books.) The historian Frederick Jackson Turner argued strongly that a new frontier was essential to the maintenance of freedom and an open society and growth. The deeper analyses of Spengler and Toynbee reinforced the same kind of conclusion. In their histories - civilization always decayed until a new culture was born, always on new soil, always as a new frontier appeared. And so, rationality demanded reaching out to space, not just as a way of increasing Utility (human happiness), but as a way of reducing the probability that civilization might decay and fall altogether. And I began to realize that the human species might well be in danger of ultimate extinction, if social decay and nuclear proliferation combined together. (Already, I learned enough technical detail that I wouldn't want to bore you with it here - but there is a lot of mathematical modeling and reality and empirical data behind all this. And histories of dynasties of China and so on.)

But finally, for the tenth grade, I really had to leave Chestnut Hill Academy to go to a boarding school, the Lawrenceville School in New Jersey, in order to take mathematics courses at Princeton suitable to where I was at that time. (In the book A Beautiful Mind, about John Nash, they mention even seeing a rather young teenager at the wonderful mathematics teas, sitting at a go board... that was me, playing with Daniel Cohen.) There was a different kind of discipline there, and many kids from New York and from the oil industry - and Texas, Arabia and China. One of my best friends in tenth grade was a guy named Chen from a prominent Kuomintang family, who gently and elegantly encouraged me to ask a few questions about my Western Utilitarian ideology. He would ask questions like: "Who are you? Really? Who are you?" And he would talk about "Watch our for your blind spot. The blind spot is important for everyone." I did not change my worldview instantly, but I remembered and respected my friend. I have often thought lately about trying to get back in touch with him—but even in the age of Google, finding a particular "Chen" is not so easy.

By the next year, my senior year at Lawrenceville, I thought I had a kind of logical argument for why we SHOULD be utilitarians. And it was influenced by Chen.

It began with a kind of argument about "who are we really, who is the greater Self." It was a kind of abstract formal philosophy, not theological at all. And I argued that of course we should strive to be truly and perfectly rational about maximizing Utility, since Utility is the bottom line, the real "score" for success in life. I read the classic book, the Theory of Games and Economic Behavior, by Von Neumann and Morgenstern, which I still believe is a fundamental part even of true spiritual enlightenment. (My goals and values have changed since then, but goals are not real and honest, and the person is a hypocrite, if the goals are not pursued with full consciousness and rationality. Later, in graduate school, I also read Raiffa's books on how to apply Von Neumann's ideas, and they too should be "must reading" for many people.) I even got an article into the Lawrenceville school newspaper summarizing this philosophy briefly, which deeply antagonized at least one of the English teachers who complained that he couldn't begin to understand what I was talking about. In the end, however, all of this philosophy still said we need to go into space (and do some other things to amplify human happiness) - and it didn't really say we needed to go into philosophy.

Early that year, I made a new friend, a wild guy named Lincoln Kaye from New York, who was a new student and declared a Certified Genius somehow as he entered the class. He didn't agree with my philosophy, but he mentioned that there was something similar already in the Upanishads, the "highest" part of the Vedas, the ancient sacred scriptures of India. And so, while waiting for my math classes at Princeton, I hung out many times in their undergraduate library, in the shelves where they had rows and rows of books of Upanishads, mostly in English. I remember sitting on a hard concrete floor with those books in my hand, and bringing at least one back to my house in Lawrenceville, where I would pore over it while sitting in a strange position over the stairs, distressing the English teacher even more. I found that the Upanishads were basically divided into two parts, a kind of older "higher" part and some later more yoga-oriented parts, both of which I read. I was astounded to see that the "higher" part was almost precisely the same as what I was saying!! And so, I added references to that in what I wrote - which did not help at all in getting people to accept or understand what I was saying. The Greater Self of the high Upanishads is a kind of universal concept of consciousness, which I hoped we could manifest and satisfy more fully throughout the galaxy and universe - bringing life and happiness everywhere possible.

I did write a few pages trying to explain this version of utilitarianism which I sent off to Bertrand Russell. I am very grateful to him for his kindness in sending me a brief but coherent reply - which did not make me happy, of course, but did influence me. And incessantly, my less moral, more existentialist friend, Lincoln Kaye, would keep asking questions like: "but what if I just simply don't FEEL like being logical and utilitarian and all that? Why should I bother? The Constitution is all well and good, but if you try to talk to some poor guy in the ghetto who is about to steal some food, why should HE obey your Constitution? He didn't write it. What's it to him?"

And gradually I put this together. I can even remember the exact moment when I was sitting on a hard (concrete?) floor in the basement of Memorial Hall in Lawrenceville, when I said to myself: "OK, Bertrand Russell was right. It is impossible to use reason and science and objective knowledge to draw any conclusions at all about what we 'should' do. But instead of asking 'what should I do?', I can ask the question 'what WOULD I do IF I were wise." And I thought: I can do this by having an operational concept of "wise," a concept which refers to a kind of end point of logic that I would find wholly satisfying, an equilibrium of my own mental processes. In effect, I could ask what ethical principles or goals would truly satisfy me. PART of the requirement would be to totally satisfy my logical, rational analysis - but there needed to be additional requirements. I immediately followed this by the thought: "This sentence CAN be approached in a logical, scientific, objective way, because it is a sentence about 'I.' Just as Chen said, this question requires a systematic effort to understand the self as a foundation for all action."

At first, I tried to hold on to a formal, traditional Utilitarian view, by saying that we should go to the higher Upanishads to understand what "I" is. But I slowly began to appreciate how this did idea not wholly fit with reality. I had conversations with other members of the Conservative Club at Lawrenceville, who had me read the book Atlas Shrugged" by Ayn Rand, who said the "I" was all about the simple (biologically based) struggle for survival. (Ayn Rand also suggested that we might consider shooting the Communist copywriters who try to outlaw use of the word "I." My father said the same, and had me read books by Rudolf Flesch supporting that view.) I didn't like that dark picture, but I could not in all honesty just ignore it. And I read the Asimov Foundation Trilogy, which, among other things, portrayed some more practical pictures about how the "I" in our brains really works. Should we give in to the reality of the "I" of the brain as we see it in neuroscience? Is that the real way to see ourselves? And - as an atheist - I gradually had to admit that the real "I" is the "I" of the brain and of biology, warts and all. And (as various females kept urging me) I had to truly accept my real, biological self. (Not that I went so far as they wanted for many years... They are also an important part of the story, but this is not the National Enquirer. Of course I remember, but one must respect people's privacy.)

In the final summer after Lawrenceville, before coming to Harvard, I had a wonderful summer intern job under a Dr. Karreman at Jefferson Hospital in Philadelphia. He had me read another classic book, the Organization of Behavior, by D.O. Hebb. I then plunged very deeply into the effort to understand the brain—in a functional way that supports useful subjective understanding of how we get things done and define our higher goals. At this point, I can claim to be a leader in this complex scientific area; for example, I was one of the first (two-year) Presidents of the International Neural Network Society, and received IEEE's neural network pioneer award for developing (among other things) the mathematical algorithm most widely used in practical applications of artificial neural networks. At arXiv.org, if you search on my name under q-bio, you can learn more about the realities of life and—perhaps more important—pointers to a very large literature, which is truly relevant to these larger issues about the meaning of life.

But for now - here is the main story about why we should go to space, as I saw it in my middle year as an undergraduate at Harvard.

I was more firmly an atheist than ever, for two reasons: (1) I could see clearly that the popular noises saying "I have feelings so I must have a soul" were simply neurotic errors, sophistry quite similar to what Bertrand Russell elegantly punctured, depressingly common among humans who delude themselves every day of the week in all areas of life because of ego-driven psychopathology; and (2) I agreed completely (and deeply) with Hebb's argument which Sagan has summarized as "extraordinary claims require extraordinary proof." And yes, folks, I read some stuff by Nietzsche and Marx that fits all this, but that's a complicated story; I certainly did not agree with everything those folks wrote.

I argued that we as humans are naturally born intelligent—which means that we do not START OUT with correct ideas, feelings, or emotional intelligence, but we have a natural desire to use all of our talents and develop them, so as to be more effective, to avoid wasting energy fighting ourselves, and to truly expand our powers, our vision and our abilities. Weak and stupid animals find it hard to think ahead even one day, and are easily killed when there is a change in the weather—but it is our nature to learn more, to establish a stronger foundation, to learn how to enhance our chances of survival across greater intervals of time and larger circles of kinship and respect and evolution. As we grow stronger and more sane, supported by greater understanding of ourselves and our potential, we give up silly illusions, and we realize... that, hey folks, we as humans are ALL in really serious danger of extinction. It is no joke. It's US, all of us, and it's not just an intellectual matter. Our lives are at stake. Space solar power and space settlement are certainly not the only critical issues here—but they are one necessary part of the equation, and we need to move if we want to stay alive.

In sophomore year, I had given up being a slave of my intellectual, symbolic brain. I knew enough about the brain to realize that the input of "feelings," of "primary reinforcement," is what drives the system. Thus a fully wise and sane person really does do just what he feels like - that is true freedom, and a powerful society empowers and encourages that (more than ours does yet). Yet such a person also has an educated and understanding self-consciousness, and appreciation and feeling for his or her own natural intelligence, which strives to see and understand his or her feelings as clearly and precisely and vividly as any scientist understands the fine points of an experiment.

There is—as Chen would say—a kind of harmony in the self. And so, I fight to stay alive because I care about living and not dying, and because I really don't want everyone human to be dead in X years. I truly appreciate in myself that everything we do will be for naught, if we allow extinction to happen. And I realize that all will be well, in the end, if we reach a new foundation for our material existence.

I look back on Aristotle's words, where he said that a sane and balanced person is moderate—and looks ahead a moderate number of generations into the future in his planning and thinking, not too many and not too few. I look back on the words of an old classmate from Chestnut Hill, George Davis Gammon, who told me: "Freud is not trying to dictate value judgments to you. His concept of sanity is to keep you from fighting with yourself, to avoid hypocrisy, to achieve whatever YOU value more effectively." They were right about these points. It is time for us to learn to be more sane, and to look forward to the hard realities we must cope with. I allow my inner self to fill itself with the healthy words of the song "Remember the future." (I was so sad when Gore replaced that with Macarena the next time around! That girl was nice, and deserved appreciation, but we had a job to do.) It is time for us to fight for the survival of humanity, in as systematic and rational and determined a way as we possibly can. Space is a key part of that.

3. New Motivations: Teilhard de Chardin, "Gaea," and Back to The Soul

Many people really should stop reading right here. To this day, I deeply respect the intellects of many people who would think exactly as I did in my sophomore/junior year at Harvard. I fully respect and understand the logic of their views, and I fund many people who believe as they do. Many of them are making great and important contributions to humanity. My PhD thesis adviser at Harvard, Dr. Karl Deutsch, was a man like that—a man of great vision, with a visibly powerful and great and benevolent soul, who did not believe at all in the soul and would be shocked if he could really see his own self in the mirror. It reminds me a bit of what Jesus said about the Good Samaritan, versus the evil Pharisees in the Temple. The stuck-in-the-mud Pharisees and suicide bombers who doubt both evolution and the truth of the Apollo program, and the rights of women, are a major threat to the spirit of America, and to the real soul of Christianity and Islam and to humanity as a whole.

But—in the end—real experience of real life has taught me greater rational understanding of what my mother really was about. I saw it all around me when I was a child, and I remember things I later blocked off from myself—but they were all around me, and I needed to open my eyes. My logical rational self needed to be brought kicking and screaming, by the pressure of unremitting personal experience, to a few rudiments of the greater spiritual reality around all of us every day of our lives. (Though, OK, there are a few dark days in government office buildings when we do need to go out and see a family member or a tree to unblur the eyes... And the same is true of many academic theologians!)

I did not know it at the time—but once I had whole-heartedly embraced the goal of Sanity and Whole-Brain Effectiveness, and focused on the goals of intense objective understanding of all details of what goes on in my own brain, as well as the goal of greater human survival... that set a certain kind of process to work. Odd sorts of things started to happen. I started noticing things. I sometimes wonder whether John Nash's sanity went when he COMBINED an ability to really put together some of what was going on around him, with an intellectual refusal stricter than mine to accept a paranormal or spiritual explanation. I do not expect others to accept all of the first person experience which led me that way—but for me it eventually reached the point where my sanity seemed equally threatened either way, either by rejecting or accepting paranormal reality of some kind. I remembered Hebb's calculation—but, in the light of extraordinary events, I felt I now had to be open-minded, 50-50. I remember the exact moment (I believe it was in March 1967) when a specific confirmed memory from the future drove me to be 50-50 open-minded... and after that, my intellect was mobilized and I learned fast.

Though I would not feel right making claims and noises about my personal inner life... maybe one story which reminds me of John Nash might be in order. In 1968-1969, when I was a first-year graduate student at Harvard, I had read about a UN conference on global environment sustainability. I had a flash of ideas about how new remote sensing technologies and parallel computing could be used to help solve some of those problems (and incidentally help the space movement). I spoke to no one about this unusual idea, which I had not heard anywhere around me. I ran to my dorm room, locked the door, and typed up a long letter on the stationary of the Harvard Center for International Relations (which had names like "Kissinger" on it), and put it in an envelope to send to the key conference people. As I walked down the cold tiles of the dormitory hallway, heading for the mailbox, a close friend of mine from India walked up to me. This friend, "Mani" Subramaniam, was very much revered by his fellow members of the Vedanta Society at Harvard, and we were in the same department. As he walked up to me, he said that an interesting idea had just popped into his head. I kept quiet, and held my sealed envelope tightly. He then proceeded to spell out almost verbatim the exact contents of what I had just written in total silence, based on putting together ideas from many very different sources and doing analysis no one else had done, to my educated knowledge. Even the choice of words was the same. If Mani had worked for DOD, I can imagine how a John Nash might have concluded that he was a CIA agent with a camera in my room; the probabilities were just 'way too low for a total coincidence. (Can you imagine living life when this seems to happen every day?) But Mani did not work for the CIA. (Hey, guys, There is enough evidence for that.) And truly, I got into deep danger at various times of my life when other folks started believing that I worked for the CIA, or the KGB, or even MI-5. (The last was by far the most fun.)

But—I certainly would not give us the quest for greater sanity and rationality.

How could I rationally explain all this? How could I reconcile things? I certainly was not ready to eat the magic mushrooms of fuzzy craziness, and depart to the mushroom islands of believing everything and nothing.

And so—various things. Of course, I read some of the work of Carl Jung, the "other" great founder of psychoanalysis, who talked about synchronicity and the collective unconscious. This was interesting - but there was no physical explanation.

And I remembered the famous French theologian, Teilhad de Chardin, whose writings were once a great inspiration to my mother, as well as to John Kennedy. In fact, I had some long talks with the Harvard janitors, who knew all the Kennedy brothers, and told me how John Kennedy had spent hours enwrapped in the writings of Teilhard de Chardin—bemusing the janitor as much as I had bemused that English teacher earlier. Teilhard de Chardin was in many ways the real spiritual father of the American space program, and it is clear that Kennedy's vision was a lot larger than most of everyone else's (and beyond my real understanding at that time). Still... there was no physical explanation.

Because I was running very low on money at a certain point in graduate school, I moved to the bad side of Boston, in the Roxbury slums—which were a one-mile walk from the Harvard Medical School Library. I read a great deal of neuroscience for my PhD thesis project, but I also devoured all the back issues of the various parapsychology journals in a few weeks, in order to look for clues. I did find a few things that seemed highly significant, given that I now had an open mind. (I still was skeptical however, despite a lot of pressure from experience.) I read all about the various SRI projects funded by the CIA and DOD and others; some of that work passed tough review at Proc. IEEE, which is the most selective, flagship journal of the world's largest professional/scientific society. I was impressed by some of the evidence I heard about, at the University of Toronto, described in the popular book Conjuring Up Philip, though I never had time to check out the details first-hand as much. And I later heard a talk by Bierman of the Netherlands which seemed quite solid. There was an article on a poltergeist associated with a Cuban boy, Julius, which did seem to hint at some possible physics - and frankly, did resonate with some of what I experienced myself in some of those years.

Once you start to consider that we MIGHT need to explain strange stuff... you naturally do your best to find a truly rational explanation.

At one point, I was intrigued by the idea that paranormal abilities might have been evolving slowly for billions of years, based on physical phenomena which are so weak that they are harder to see than ordinary light, and yet with enormous long-term potential to be useful in exchanging information. After all, if we did not have first personal experience of using eyes in our own heads, what would our "rational mainstream science" be telling us today? Wells (who also write The Time Machine) wrote a story about that, called "In the Nation of the Blind." (I wonder why Disney never made a movie of that. They should.) We probably would not even believe that the stars and planets exist! We would not have made telescopes. The warm and fuzzy feelings we get form the sun would have been dismissed as a childish hallucination. And I wondered: did Wells himself have eyes? Have other science fiction writers had eyes they were too cautious to tell us about? (And to this day, I wonder that about Orson Scot Card, Modesitt, Eric Temple Bell, Jules Verne and even Asimov and Bear sometimes. And others.) Could it be that the odd molecules in the pineal gland (which are wired up to the epithalamus in a way very similar to the way that the pituitary gland is wired up to the hypothalamus, which does have sensory functions) provide the "receiver" in this system?

But as time went on, I realized this simply was not good enough to explain the experience. Not even close. As an example - how could your pineal gland sort out all the multitude of signals from other people? Some folks say "oh, that's easy, using quantum theory." But in all fairness, there are folks who understand quantum theory a whole lot more than you can believe, who would love to build systems that act as well as remote viewing, and it's not easy. There are lots of reasons to believe that this simply does not work (is not sufficient) as a physical explanation. (Of course, I am not deducing here that we should simply chuck out quantum theory!! That's an extremely complicated subject, and I would refer you to the physics part of arXiv.org for that.)

If quantum theory is not enough - what is? What is the minimum change we would have to make in our basic assumptions about physics to accommodate such empirical evidence? (This is not "science" here and now as Kuhn defines it, and I apologize again to those who do not accept this empirical evidence.)

OK, folks, it's a big swallow, but here is the best I can do. It's the "theory" most consistent with Occam's Razor, a basic principle of rational, scientific thinking. I think of it as "the standard model of the soul." I don't have religious faith in it, any more than a good physicist has religious faith in today's standard model of physics, but for myself, I can't get by without it. And frankly, I would argue it has a better chance of being Completely True in the end than the standard model of physics does.

The idea is—it wouldn't take a very fundamental change in our understanding of the laws of physics to allow for the possibility of a couple of extra force fields as yet observed, which give our universe the property that it is a nonlinear dynamical system supporting pervasive life as an emergent property. Not just the life we see everyday with our mundane eyes, but an additional kind of life. And then "we"— our whole selves—are a kind of symbiosis of two life forms, a "body" and a "personal soul" which is in turn a kind of cell or region with a kind of Earth-spanning life-form. A life-form in a kind of childhood state of development, as is necessary to explain a lot of what we see.

And—of course—this is essentially the same viewpoint that Teilhard de Chardin was trying to teach us! And it fits nicely with ideas about "the Gaia hypothesis," the collective unconscious and the "noosphere" and so on. It fits with Taoist and Native American feelings about the living Earth. It is actually more consistent with the later, more yoga-based parts of the Upanishads than with the parts I used to like. Maybe some people in India learned a few things over time, just as I did. For true sanity, we need to be true to our entire selves, and seek harmony of all parts of ourselves, a harmony which ancient mystics like George Washington used to call an "alchemical marriage." (Are you skeptical about that last part? Well, in the open museum in the George Washington Masonic Memorial, a few minutes from my house, you can see a picture of him in his Scottish Rite apron, and many related materials. Loren Eiseley's introduction to the science fiction Voyage to Arcturus, by Lindsay, echoes a view very similar to this standard model of the soul - and the book itself clearly reflects one strand of the Scottish Rite Masonry. Those interested in authentic esoteric history might read Colin Wilson's book on Lindsay - but not the more recent edition captured by the fuzzies.)

This is just a small opening to a huge and important subject. But what does it tell us in the end about space? To begin with, it says that Teilhard de Chardin was right. In addition to the rational mundane reasons to want to move out into the galaxy, there is a strong natural spiritual impulse as well. Tsiolovsky said: "Earth is the cradle of mankind - but we can't stay in the cradle forever." It would be grossly unhealthy for ANY intelligence to spend all of eternity talking to and seeing nothing but itself. Thus those who have real sensitivity to the spirit which moves us all on the winds should be able to feel directly how this spirit, like the birds, does want to soar. And of course Teilhard de Chardin, a Catholic priest, could see clearly how this fits the true inner spirit of true Christianity.

There is another, more grim side of this. Some people imagine that we can simply let the whole Earth go to hell. Even if all bodies on Earth die, the Elect can just go to heaven. They imagine a strange kind of God the Father who prefers unicellular babies who scream and ask for favors all day, never really listen to anyone else, throw their talents away in a way that Jesus deplored, and stand idly by as their brothers and soul-mates die by the billions and as they cheerfully smash up a very expensive billion-year-old toy. Sorry, kids, that's not the way it works. We are all in this together. Smash up this toy - and there are no guarantees at any level of existence, mundane or spiritual.

One never knows what to make of dreams. Sometimes they are simply effective tools of our subconscious mundane mind to make us more fully aware of realistic possibilities we need to take seriously. Sometimes, in my experience, they foreshadow real possibilities - perhaps from our collective intelligence or perhaps from some other physics we do not yet know. Whichever it is, they are a proper part of our thinking, and a truly sane person does not just repress them. (My apologies to the Freudians who would prefer to write a book to clarify this.) For myself, I am heavily influenced by different dreams, which have shown very detailed and very scientifically defensible alternative scenarios for the human future. In one scenario, I do indeed see the faces on the billions of dead bodies all over the entire Earth, so many of them barely and stupidly dawning to a realization that "all of humanity going extinct" was a literal statement including them personally despite all of their good works and so on. In another... long before I met Ray Chase, I saw his vehicle flying (yes, the exact same weird angles with special CFD properties)... and I saw humans truly reaching the stars. There were great struggles in both tracks or books... but we really do face a crossroads, here and now. I hope we can get our act together to drive down the right fork.

4. An Ecumenical Postscript

Section 3 really gives my final message, but a few caveats and clarification are in order.

First, rationality does demand that we try to stay open-minded, at least towards alternative models that make sense. And of course, there is good reason for us to do so. Our experience and even our understanding of mathematics are all at an immature stage and we have a lot to learn.

Second - I am often very amused by a silly expression of that uncertainty. Some followers of the Gaia hypothesis can become a bit degenerate about it, and depict "Gaia" as a literally female humanoid goddess. For some folks, that can be useful, in much the same way that a rosary can be useful for some people, but it's important not to confuse the symbol with the reality. For us to speculate about the gender of the "noosphere" is a bit like a couple of twin fetuses giggling about their fantasies of what sex might be like. Maybe they have better ways to learn about reality in their present state of impoverished input.

Third - as I speak of what learned from my mother, I really must give equal time to my wife, Dr. Ludmila Dolmatova Werbos. This very day the spiritual sustenance and palpable feelings of love which allow me to put all this on paper certainly trace back to her. I will not say as much about her as I did about my mother here, mainly because it would be inappropriate to go too high in levels in a paper for the public.

Fourth - the standard model of the soul certainly does not say that intelligence or soul is unique to Earth, or that maturity exists nowhere in the universe. Almost all of our spiritual life properly revolves around our connectedness on Earth - but, like Teilhard de Chardin, I suspect we should have a bit more literal respect for the specific words of Jesus when he refers to "Our father." Jesus and Lao Tzu both made it clear they wanted to be listened to, not worshipped as if they were gods, just as Mohammed did, and it is very sad that so many have come in their name who tried to undo their work, all for the sake of petty things like personal ego and vanity.

Fifth - like Isaac Newton, I believe that the Greek Bible used today was in great part a twisted document due to Constantine, whose role as the head of Christendom was essentially the same as that of Lenin when he nationalized the Russian church. (Is it truly a miracle when you get a unanimous vote of the Politburo?) The neo(?)-pagan pseudo-Christian works from those days are no better and no more authentic.

Sixth - there is no time here to recount long and illuminating conversations with people from many deep and sensitive cultures around the Earth. Such conversations are an essential part of our common, indivisible spiritual growth. But I can relate one thing.

A true Sufi leader who once asked "Why a mathematical model?" At one level, experience is enough by itself. That is what a mouse would tell you, as he questions the weird perversity (in his view) of the human way of life, using words and ever so self-conscious. The spiritual leaders of the past have always been right to struggle with the oppressive kind of rational ego, which does not open itself up to the greater realities of life. But it is not natural give up our total being, our intelligence, our vitality, and even our capacity for higher pleasure and pain, as the price of doing so. Is it really possible that a mathematical model could "explain" or fit absolutely all of existence in a coherent way? Ultimately, we cannot be sure. But it is in our nature to try, and to rejoice as we see that we really can continue to grow in understanding, now as in the past, and hopefully forever.

And finally – the experiences I have described here are not so unique and not so alien to the human spirit. Many humans today may never have the experience of walking out on a cool night, and seeing a sky so clear and full of stars that they can see the Milky Way and feel the real size and majesty of it. Many even believe that "feelings in the heart" are just a metaphor. But we all have the capacity to see and appreciate the larger objective reality that we are part of. If we feel safer in the light than in the darkness, we will not shy away from this larger reality. Years ago, Andrew Greeley reported an intensive NSF-funded survey of basic values in American life – and he reported that fully 70% of PhDs in their prime in the US have had the same kinds of experience I have hinted at here – but mostly shied away in fear. ("Are We a Nation of Mystics?", New York Times Magazine, Jan. 26, 1975.) Perhaps it is time to stop hiding in fear – and to walk forward with big strides into the light.

About the Author

Prior to arriving full-time at the National Science Foundation (NSF) in 1989, **Dr. Paul J. Werbos** worked since 1979 at the Energy Information Administration (EIA) of the Department of Energy. He holds four degrees from Harvard and the London School of Economics. Dr. Werbos has core responsibility for the Adaptive and Intelligent Systems (AIS) area within the Controls, Networks and Computational Intelligence (CNCI) Program of ECS. He is also leading the development of a new Cybersystems thrust within the Integrative Hybrid and Complex Systems (IHCS) program of ECS. He is the ECS representative for the CLEANER initiative, for biocomplexity (MUSES), and for Collaborative Research in Computational NeuroScience. He is one of the two ECS representatives for cyberinfrastructure. Much of the specific research alluded to here is described in detail at www.werbos.com.

Dr. Werbos is an elected member of the Administrative Committee (AdCom) of the IEEE Computational Intelligence Society, which he represents on the IEEE-USA Energy Policy Committee. (See www.ieeeusa.org/policy/energy_strategy.ppt.) He also serves on the AdCom of the IEEE Industrial Electronics Society, and the Governing Board of the International Neural Network Society (INNS). He is a Fellow of the IEEE, and has won its Neural Network Pioneer Award, for the discovery of the "backpropagation algorithm" and other basic neural network learning designs. In 2002, he and John Mankins of NASA initiated and ran the NASA-NSF-EPRI initiative on enabling technologies for space solar power (search on "JIETSSP" at www.nsf.gov). In 2003, he participated on the interagency working group for the Climate Change Technology Program. At the 2005 Space Development Conference in Arlington, he was invited to present a new strategy for sustainable exploration and development of space, drawing in part on previous work funded by NSF.

Chapter 22

Space and Humanity's Evolution

By John Stewart

"The most meaningful activity in which a human being can be engaged is one that is directly related to human evolution. This is true because human beings now play an active and critical role not only in the process of their own evolution but in the survival and evolution of all living beings. Awareness of this places upon human beings a responsibility for their participation in and contribution to the process of evolution. If humankind would accept and acknowledge this responsibility and become creatively engaged in the process of metabiological evolution consciously, as well as unconsciously, a new reality would emerge, and a new age would be born."
Jonas Salk, 1983

Movement out into space has the potential to deliver clear benefits to humanity—it has the potential to improve the quality of life for those who remain on Earth, and to also provide enriched experiences for those who go out into space. By themselves, these benefits make space colonization almost inevitable. But the expansion of humanity into space is also of great evolutionary significance, and recognition of this has the potential to substantially increase support for the space effort. Awareness that space settlement is one of the next great steps in evolution on this planet and that contributing to this amounts to participation in the advancement of the evolutionary process has the potential to strongly energize support for the expansion into space.

At present, this is not a major factor - very few humans see their lives as part of a broader evolutionary trajectory, and even fewer draw on the evolutionary process to define their goals and values.

However, this is changing. Increasingly, individuals are beginning to see that their lives and actions are an important part of the great evolutionary process that has produced the universe and the life within it. They realize that they have a significant role to play in evolution on this planet, and that the eventual movement of humanity out into space is a key step in this evolution. Some of these individuals are choosing to dedicate their lives to consciously advancing the evolutionary process. They know that if evolution on Earth is to continue to fulfill its potential, it must now be driven consciously, and their responsibility and destiny is to contribute to this.

At the heart of this evolutionary awakening is the understanding that evolution is directional. This is the unmistakable conclusion that can be drawn from a growing body of interdisciplinary work (for example, see Huxley, 1957 [3]; Maynard Smith and Szathmary, 1995 [5]; Stewart, 2000 [6]; Chaisson, 2001 [1], and Gardner, 2003 [2]). The emerging picture is that evolution is not an aimless and random process, but that it is headed somewhere. The realization that evolution is directional has major implications for humanity—once we understand the direction of evolution, we can identify where we are located along the evolutionary trajectory, discover what the next great steps in evolution are, and see what they mean for us, as individuals and collectively.

Where is evolution headed? Contrary to earlier understandings that emphasized the evolution of selfishness, an unmistakable trend is towards the formation of cooperative organizations of living processes of larger and larger scale.

The significance of this trend towards larger-scale cooperatives is well illustrated by a short history of the evolution of life on Earth. For billions of years after the big bang, the universe expanded rapidly in scale and diversified into a multitude of galaxies, stars, planets and other forms of lifeless matter. The first life that eventually arose on Earth was infinitesimal—it was comprised of a few molecular processes. But it did not remain on this tiny scale for long. In the first major development, cooperative groups of molecular processes formed the first simple cells. Then, in a further significant advance, communities of these simple cells formed more complex cells of much greater scale.

A further major evolutionary transition unfolded after many more millions of years. Evolution discovered how to organize cooperative groups of these complex cells into multi-celled organisms such as insects, fish, and eventually mammals. Again the scale of living processes had increased enormously. This trend continued with the emergence of cooperative societies of multi-celled organisms, including bee hives, wolf packs and baboon troops. The pattern was repeated with humans—families joined up to form bands, bands teamed up to form tribes, tribes joined to form agricultural communities, and so on. The largest-scale cooperative organizations of living processes on the planet are now human societies.

This unmistakable trend is the result of many repetitions of a process in which living entities team up to form larger scale cooperatives. Strikingly, the cooperative groups that arise at each step in this sequence become the entities that then team up to form the cooperative groups at the next step in the sequence.

It is easy to see what has driven this long sequence of directional evolution—at every level of organization, cooperative teams united by common goals will always have the potential to be more successful than isolated individuals. It will be the same wherever life arises in the universe. The details will differ, but the direction will be the same—towards unification and cooperation over greater and greater scales.

Life has come a long way on this planet. When it began, individual living processes could do little more than influence events at the scale of molecular processes. But as a result of the successive formation of larger and larger cooperatives, coordinated living processes are now managing and controlling events on the scale of continents. And life appears to be on the threshold of another major evolutionary transition - humanity has the potential to form a unified and inclusive global society in symbiotic relationship with our technologies and with the planet as a whole. In the process, "we" (the whole) will come to manage matter, energy and living processes on a planetary scale. When this global organization emerges, the scale of cooperative organization will have increased over a million, billion times since life began.

If humanity is to fulfill its potential in the evolution of life in the universe, this expansion of the scale of cooperative organization will continue. The global organization has the potential to expand out into the solar system and beyond. By managing matter, energy and living processes over larger and larger scales, human organization could eventually achieve the capacity to influence events at the scale of the solar system and galaxy. And the human organization could repeat the great transitions of its evolutionary past by teaming up with any other societies of living processes that it encounters.

Further extrapolation suggests that the great potential of the evolutionary process is to eventually produce a unified cooperative organization of living processes that spans and manages the universe as a whole. The matter of the universe would be infused and organized by life. The universe itself would become a living organism that pursued its own goals and objectives, whatever they might be. In its long climb up from the scale of molecular processes, life will have unified the universe that was blown apart by the big bang.

As life increases in scale, a second major trend emerges - it gets better at evolving. Organisms that are more evolvable are better at discovering the adaptive behaviors that enable them to succeed in evolution. They are smarter at finding solutions to adaptive challenges and at finding better ways to achieve their goals.

Initially living processes discover better adaptations by trial and error. They find out which behaviors are most effective by trying them out in practice. Initially this trial and error search occurs across the generations through mutation at the genetic level. An important advance occurs when this gene-based evolution discovers how to produce organisms with the capacity to learn by trial and error during their lives.

In a further major transition, organisms evolve the capacity to form mental representations of their environment and of the impact of alternative behaviors. This enables them to foresee how their environment will respond to their actions. Rather than try out alternative behaviors in practice, they can now test them mentally. They begin to understand how their world works, and how it can be manipulated consciously to achieve their adaptive goals.

Evolvability gets another significant boost when organisms develop the capacity to share the knowledge that they use to build their mental representations. Imitation, language, writing and printing are important examples of processes that transmit adaptive knowledge. These processes enable the rapid accumulation of knowledge across generations and the building of more complex mental models.

Eventually organisms with these capacities will develop a theory of evolution - they will acquire the knowledge to build mental models of the evolutionary processes that produced the living processes on their planet, including themselves.

In fact, on any planet where life emerges, this trend is likely to eventually produce organisms who awaken to their evolutionary history and its future possibilities. They will begin to understand the wider-scale evolutionary processes that have produced them and that will govern the future of life on their planet. The organisms will begin to see themselves as having reached a particular stage in an on-going and directional evolutionary process. They will know where evolution is headed, and what they must do if they are to advance evolution on their planet.

On any planet which reaches this stage, some individuals will begin to undergo a critical shift in consciousness. Increasingly they will cease to experience themselves primarily as isolated and self-concerned individuals. Instead, they will begin to see and experience themselves as participants and actors in the great evolutionary process on their planet. The object of their self-reflection will change. When they think of themselves, they will tend to see themselves-as-part-of-the-evolutionary-process. Their conscious participation in evolution will increasingly become the source of value and meaning in their lives. Key realizations that will contribute to this shift in consciousness are:

* they have the opportunity to be conscious participants in the evolutionary processes that will shape the future of life on their planet and its movement out into space. They can play an important role in the actualization of the next great steps in evolution;

* the successful future evolution of life on their planet depends on their conscious participation. The most successful planetary societies will be consciously designed and envisioned, as will their successful expansion out into space;

* their actions can have meaning and purpose insofar as they are relevant to the wider evolutionary process. To the extent that their actions can contribute positively to evolution, they are meaningful to a larger process outside themselves that has been unfolding long before they were born, and that will continue long after they die;

* the evolutionary perspective therefore provides them with an answer to the great existential question that confronts all conscious individuals: What should I do with my life?

* their awakening to the evolutionary perspective and the awakening of others like them is itself a critically important evolutionary event on their planet.

The emergence of individuals who undergo this shift in consciousness equates to the evolutionary process on the planet becoming aware of itself. Through these individuals, the evolutionary process develops capacities for self-reflection, self-knowledge, and foresight. It will use these abilities to continually redesign itself to accelerate its own advancement.

Individuals that embrace the evolutionary perspective will set out to align their personal goals with evolutionary objectives. They will attempt to free themselves from pre-existing motivations and needs that conflict with evolutionary goals. They know that this will be essential if their species is to continue to contribute to the advancement of the evolutionary process—the organisms that play a significant role in the future evolution of life in the universe will not be those that continue to stay on the planet on which they emerge, masturbating stone-age desires forever.

Individuals that develop the psychological capacity to transcend these motivations and needs will actualize a further major transition in evolvability. They will be self-evolving beings - organisms that have the ability to adapt in whatever directions are necessary to advance the evolutionary process, unrestricted by their biological and social past.

Humanity has reached this major evolutionary threshold. The next great steps in social evolution on Earth are the formation of a unified, sustainable and creative global society and the expansion of this society into space. On Earth, individuals and groups are beginning to emerge who have decided to consciously contribute to the evolutionary process by doing what they can to actualize such a global society and its expansion. They are energized by the realization that their evolutionary awakening and activism is part of a significant evolutionary transition on Earth.

When larger-scale cooperatives of living processes have emerged previously in evolution, they have undergone a process of individuation. The unified and expanding human society can be expected to follow a similar evolutionary path. It will progressively develop internal processes that enable it to act, adapt and relate as a coherent whole—eventually humanity will be able to speak with one voice. For the first time, there will be an entity that other planetary societies could relate to and interact with. There will be an entity at the same level as other planetary and trans-planetary societies. If humanity is successful in reaching this level, a new universe of possibilities and experiences will open up to humankind.

As documented throughout this book, some kind of movement of humanity out into space seems inevitable. But this great step is likely to be far more successful and meaningful if it is guided and energized by awareness of the wider evolutionary trajectories that will eventually determine the significance of humanity in the universe.

References

(1) Chaisson, E. J. (2001) Cosmic Evolution: the rise of complexity in nature. Cambridge, Massachusetts: Harvard University Press.
(2) Gardner, J. N. (2003) Biocosm: The new scientific theory of evolution: Intelligent life is the architect of the universe. Makawao, Hawaii: Inner Ocean Publishing.
(3) Huxley, J. (1957) New Wine for New Bottles. London: Chatto and Windus.
(4) Maynard Smith, J., and Szathmary, E. (1995) The Major Transitions in Evolution. Oxford: Oxford University Press.
(5) Salk, J. (1983) Anatomy of Reality: Merging of intuition and reason. New York: Columbia University Press.
(6) Stewart, J. (2000) Evolution's Arrow: the direction of evolution and the future of humanity. Canberra: The Chapman Press.

About the Author

John Stewart is an Australian evolutionary theorist and is an external member of the Evolution, Complexity and Cognition (ECCO) research group of the Free University of Brussels. A number of his theoretical papers on evolution have been published in key international science journals. He is particularly interested in the implications for humanity of an evolutionary understanding of the universe and life within it. This is the major focus of his book Evolution's Arrow: The Direction of Evolution and the Future of Humanity. The ECCO website is at http://pespmc1.vub.ac.be/ECCO/, and the Evolution's Arrow website is at http://users.tpg.com.au/users/jes999/

PART III.

SCIENCE, TECHNOLOGY, ENGINEERING, AND MANAGEMENT FOR SPACE

The totality of existing sciences, plus new sciences that are already being created and others yet to be created, will be needed to provide the "How to ...." for the successful exploration and permanent human habitation of space. The authors of Beyond Earth are optimistic that this can be done, but they also understand that doing so will require a unique mix of research, planning, and political will. Our belief is that *"When in doubt, choose optimism, then manage wisely to achieve a self-fulfilling prophecy."*

In Part III, 15 authors examine some of these critical technical and managerial challenges, including topics ranging from space settlements to extra-terrestrial intelligence, mission operations, spaceports, resource management, risk management, an overview of technical challenges in many fields, and the oasis in space.

Chapter 23

Strategic Thinking for Space Settlements: Energy from Space

By Paul J. Werbos

1. Defining the Target: Sustainability

Forty years after the Apollo program, we now recognize that the dream that inspired us was not the dream of funding a few short-term visits to the moon, leaving behind only footprints in the sand, at a cost of millions of dollars per footprint. If the space program continues to cost a million dollars per footprint, and if the taxpayer on Earth pays the entire bill, we haven't done what we set out to do. But what did we set out to do? How can we translate our broader vision of humans settling space into a more concrete, operational target that we can use to guide concrete engineering and management efforts?

President Bush's vision for space exploration rightly stresses sustainability as a key goal in space. Sustainability got into the plan as a result of political considerations, since using 1960's technology a voyage to Mars would cost many times more than the entire NASA budget, and Congress would not support that. Therefore, we know we need to develop new technology to reduce the cost of space exploration, so as to make it sustainable as a larger-scale activity within a fixed real budget.

This vision was a good start—but we need to take it further. Human presence in space will not be truly sustainable until it is capable of independent economic growth, without needing any net subsidy at all from government, and without being limited by the current rate of growth on Earth itself. There are three main requirements that need to be met:

* Money earned by space (revenue) must grow enough to be able to pay for the entire activity.
* Activities in space must be large enough and diverse enough that they lead to "multiplier effects," where people will invest ever more money to support new activities in space paid for by supplying existing activities and people in space.
* The underlying technology for productive activities in space must be efficient enough that we can "close the loop" economically.

I would propose that the number one goal for human activities in space should thus be to maximize the probability that we eventually meet all three requirements. Once the goal is accepted, everything else we do should be part of a rational, adaptive strategy to meet the goal - to maximize the probability. This chapter addresses the question: How do we actually do that? But I need to say more about the goal itself before moving on.

First, the three requirements here are really just an English-language version of a more precise mathematical criterion. Economist Eugene Rostow wrote a famous book decades ago on "take-off economies" in which he noted how some poor nations seem to go from bad to worse; others become "banana republics," earning money in proportion to the current market for bananas in rich nations, but unable to grow any faster or to catch up; and yet others reach a "takeoff point." Hence, the real goal here is for the human economy beyond the Earth to reach the economic take-off point.

In mathematical terms - there is a kind of matrix XS, the portion of the input-output matrix for activities in space which can be economically supplied from space, augmented by considering humans in space as an activity; we want this matrix to have a real eigenvector with eigenvalue greater than one. Because XS is actually a matrix of derivatives, it depends both on existing infrastructure and activity levels in space, and on the characteristics of the technology. In practical terms, this means that we want the infrastructure in space, the revenue-generating activities, and the underlying technologies to reach the point where space is no longer just a banana republic cum subsidy farm.

Second, it is very hard to assess the total probability that humans will someday meet all three requirements, across all future history. Thus, as a practical matter we may approximate this goal by trying to minimize the expected delay time between now and the time when all three requirements are met. The longer we wait, the greater the chance that world events will make it harder to keep going, particularly as we do not reap the economic benefits that space could give us in the meantime. There are critical technologies we have inherited in the U.S. which could become lost (and extremely difficult to re-invent) if we do not use and update them.

Some advocates would argue that this approximation is not so good after all. They would say: "The probability of someday meeting these requirements may actually be greater if we focus all our efforts for now on the goal of humans walking on Mars. Footprints on Mars would energize the public, and provide a growth in budgets which makes it easier to reach your requirements." In effect - this advocates for a gigantic, expensive public relations effort.

But footprints on the moon during Nixon's presidency were followed soon after by the largest percentage cut ever in NASA's budget. Still, there is a legitimate argument here for accelerating the Terrestrial Planet Finder (TPF) mission, drawing on the new minimum-cost technology initially developed by Prof. David Hyland (of the TPF advisory board) under a small grant from NSF. TPF, if fully funded and focused on using advanced technology, will probably be able to see decisive signs of large-scale plant life on Earth-like planets in other solar systems.

I would guess that such a discovery would galvanize public interest in learning more and in strengthening Earth's technology far more than any rerun of Apollo would do (even on Mars). There is room in NASA's budget to accelerate a few key activities like TPF, even if the bulk of the budget is redirected to the goal I propose here - to minimize the expected delay time between now and the time when the three requirements are met. There is also room for a few lower-cost strategic efforts to encourage the science and technology that may be needed, someday, to maximize our chance of someday reaching beyond our solar system.

The greatest convulsions in the history of US space policy can all be traced back to people chasing the wrong targets. For example, the three requirements here are a goal, not a present reality. If we try to turn over all of space to truly commercial market companies today we may never reach the goal. President Bush's vision does not require that we turn over all of space to private companies in the short term, in a way that leads to the elimination of NASA and enormous loss, pain and liquidation of America's strongest technological capabilities. Yet some have interpreted it that way.

Likewise, it does not call for us to create a vast new government program with its own autonomous goals, disconnected from the natural growth of American industry and from the positive, egalitarian spirit of the frontier. The target proposed here would require a new kind of collaboration between NASA, space enthusiasts in other agencies, private industry, and other nations. The goal is far too big and complex for any one group to do it all alone.

Finally, the greatest risk to NASA today comes from paying too much attention to the goal of minimizing the time and cost between today and the time when we are certain that American footprints will appear again in the dust on the moon. We need instead to minimize the cost and time between now and our average best guess of the time when our enabling technologies, discoveries, and infrastructure will truly allow profit-making entities aimed at real pubic market to "take over" - to operate on a large enough scale, with diverse enough activity, that they truly possess self-sustaining growth in service to humanity.

Notice the big difference between a goal which demands certainty and a goal which aims at the average expected time delay! There is a huge difference between pursuing a difficult longer-term goal, where success cannot be guaranteed, and pursuing a more certain near-term goal. Many prefer the goal of "new footprints in the sand" simply because it can be guaranteed, technologically, by using technologies NASA proved many years ago. There is an analogy here to the large programs on lead-acid batteries at DOE years ago, at the time of long gas lines; everyone knew that lead-acid batteries would never be good enough for real market-worthy electric cars, but they were funded anyway, because they were a "low risk" technologically.

We need to remember that there are two kinds of risk - technology risk (when you might not meet your technical goal) and market risk (when the product may not have market value). Re-inventing Apollo would have minimum technology risk, but without a market benefit in sight, the risk of total bankruptcy is very serious. Hence, we need to adopt a more balanced attitude to risk, like that of the entrepreneur who realizes that he must make a trade-off between technology risk and market risk. He must be bold enough to aim directly at a real market profit, and accept the fact that some degree of risk is unavoidable. In the end, the risk that matters is the risk that we never get to our true target. Advanced technology is a way of reducing the risk that really matters. We need to do it right or not do it at all.

2. Energy from Space: New Options and a Critical Need on Earth

The number one requirement for sustainability in space is a huge new revenue stream, aimed at markets on Earth, but requiring diverse new activities in space. The most serious candidates right now, in my view, are energy from space ("Space Solar Power," SSP), space tourism, space manufacturing, and materials from the asteroids. In order to implement the strategy of section 1, our civilian space program should be centered on the goal of bringing one or more of these four candidates all the way to commercial fruition. Our efforts must be highly adaptive, because our knowledge about these four markets and the required technologies will change a lot from year to year, if we do our jobs right. And if we do our jobs right, we will develop many new technology options and push hard to improve our knowledge by addressing the most difficult challenges and uncertainties surrounding them.

The four big new markets are not alternatives, as at the level of nuts and bolts, efforts in the four different areas can all help each other. But in this section I will write just about SSP, for four reasons: (1) thanks to new progress, SSP shows the most promise of meeting our requirements as soon as possible; (2) this is an area where I personally have special knowledge, as a result of co-managing the NASA-NSF-EPRI joint effort in SSP funded in 2002; (3) SSP can play a vital role in helping us achieve a sustainable global energy/environment system on Earth—a crucial high-risk challenge which is a matter of life or death in its own right; and (4) new energy sources motivated by the needs of Earth can also play a crucial role as an enabling technology in achieving sustainability in space. In section 3, I will discuss how SSP fits into a larger concrete strategy for sustainable space settlement.

Years ago I was far more skeptical about SSP as a flagship market for space, and I still have great respect for those people who remain skeptical today based on the options developed back in the 1970's. In the 1970's, NASA and the Department of Energy (DOE) were both funded by Congress to evaluate SSP. NASA funded two major studies to develop "reference designs" for SSP which, they estimated, could produce 24-hour ("base load") electricity at 5.5 cents per kilowatt hour (kwh). That is roughly competitive with coal and nuclear power, whose total cost runs at about 4-8 cents per kwh today.

The DOE study, led by Fred Koomanoff, of the Germantown Office of DOE, was more pessimistic. Fred brought in assistance from several places—including me, since I was then the person in DOE Headquarters (the Office of Energy Information Validation, located in the Energy Information Administration) responsible for evaluating all aspects of long-term energy projections and their assumptions. Our report was far more skeptical than the NASA report. It questioned the cost estimates, and pointed to major unresolved uncertainties about whether the technology would work as advertised. Unfortunately, Ralph Nader and others misrepresented this report as an attack on the idea of exploring SSP at all; in fact, that was not the intent. A rational and honest assessment of uncertainties and realities is an essential first step to making a technology real—if there is any hope of making it real at all. Koomanoff and I both tried to be clear on that point, but Nader's voice was of greater interest to the media.

In the 1990's, thanks to Congressman Rohrabacher and NASA's John Mankins, research on SSP was revived at NASA. Mankins' "SERT" program achieved a number of very solid accomplishments, essential to the new hopes we see today. He did verify that some of the problems we were worried about would have been show-stoppers in the original reference system designs, but then he moved on to develop new solid, verified technologies that would overcome those problems. He obtained detailed reviews and recommendations from the National Academy of Sciences, whose report is still an important document on SSP, and he also developed serious cost estimates for the best new designs which resulted from this effort. Unfortunately, by about 2002, the best serious cost estimates came in at just under 20 cents per kwh—good enough for some niche markets, but not good enough to compete with coal or nuclear fission for the big baseload power markets on Earth.

In 2000, two important new developments arose. First, we led a new workshop, jointly with John Mankins, on how we might use computational intelligence, learning, and robotics to reduce the cost of SSP. That workshop report, still available on the web site of Prof. George Bekey at the University of Southern California helped set the stage for our new joint effort in 2002.

Second, the Millennium Project (MP) of the American Council for the United Nations (ACUNU) received a grant from DOE to explore the needs and opportunities for science and technology in general, all across the Earth. That study asked key decision-makers and policy experts all over the world what contributions science and technology could make to improve the human condition in general; the number one answer was a non-fossil, non-fission source of affordable baseload electricity on a scale large enough to meet all the needs of Earth. (See www.stateofthefuture.org .)

SSP is one of the very few serious candidates to meet that need. And in 2001, my current employer - the Electrical and Communication Systems Division of NSF - was approached by one of the leaders of the microwave communication community, Prof. Michael Steer of North Carolina Sate University, whom we funded to do a quick reassessment of the microwave power beaming issues associated with SSP.

Leaders in microwave communication suggested that the challenges of power beaming and its supposed hazards were all basically matters of engineering and verification - but that little if any objective risk is present in the underlying technology: We can do it. A subsequent study by Frank Little of Texas A&M University provides the best current information on this aspect.

Hence, all three sources gave us very strong encouragement to initiate a funding activity in this area, and so in 2002 James Mink and I from NSF approached John Mankins of NASA to propose a 50-50 new joint venture, to be routed through the NSF funding for universities and small businesses, with an open call for proposals distributed across the US.

This solicitation was published as NSF publication "number" NSF 02-098, still available on the web, with references to prior work and to the funding priorities as we saw them in 2002. We announced that we only had $3 million to hand out - but we still received a very diverse and impressive collection of 98 proposals. After very intensive peer review, demanding that we only consider truly original and unique work with a major potential impact on the economics of SSP, our panels recommended funding of about 50 proposals, amounting to $21 million. There is a major unmet opportunity here, unmet to this day, and we deeply regret the very valuable work proposed that we could not fund as yet.

But even so, we learned a great deal from this exercise. John Mankins and I jointly ran the exercise, and we both had very intensive discussions with the various groups as their work progressed, and we learned a great deal about the realities. With help from NASA's Jet Propulsion Laboratory, we jointly wrote a "technology profile" on SSP included in the 2003 technology book of the interagency Climate Change Technology Program - the only relatively official summary of the overall (interim) outcome of the SSP effort.

So where do we stand today on actually making this work at an affordable price? now have what I would call "four new design concepts," all much more promising than the 17-cent SERT options. These are:

* A SERT-plus design, a concept from John Mankins, similar to some previous designs, but embodying new "solar sandwich" ideas from Neville Marzwell of the Jet Propulsion Laboratory, and other innovations.

* The "spinal cord laser" design that Richard Fork (University of Alabama - Huntsville) and I presented at the 2005 Technical Interchange Conference hosted by the Ohio Aerospace Institute.

* A hybrid laser/D-D design that I first proposed in the State of the Future 2003 (CD ROM appendix).

* The not-so-new or not-yet-new-and-ready concepts for using nonterrestrial materials, as proposed by Gerard K. O'Neill, David Criswell and others.

None of the four now offer such a credible price tag as the SERT designs, simply because we have not had the support yet to detail them sufficiently, evaluate the uncertainties, or develop sharp cost estimates. But in at least three of the four cases, we have excellent reason to expect that costs will be much less than 17 cents. There are important risks - but when we consider all four options together, I believe that the objective risk of not being able to beat coal or fission on cost is far less here than with any other technology that is large enough to meet all the world's energy needs.

For an updated version of the global energy story, see www.werbos.com/energy.htm.

This does not mean that we should not fund those other options, but nor should we count on them. Earth cannot afford that we neglect our best option for baseload alternative power.

SERT-plus

The key concept in SERT-plus is the use of lightweight mirrors, like those proven out by Entech, to focus light onto "sandwich cells," which produce electricity that is then beamed to Earth by microwave.

On the whole, I would guess that SERT-plus would end up around ten cents per kwh, if we meet certain assumptions built into the SERT cost estimates. The most worrisome assumption, in my view, is that the cost of access to low Earth orbit (LEO) will be cut to $200 per pound. At $2,000 per pound or more, the best we can realistically hope for with expendable rockets, that gets raised to $1.00 per kwh - strictly a niche market.

Spinal Cord Laser

In the "spinal cord laser" (which Richard Fork initially called the "backbone laser"), there is almost no flow of electricity at all in space. Lightweight mirrors concentrate light onto disks of semiconductor material, which are excited by sunlight in the same way that photovoltaics are excited. These disks also exhibit similar efficiencies and weight, but emit energy in the form of coherent light that can be beamed directly down to small receiving stations on Earth. The same laser can easily punch through clouds on cloudy days.

Concentrator cells like Marzwell's could be used to extract electricity on the ground. Because of the similarity to photovoltaic designs, and because we avoid the weight needed for electric power distribution (about half the weight of the SERT designs), it is reasonable to expect costs in the 10-20 cent range. One beauty of this design is that it can be tested and used at a much lower minimum power level than the SERT designs.

Hybrid laser/D-D

My hybrid concept is riskier than the other two, but still has an excellent chance of working out, and promises much lower costs than the other two. The idea is to start with a different light-to-light laser, analogous to the spinal cord laser, but harder to design. We need a laser that can emit energy in the high-speed megajoule pulses needed to ignite nuclear fusion in the deuterium-deuterium (D-D) pellets designed by John Perkins of Lawrence Livermore Laboratories. One way of looking at this is that Perkins' pellets augment the power production by a factor of 100 or more, beyond what comes out of the laser. Thus, if the required lasers end up weighing three times as much as the spinal cord laser for the same output of light, we still end up increasing power output per pound in orbit by a factor of (100/3), or 33. That implies a cost of well under 1 cent per kwh, for electric power available at a central point ready to be beamed down to earth.

From another viewpoint, the hybrid design has major advantages over nuclear fusion on Earth. There are two mainstream versions of nuclear fusion for use on Earth - fusion in magnetic bottles, and fusion based on lasers hitting pellets. Both versions are expected to cost more than today's nuclear fission because the energy comes out as energetic particles which turn into heat and radioactivity very quickly. They require the same complex heat systems as today's fission and coal plants, and the raise the same kinds of issues in handling radioactive materials.

Building lasers on Earth imposes many additional costs, but by doing this on space, and using D-D pellets, we end up with 80-90% of the fusion energy coming out as electric currents - protons moving through the vacuum of space. Instead of a heat chamber, we only need lightweight magnets to transform the electricity into a form that can be beamed to Earth as microwave energy. Several credible leading laser laboratories say that they know how to design such a laser relatively quickly, using new concepts such as photonic bandgap materials, fiber lasers and variations of Raman backscattering. But certainly the characteristics of the new laser are the main objective uncertainty here, and we need to clarify this as soon as possible (as constructively as possible).

To make this option certifiably real, we have two urgent needs: (1) to work on the access to space issues; and (2) to create a small competitive funding venue to develop, simulate, and evaluate competing laser designs from universities, small business and laboratories all over the Earth, if possible. Even at $2,000 per pound, however, this option looks as if it will come in well under 10 cents per kwh. If the laser costs meet our present best guess, then the costs of delivered power on earth could be competitive with existing baseload costs – i.e., about 5 cents per kwh—if the cost of power beaming itself is 4 cents per kwh (as estimated today for off-the-shelf designs) or lower (as many believe could be achieved with further research on power beaming as such.).

Nonterrestrial Materials

Finally, the fourth option - SSP based on nonterrestrial materials - is not as well explored as these others. Because of the hope of deep cost reductions, the need for nonterrestrial materials in the long-term anyway, and because of other technology benefits, this option cries out for much more effective strategic analysis and pursuit. The cost estimates available today are far less reliable than the crude guesses here for the other three - but we need to work to change this. The costs will still be proportional to the Earth-to-LEO cost, because the main cost is still the cost of lifting a lot of material into space. Also, a great deal of testing will be needed to clarify the options and make it possible to estimate costs.

In summary, to approach the normal costs of baseload electric power in the big markets on Earth today, we will either need to develop an advanced form of SSP (options 3 or 4) or develop $200/pound access to space, or both. If we develop both, energy from space could become a true bonanza for both Earth and space.

This book mainly addresses the goal of developing space - but SSP is equally essential to the strategic picture on Earth. While there is not sufficient space here to do justice to that complex issue, here are some overview thoughts and citations. In 2003, at a workshop in Mexico linked to the ACUNU MP, scenarios for the future of the world were discussed that were based on present trends in all sectors all over the world. The scenarios were very, very worrisome, as there is an emerging intersection of growing world dependence on oil and natural gas in sensitive areas, together with intense world tensions, and acceleration of nuclear proliferation linked to civilian programs like those of Iran. It is thus now quite rational to worry about the possibility of human extinction.

At present, hybrid cars (up to 60mpg "city") and plug-in hybrids (using half as much gasoline, but also using overnight electricity) are the only force actively moving us towards independence from OPEC oil - but they aggravate the already-serious problem of baseload electricity supply. Earth-based solar farms are the only renewable energy source which - like SSP - could easily meet all the world's energy needs, but they only work in the daytime; storage to provide energy at night adds a lot to the cost, and the going market cost today is already about 12 cents per kwh in the daytime (though we have ways to reduce that).

From a purely technical viewpoint, I see no reason why SSP could not provide tremendous amounts of electricity as soon as 10 years from the start of a massive, focused effort; this would be soon enough to prevent the kind of disaster we are now on track to encounter by about 2025. For more information, see the State of the Future (www.stateofthefuture.org), or www.ieeeusa.org/policy/energy_strategy.ppt. (These slides include text that can be viewed using "View - Notes" option in powerpoint, or by printing out the slides with the "notes" option selected.)

3. What Would an Optimal Space Policy Look Like?

According to the goals of section one, the "optimal" space policy, the best possible space policy is the one that enables us to fulfill the three requirements in the shortest possible time, subject to budget constraints. To devise such a policy we need to consider several key principles.

First, we should not leave it all to NASA. Space settlement is not just a goal for NASA, but conversely, humanity also has other important goals - such as energy sustainability and security - which should concern us all. There are exciting opportunities to "kill two birds with one stone" by expanding collaboration between government agencies and others. For example:

* Commercial enterprises can provide guidance to NASA on how to do a better job in supporting the four key markets discussed in section 2.
* The military and intelligence agencies have great needs, essential technology, and funds for improved access to space and related goals.
* The National Science Foundation - in partnership with NASA, as in NSF 02-098 - can effectively solicit ideas from a wide spectrum of universities, disciplines and small businesses, and provide more novelty and feedback than NASA could manage on its own, as well as an agile funding mechanism.
* Lawrence Livermore and other DOE labs could play an important role in implementing concepts of energy from space.
*And international partnerships could expand all this in many directions; for example, robotic and telerobotic technology from Japan could be married with new intelligent systems technology from NSF to meet many crucial needs in space in a cost-effective way.

Second, while these programs must be led by champions who truly care about long-term success, they should not be led by advocates who believe in tilting their message to create a less than accurate image in order to elicit support. When programs are not led by sincere champions, funds have a way of disappearing into activities which support the leader's constituents, other goals, or nicely packaged but insufficient efforts. On the other hand, when programs are led by advocates or true believers there is an equally long history of ultimate failure. When "team players" filter information and do not face up to key difficulties early on, it is common that difficulties surface too late to explore alternative pathways and save the program. Advocates, publicists and honest worriers should always be "in the loop" because of their creativity, but they should not be in charge.

Third - no matter which big markets for space we focus on, the biggest single barrier today is still the cost of getting to Earth orbit. If present trends are continued, I worry that all versions of space development will be doomed to failure in the end. The private sector can help, but it cannot work miracles. There are difficult technical barriers that require the very best enabling technologies to drive the costs down enough. Sheer scaling laws tell us we need a big vehicle to get the best dollars per pound - and we cannot afford anything less than the best. Furthermore, we need to remember that the market for access to space has two very different segments - the segment which can get by with payloads of ten tons or less, and the segment which requires heavy lift.

For the ten-ton market - NASA has known for a long time that operations costs are the dominant cost in getting to space, and that aircraft-style operation will be necessary. (For example, see the 1960's reports from Mueller at Goddard, which inspired the idea of a space shuttle - a great idea, ruined by political confusions and wrong metrics from the bean-counters.) This then requires horizontal-takeoff reusable rockets, at least until more exotic alternatives become available. In fact, at NSF I funded some very aggressive research on some advanced alternatives, through Ray Chase of ANSER and Miles of Princeton University. A startling outcome was that really advanced alternatives can become real only if we first resurrect and enhance some endangered legacy technologies. Ideally this means that we would develop a major program to insert those new technologies into a new program for a 1.5-million-pound horizontal-takeoff rocketplane.

The most crucial endangered technologies are "hot structure" technologies that would allow such a rocketplane to return safely to the Earth without using the weird and problematic tiles, or ablative structures, or slush hydrogen that other approaches call for. Ray Chase has developed a detailed plan for how to do this, using off-the-shelf technology all the way, with a project planning chart that goes out just 4 years - though we would be better off starting from a detailed one-year tradeoff/design-refinement study for $900,000, as Chase has also proposed. Some have claimed that space elevators would be better than any kind of spaceplane - but before we can afford to build space elevators, we will need affordable transportation to carry the required material into space!

In an ideal world, competitive launch services contracts from NASA, DOD, and the intelligence agencies would give the big aerospace companies the resources they need to go to Wall Street and finance a new vehicle on a private sector basis. In addition, launch service contracts for heavy lift capacity at 60 tons payload would allow companies like Space Island Group to build a shuttle-C vehicle enhanced by bringing the expended fuel tanks into orbit. This would cost much less than the presently planned heavy lift vehicles, and help NASA redirect funds to the development of the rocketplane and other key needs. "Meyer's ark" is an exciting concept, although it (like Criswell's) still requires a bit of fine-tuning.

President Bush's goal of a permanent human presence on the moon certainly plays a critical role - along with the asteroids - in opening up a source of lower-cost materials which will be essential, sooner or later, to the full economic development of space. But development of the moon must be part of the larger development of space described in section 1, in order to contribute the most it can to the long-term goal.

More developed infrastructure will be needed both in Earth orbit and in low lunar orbit (LLO) to make this possible. We need to do it right as soon as possible - and that means doing it with efficient infrastructure, and not "saving time" and wasting money by doing things the wrong way. If 10-ton payloads end up costing $200/pound (and lower, with time), and if heavy lift costs ten times as much or more, then it will be essential to economize on how we use that heavy lift capability. Reynerson of Boeing has described concepts of a totally reusable rocket (RLAV) to go to and from the moon, and he reports that an RLAV can be quite efficient even weighing only 60 tons if it is launched dry and unmanned. (Fuel and people can be carried aboard the cheaper rocketplane.) Kistler's new lunar rocket company has proposed similar ideas. Long-term fuel depots in LLO are well within today's best technology. Ion rockets have been proven in deep space by Japan, and numerous other concepts for deep space transportation look very exciting and very credible as a way to round out this infrastructure.

The technology is there, and the challenge to us is to go ahead with it.

About the Author
Paul J. Werbos, see Chapter 21

Chapter 24

The Intelligence Nexus in Space Exploration: Interfaces Among Terrestrial, Artifactual and Extra-Terrestrial Intelligence

By Joel D. Isaacson

Introduction

Space exploration and habitation will surely tax our natural intelligence, stretch our ability to invent intelligent artifacts, and will possibly put us eventually in contact with extraterrestrial forms of intelligence. Hence, many different forms of intelligence will, at some fundamental or generic level, be at the core of all space-related endeavors. My journey, my long journey, has been in quest of roots of "intelligence;" I have arrived at an ultimate simplicity that explains a great deal about intelligence, and this I wish to share with you.

This chapter speculates about generic aspects of "intelligence" as would be shared by all intelligent entities, and suggests modes of interface among all three types of intelligence, i.e., terrestrial, artifactual, and extraterrestrial. It also suggests that intelligence permeates Nature; and that it has no central focus because it is distributive, dissipative, cooperative, and emergent. Questions relating to Intelligent Design are raised, and some future research directions are proposed.

Elements of Perception

Consider an array of sensors that detect local differences in certain attributes of signal patterns. Such distinctions are recorded and form new signal patterns in their own right. Since a fundamental operation in perception is distinction-making in patterns, another iteration of distinctions-of-distinctions is then performed, and this kind of repeating process is applied recursively. To summarize, there are two fundamental elements in perception: (a) distinction-making and (b) the indefinite recursion of distinction-making.

The importance of distinction-making is not new. For example, G. Spencer-Brown focused on distinction-making in his Laws of Form [1], resulting in a very rich formal system. However, these distinctions require a full-blown human observer and the use of writing modes to record them on a piece of paper. What would be a logic of distinctions at the level of single receptors that could not write?

Imagine a linear array of receptors that processes signals reaching it, whose basic function is to determine local differences among adjacent signals. The recording of differences would require at most four types of tokens. A "token" is a kind of sign or symbol. Here it stands for the relations of distinct-from/not distinct-from that hold between a signal and its two adjacent signals. There are exactly four possibilities when a signal is compared with its two neighbors — see Ref [2]. The next pass of distinction-making re-encodes a prior 4-token array with a newly-formed 4-token linear array, and this repeats recursively.

Traces of such processing show an indefinite succession of 4-token strings.

The structure of such traces has been studied in great detail in (2). The main characteristics are:
* autonomic mode of processing (self-organization)
* autonomic error-correction
* autonomic mode of 3-level memory

(See Ref [2], where the 3 levels are "Subsurface Memory," "Intermediate Memory," and "Deep Memory." These levels develop spontaneously, of their own accord, in the course of BIP-processing. The levels correspond roughly to "short-term memory" (STM) and "long-term memory" (LTM) as commonly understood in cognitive science.)

* dialectical patterns
* autonomic syllogistic inferences
* limit cycles or attractors (Hegelian cycles)
* autonomic generation of palindromes
* complementarity of 4-letter strings

Some of these features are related to Douglas Hofstadter's "strange loops," as presented in his discussion of the natural intelligences of Gödel, Escher, and Bach in [3].

One subtle consequence of this model is that actual raw patterns of signals are not required for perception. Rather, it is their patterns of internal/local distinctions that matter. Thus, ordinary sensory detectors may be bypassed, and the first stage of processing be applied to patterns of distinctions if captured by some means. To get quickly to the main point, for cognitive subjects—entities capable of cognition—of diverse sensory modalities, common ground lies in being able to process "streaks" of patterns rather than just raw sensory patterns

Patterns are not innate "information," but they become so only when they enter into a certain relationship with a cognitive subject. So it's meaningless to discuss information without regard to a cognitive subject interacting with it. A given chunk of information, when considered with respect to a given cognitive subject, need not be physical with respect to that cognitive subject! It may be non-physical (at least in relation to the physical world in which the cognitive subject is immersed). How could this be? In hard science and engineering we normally think of "patterns of signals" as the physical carriers of information. However, the physical signals, per se, are not always important to the cognitive subject. It is rather local signal-differences — a derivative property of raw physical signals — that matters. This allows the unexpected introduction of information carriers that are non-physical signals, referred to here as "fantomarks."

What are "fantomarks"? Well, let's first examine what "marks" are. In physical-symbol systems (including conventional digital computers), "mark" is a generic name for physical signals of all kinds, and for physical symbols or their representations in term of physical bit-signal patterns. In short, "marks" are physical carriers of information that is being manipulated. "Fantomark" stands for phantom-mark. Fantomarks are information carriers that cannot be sensed or detected or recorded by human beings, and/or other living things or systems, and/or instruments, devices, or systems made by human beings. Patterns of fantomarks, while not physically perceivable to a cognitive subject, also have local differences, just as patterns of physical marks have differences. These local differences constitute patterns in their own right (dubbed "streaks") that correlate with the underlying fantomark patterns, but are not coincident with them. Streaks are derivative entities, and are physical. If a subject has access to a given streak, it can process it as a source of information based on a basic principle which states that:

WITH RESPECT TO ANY SPATIAL OR SPATIO-TEMPORAL SPACE OF ANY DIMENSIONALITY THAT EMITS PATTERNS OF SIGNALS, ABSOLUTELY NO COGNITIVE FUNCTION IS POSSIBLE UNLESS A COGNITIVE SUBJECT HAS THE CAPACITY TO LOCALLY MAKE AND RECORD SIGNAL-DISCRIMINATIONS. IF THE SUBJECT DOES POSSESS THIS CAPACITY, THEN SUCH CAPACITY IS THE ULTIMATE GENESIS OF ANY COGNITION, OR LEVEL OF COGNITION, THAT MAY DEVELOP IN THE SUBJECT. (Isaacson, 1987, see [4])

Namely, it recognizes a streak as a physical signal-pattern in its own right, and attempts to determine and record its local differences. The resulting pattern is the streak pattern of the previous streak. This activity may proceed recursively, which sets in motion a pattern-processing cellular-automaton.

Cellular automata are special kinds of computational models that are used in dynamical systems and share some properties with fractal and chaotic systems. One important aspect of cellular automata is that global patterns resulting from their computations are emergent (essentially non-algorithmic) from simple local interactions. It turns out that the emergent behavior of the cellular-automaton in question is dialectical, more or less in the classical Hegelian sense.

Why introduce "dialectical" elements into the discussion? The simple answer is that we have no choice in the matter. Dialectical behavior and features are emergent from the underlying cellular automata, which, in turn, are governed by the basic principle stated above. Further, cellular automata can be represented in terms of certain neural networks that are functionally equivalent to them. This provides us, then, with working models of neural mechanisms that give rise to Hegelian-type dialectics inside the brain. In effect, the model—of its own accord—offers a neural correlate of the Hegelian dialectics.

This was documented in a technical report titled "Dialectical Machine Vision," prepared for the Strategic Defense Initiative Organization and the Office of Naval Research in 1987. Following is a summary of some relevant points.

The report explores four different neural structures that are necessary for the existence and functioning of "vision":

* Sensory-level,
* Sensation-level,
* Perception-level, and
* Recognition-level.

These levels are essential to a cognitive function, as dictated by the Basic Principle and the recursive process engendered by it, and they map readily to anatomical structures in the neurological system's visual pathways. The sensory-level is in the retina; the sensation-level corresponds to the lateral geniculate nucleus (LGN) — this is the standard name for this anatomical structure in the visual pathway —; and the perception/recognition levels are mapped to the V1 area (and beyond) in the visual cortex of the brain, through bidirectional interaction with the LGN.

Visual awareness is postulated in virtue of a "sensation-resonance" process, whereby incoming imagery is held in short term memory in the LGN, and resonates with selected reverberatory circuitry from a massive neural network imagery-bank held in the visual cortex. As the communication between the LGN and V1 is bi-directional, visual memory turns out to be not so much storage of information patterns, as storage of sensations associated with previously experienced patterns of visual information. Perception and recognition are accomplished via "sensation-resonance," namely, resonance between current and previous sensations.

Among other things, this permits us to work out the binding problem for two-dimensional shapes, and also offers a scenario for face-recognition in infants that is based on the theory developed in the report. That theory contains fairly detailed sketches for possible solutions to both problems, with actual demonstrations. The "binding problem" relates to mechanisms whereby individual pixels (which fire/not-fire) are bound together into unified objects that are subject later to perception and recognition.

At the core of "dialectical image processing" some cellular automata (CA) perform recursive-edge-detection on imagery. If we translate those CA into a functionally equivalent neural-network, we can fill in much detail that is missing so far in the art, and is relevant to a stage we call "very-low-level-vision" (VLLV), where VLLV is even below David Marr's "early vision" [5].

The cellular automata in question are called "Dialectical Image Processors" (DIP). A DIP finds boundaries of objects in a recursive fashion. The first step, comparable to the retina proper, finds the boundary of a "raw" external image. The output of this step, comprised of the boundary, is an image in its own right. The next step finds the boundary of the current boundary. (Note: This kind of image processing, i.e., finding the boundary-of-a-boundary, is not intuitively obvious to most people and constitutes one of the innovations in this approach.) If we have an image that is a boundary (of some object), the only way to perceive that boundary-image is to determine its own boundaries.

The subsequent step defines the boundary-of-the-boundary-of-the-boundary, and so on, as an indefinite number of steps recursively follow. Each step results in a new image comprised of the boundary of the previous boundary. So, in effect, each step is like having a new ("virtual") retina operating on imagery received from a previous retina. This sequence of "retinas" can be thought of as belonging to a train of "virtual homunculi," where each homunculus attempts to interpret what the previous homunculus' retina has just processed.

However, old theories based on homunculi are notoriously naive, as they can't escape the obvious problem of "infinite regress." Why bother, then? Well, DIP is able to escape the curse of infinite regress because it is (mathematically) guaranteed to converge on a limit cycle. "Limit cycle" is a term from dynamical systems theory, including chaos and cellular automata. A process that converges on a closed/fixed loop is said to have reached a limit cycle. Once in the limit cycle, at the LGN stage of processing, the image is afforded a (dynamic) short term memory. In addition, the cycling activity causes periodic firing of neurons in the LGN over the entire area of the mapped image, giving rise to some primitive sensation in the cognitive subject. That "sensation" is the basis for a theory of visual awareness that follows in the report. That "sensation" element also builds a bridge towards ideas about the role of sensation in consciousness, proposed by Nicholas Humphrey in the early 1990s.

Ties to particle physics

When this recursive process is applied to a single signal (against a backdrop of the 'void') and linear patterns are allowed to expand in both directions, (that is, along a one-dimensional line where things are propagating on both sides of active segments - see figures in Ref [6]) from step to step, the entire pattern that emerges is self-similar fractal. It can be shown that these self-similar configurations subsume the structure of elementary particles, at the quark-level description, known as the "baryon octet" [6]. This indicates that, hidden deep in elementary acts of perception are structures that mimic the organization of elementary particles of matter.

In other words, the observable universe, down to the level of quarks, is consistent with the structure of elementary perception. Or, put another way, physicists who have discovered the structure of the baryon octet may have reported about their own fundamental perceptual apparatus as much as on fundamental particles of matter. Is it possible that the structure of perception and the structure of matter are coincident?

I believe that this is the case, that perception (observer) and matter (observed) are inseparable, and that both are processes involving recursive distinction-making.

Intelligent Design

The picture emerging here is that intelligence is distributed at all scales across vast, interlocking networks throughout the universe, and that there is no particular explanatory advantage to be gained in localizing intelligence in a single agency, such as a "designer" or a deity.

Following the logic of Occam's razor, which modern science defines as: *"Given two equally predictive theories, choose the simpler"*, when we suppose that the universe has a single designer we introduce a very complex (and unknown) entity, and this only complicates the scientific explanation of intelligent patterning throughout the universe. This implies that as a matter of science, the notion of a designer-less universe provides a more consistent and simpler explanation of "reality." This is, however, a matter of economy and simplicity, and as matter of deduction it cannot be settled by decree one way or the other. Hence, people who believe in deities are entitled to their non-scientific beliefs even though their beliefs would not meet Occam's criteria.

In this preliminary report, analogous to the tip of an iceberg, I postulate that recursive distinction-making is a generic process that occurs across all types of perception and intelligence, and is independent of the particular sensory modalities deployed. Hence, I conclude that raw sensory signals may be bypassed, and fantomark patterns (or their streaks) may turn out to be the preferred modes of intelligent communication among species. The relevance to our venture into outer space, of course, is that such a possibility may be crucial for our ultimate success in communicating with extra-terrestrial intelligences.

This gives us, of course, a rather different view of interspecies communication than we might get from the cartoons wherein aliens land and tell some innocent to "Take me to your leader." Nor are we talking about Star Wars or Star Trek, where humanoids from different planets, systems, or galaxies converse over a glass of space beer via the convenience of Gene Roddenberry's marvelous concept of a "universal translator." This is a much more plausible scenario involving the sort of encoded messages we could expect will be sent, received, and understood across the vastness of space by intelligent forms of organic or non-organic life. As we venture further beyond Earth, these are the types of messages that we can anticipate and that we must be prepared to deal with.

✳✳✳

Future research directions

The concept of Panspermia relates to the hypothesis that the seeds of life are prevalent throughout the universe, and that life on our planet was initiated when such seeds landed from outer space and began propagating themselves.

Francis Crick (with Leslie Orgel) suggested in 1973 a theory of directed panspermia, in which seeds of life (such as DNA fragments) may have been purposely spread by an advanced extraterrestrial civilization [7].

Critics, however, argued that this was implausible because space travel is damaging to life due to radiation exposure, cosmic rays and stellar winds.

However, the principles of intelligence described here permit us to introduce now the notion of tele-panspermia, which postulates panspermia guided by means of coded fantomark patterns (or their streaks). According to this concept, diffusion of life does not necessarily require the physical transport of actual "seeds" via meteors, comets, and the like.

Telepanspermia may be guided by means akin to pilot waves in Bohmian quantum mechanics. So, work on defining such guiding mechanisms in telepanspermia may converge with non-local hidden variable theories in fundamental physics.

Development of an information theory that is extended to fantomark-coded messages and streaks would facilitate the invention of superior intelligent artifacts. It could also hold a key to communication with extraterrestrial modes of intelligence, and eventually help us understand our cosmic ancestry and the relationship between the implicate and explicate orders as outlined by David Bohm [8].

References

(1) Spencer-Brown, G. Laws of Form. London: Allen & Unwin. 1969

(2) Isaacson, J. D. Autonomic String-Manipulation System, U. S. Patent No. 4,286,330, Aug. 25, 1981; accessible via http://www.isss.org/2001meet/2001paper/4286330.pdf

(3) Hofstadter, D. Gödel, Escher, Bach: an Eternal Golden Braid. Basic Books, 1979

(4) Isaacson, J. D. Dialectical Machine Vision: Applications of dialectical signal-processing to multiple sensor technologies, Tech. Report No. IMI-FR-N00014-86-C-0805 to SDIO, Office of Naval Research, 31 July 1987 (limited distribution).

(5)Marr, D., "Early processing of visual information", Philosophical Transactions of the Royal Society (London) Series B, 275:483-524, 1976

(6) Isaacson, J. D. "Steganogramic Representation of the Baryon Octet in Cellular Automata." Archived in 45th ISSS Annual Meeting and Conference: International Society for the System Sciences, Proceedings, 2001. Online version: http://www.isss.org/2001meet/2001paper/stegano.pdf

(7) Crick, F. H. C., and Orgel, L. E. "Directed Panspermia," Icarus, 19, 341,1973

(8) Bohm, D. Wholeness and the Implicate Order. London: Routledge. 1980

About the Author

Joel Isaacson has pioneered Dialectical Cellular Automata (DCA; per example: BIP/DIP) for well over 40 years, where DCA underlie generic intelligence. Isaacson is Professor Emeritus of Computer Science, Southern Illinois University, and principal investigator, I M I Corporation. His research has been supported over the years by DARPA, SDIO, NASA, ONR, USDA, and a good number of NIH institutes.

Chapter 25

Efficient, Affordable Explorations Operations: Crew Centered Control via Operations Infrastructure

By Thomas E. Diegelman, Richard E. Eckelkamp and David J. Korsmeyer

A significant challenge facing NASA's new explorations initiative to the moon and beyond [1] is to design a program of science and engineering exploration objectives that are both meaningful and affordable. This challenge is significantly different from that of the 1960's Apollo program wherein the object was to land an American person on the moon first and within nine years.

Apollo was accomplished with excellence. Talented teams throughout America set out to achieve an unparalleled technical marvel and to establish our country as a leader in spaceflight. Many new innovations in program management, space operations, and space hardware and software were developed methodically in the presence of enthusiasm and national competition. We Americans impressed ourselves and the rest of the world with our accomplishments.

However, low program costs were not primary drivers in these pursuits. Focus was placed on the end objective, winning the race to the moon. Some considerations were given on how to use or adapt the Apollo hardware on future space missions. Such thought helped produce the subsequent Skylab missions as well as helped develop concepts for follow-on activities on the moon, later called Apollo Applications, and for planned traverses to Mars.

Post Apollo, these relegated-to-secondary place drivers began to have unanticipated affects. In the heat of accomplishing the marvelous Apollo feats NASA had produced a large skilled work force and an extensive infrastructure. A logical follow-on activity for NASA as it would have pursued its next challenging project would have been to transform the workforce and infrastructure into forms and sizes that could be sustained. It was anticipated by NASA at the time that the next space project would be quick in coming and would be a natural extension of Apollo efforts leading to ever greater progress in the field of space exploration. Not yet conceived was the idea that the space efforts could become self-energizing through emerging economic benefits to commercial space enterprises.

But that anticipated quick next challenge did not occur. The American people, partially pushed by the news media, moved to other concerns - significant environmental issues and the quagmire of Vietnam. Missing from the American prospective were realizations of the importance and long-term significance of making additional progress on the space frontier, the importance of the strategic and eminent economic benefits of space work, and of the importance of setting and accomplishing worthwhile goals as a society. After having initial successes with the Apollo Soyuz Test Program and Skylab, the first U.S. space station, NASA struggled to outline a space future. In fact, NASA did not even offer a convincing case for a space future.

NASA, instead, found itself in organizational survival mode, concentrating its efforts on planning and selling some new space program that Congress and the American people would support, an so the Space Shuttle Program was born.

Money available for space was scarce, however. Due to ensuing yearly budget cycle pressures and schedule pressures NASA concentrated on vehicle hardware and software development, leaving to the unspecified future any concentration on providing an efficient operational system and associated infrastructure that would lower total lifetime costs.

The results have been quite costly and disappointing. As a result of high operations costs (see Tables 25.1 and 25.2). [2] NASA has cancelled or not sponsored many worthwhile and well-defined science and explorations programs. It should be emphasized that these consequences have not sprung from, lack of public support, lack of planning, or ignorance. Rather, they have come as a result of building subsequent programs upon the Apollo infrastructure and techniques, none of which were designed to conduct long-term operations. The resulting cost in operational dollars and human capital of both the shuttle and the space station programs has been and is enormous.

Cost Qualitative Metrics	Apollo	ASTP	Skylab	NSTS	ISS	Exploration
Vehicle Development	High	Low	Moderate	High	High	High?
Ops Systems Development	High	Low	Moderate	Moderate	Low	High?
Vehicle Sustaining Eng. per year	Moderate	N/A	N/A	Moderate	Moderate	?
Ops Systems Sustaining Eng./yr.	Low	Low	Low	Low	Low	?
Ops costs per year	High	High	High	High	High	?
Number of years	2	1	1	31 +	5 +	50 +

Table 25.1 U.S.A Manned Space Program Relative Costs

Cost Qualitative Metrics	Viking	Hubble	Pathfinder	MER	Exploration
Vehicle Development	High	High	Moderate	High	High?
Ops Systems Development	Low	Low	Low	Low	High?
Vehicle Sustaining Eng. per year	n/a	High	n/a	n/a	?
Ops Systems Sustaining Eng./yr.	Low	Low	Low	Low	?
Ops costs per year.	High	High	High	High	?
Number of years	2 years	10 +	90 days	1-2 years	50 +

Table 25.2 U.S.A. Robotic Space Program Relative Costs

NASA has not been alone in falling into this negative capital pit thought process. All the other space agencies: of Russia, of Europe, and now it appears, of Asia have suffered or will face these same high costs and accompanying limited results. It, therefore, is essential that a sustainable exploration program should allocate funds from the onset to develop an efficient operational exploration system. Mars and beyond exploration requires new techniques and organizational methods due to the sheer distances involved. In addition, both program and economic risks must be reduced by accomplishing early technical, much like Mercury, Gemini, and Earth orbit Apollo missions were use as stepping stones to the lunar landings. Accompanying these technical milestones were a series of infrastructural and managerial developments that enabled Apollo to succeed. We must do this again for the exploration era.

Change within NASA today is not only necessary, but mandatory. To have a successful and flourishing exploration program, we must make an "Earth-shaking change" in how we perform command and control. What is needed is to build a new robust human and robotic exploration operations command and control architecture and associated evolving infrastructure, one such version we name "Crew-Centered Control". The word "new" could sound frightening to some who might imagine huge expenditures. But how many of us, for example, still use old energy-inefficient machinery or computers when modern versions exist that are far superior in performance and are cost effective to boot? In the instance of ground control of space operations, the inefficiency lies in the large number of people required to perform tasks that now can be automated, in the toll on human health of controllers doing repetitive tasks manually, and in the accompanying historically-rooted management infrastructure.

Keeping in mind that, "They who begin a program with a large marching army continue that program with a large marching army," we propose to move from the current ground-centered control to this crew-centered control method (see Figures 25.1 and 25.2) (2), (3), (4). (We will discuss how this might be accomplished later in this chapter.) Using this method missions are designed to be as self-sufficient as practical. Operational control is pushed into the field, whether that control be in free space or on planetary surfaces. Distributed ground support teams and centers are used on a non-continual basis to augment what are essentially crew-controlled and crew-led operations.

Figure 25.1 shows a mission operation model that relies on nearly flawless communication, adequate reliability of a minimalist vehicle, and a relatively and a large staff of technical talent on the ground, serving as a human-powered archive of the necessary and sufficient information and knowledge to execute the mission. In 1969, when the first lunar landing took place, it was not possible to deploy additional technology, because we were employing all the technology that we had. Automation was minimal. The mission, to get to the moon before the Russians, was done using the only system available and capable to do so - the world's first and still best computer, man.

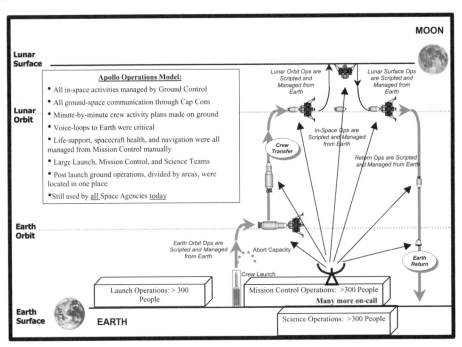

Figure 25.1 Apollo Style Mission Scenario

Considering the fact that the Apollo missions were relatively short duration (about a week), the Apollo model for organizing the mission was not only quick and effective in the short term, but was probably close to being the most cost effective model, since the infrastructure consisted only of hardware systems and the personnel. Little thought, time or process development other than the minimum needed to rush this to success were invested.

For subsequent programs, continuing to perform tasks manually began to take its toll in human capital, as may be seen in the "stress cracks" that appeared in the Skylab flight controllers and the Skylab crews. More recently, the experience of flight controller "burn-out" has been a very noticeable aspect of the International Space Station Program.

The operational scenario shown in Figure 25.2 reflects the relatively large number of personnel that will probably still be required for the early exploration activities - the Crew Excursion Vehicle (CEV), Constellation, and other launch vehicles. Fewer people are involved than for shuttle and station support, and likely could be reduced further as the systems gain maturity. But the striking change is that the crew is now the epicenter of the information, knowledge and first-line direction of the flight, while ground personnel take on more the roles as offline system experts, longer-term planners, risk assessors, and data brokers.

Further, the traditional location of science and systems experts at a single control center, such as the JSC Mission Control Center for manned missions, gives way to experts residing at various geographic locations that are linked together electronically. Today's communication technology of "e-access" can enable collaborative productive environments that are geographically, culturally, and organizationally diverse. This is quite impressive. A significant challenge is to devise human interactive processes that enhance versus slow down information flow. The slowing of information flow is technically risky and, in the past, has been blamed for accidents. As Alvin Toffler put it, "Money moves at the speed of light. Information has to move faster." [9]

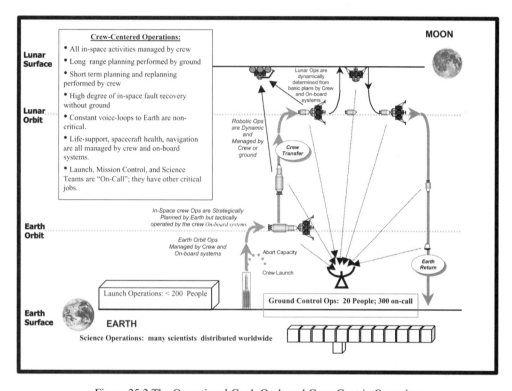

Figure 25.2 The Operational Goal: On-board Crew Centric Operations

Duty Assignments

What are the "duties as assigned" in this new operational scenario? In the crew-centered concept, the exploration crew will, as a minimum, be required to execute the following functions:

1) be the center of communications, command and control activities,

2) travel in vehicles with autonomous navigation, guidance and control and automatic system health monitoring,

3) perform all weekly task planning and daily scheduling,

4) deploy scientific and engineering equipment as scheduled,

5) perform scientific, exploration, and engineering activities in free space and on planetary surfaces,

6) control or manage resident robots in pursuit of the above activities,

7) manage in-space and planetary surface facilities,

8) perform the majority of system error recoveries using in situ capabilities,

9) manage in-space and planetary surface facilities, and

10) perform the majority of system error recoveries using in situ capabilities.

Similarly, there is a refocused set of tasks for ground personnel. The new tasks will require the execution of at least these defined tasks:

1) perform strategic and tactical monthly and yearly planning,

2) perform permission contingency analyses and produce associated execution products,

3) conduct permission crew training on the ground,

4) perform interplanetary and trans-lunar trajectory design,

5) provide launch vehicle checkout and preparation support,

6) provide traffic control for launching and landing vehicles, much analogous to the FAA for aircraft,

7) provide on-call onboard and field system expertise,

8) perform long term trending and correlation,

9) provide on-call failure analyses and trouble-shooting,

10) perform needed lengthy reconfiguration procedures development,

11) develop robotic execution scripts,

12) provide remote telerobotic robotic control when practical and cost-effective,

13) develop crew training and instructional programs and associated uplink material,

14) manage and control ground data facilities, and

15) use feedback to improve the efficacy of operational process and programs.

Measures of the success of the crew-centered operational control will include the following tactical measurements:

1) high number of exploration objectives met,

2) low cost per accomplished exploration objective,

3) high number of crew tasks and robotic tasks performed in a work day,

4) efficient rate of resource use,

5) low number of planning personnel,

6) low number of ground control personnel required,

7) low number of uplink commands from Earth,

8) high number of parallel activities accomplished per unit time,

9) short time required to train crew and robots,

10) routine ability to field new equipment and experiments without requiring crew training on Earth,

11) routine ability to manage multiple tasks at the same time,

12) crew has situational and contextual understanding of activities being performed,

13) high safety of crew maintained during EVA / IVA,

14) short time to diagnose common problems and modify plans accordingly,

15) short systems down time due to problems or maintenance,

16) low amount of logistics required to maintain operations (reuse and recycling decrease the amount of resource planning on the ground.),

17) low amount of Earth human intervention time required,

18) number of replacements versus repairs required,

19) high amount of information and context recorded for every exploration activity, and

20) short time needed to understand what occurred in the activities by third parties[5] .

To accomplish the process of vectoring toward more efficient operational control structure requires:

1) substantial advances in some areas of autonomous systems and robotic control,

2) development of multi-sensor field exploration robots, human assistant robots, and other construction robots,

3) robust space-qualified computers and advanced operational software,

4) first-time development of methods to perform crew-centered control and integrated robot-human operations,

5) distributed interconnected ground support teams and control centers, located at natural centers of talent base, used to provide crew support services that are infeasible to locate in the exploration field (the exception rather than the rule),

6) a robust web-like Earth-moon-planetary communications evolvable infrastructure, and

7) rapid crew access to large multi-medial databases for both exploration, maintenance, and repair tasks[4,5]

Shifting to the New Model

These changes can theoretically be accomplished in stages as the exploration program develops at a moderate rate, first with the moon and then beyond. A reduced manage approach is to begin is now, as shuttle operations wind down. The development of exploration command and control for lunar exploration could go as follows:

1) Earth-based control of multiple reconnaissance robots

2) Earth-based control of teams of infrastructure-building robots

3) Astronauts arrive and work in cooperation with and exercising shared control of robots.

   a. building infrastructure -robotic assembly

   b. exploring in the lunar environment

   c. Astronaut mission duration measured in months and years,

4) The lunar surface base comes on-line with near autonomous operations and daily-weekly planning capabilities.

5) Remote site sorties begin, supported by a substantially capable surface base [2], [5], [6]

A logical time line for the evolution of the operations infrastructure development to support this crew-centered command and control for lunar exploration and beyond could be proposed as follows:

1) 2006 First draft of operations processes and NASA centers' tasks functional allocation trades

2) 2007 Operations processes and functional allocation trades finalized

3) 2008 Initial Centers online for vehicle & robotic control

4) 2009 NASA distributed command and control operational plan in place

5) 2011 NASA Inter-control center systems fully operational

6) 2011-2015 International command and control plan development

7) 2019-2025 Multi-national control centers online

8) 2018-2023 International Martian command and control plan development

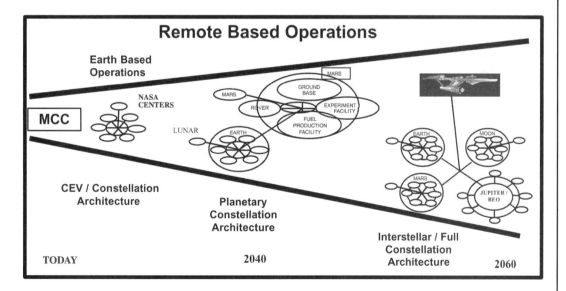

Figure 25.3 The Long Term Future of Crew Centric Operations

Given this proposed timeline, a summary extrapolation for the near term future can be produced (see Figure 25.3). Although the future may not unfold as depicted, the extrapolation presents a vision that can be used to determine required technological end products. A good reference of how end-space systems can be described using contemporary technologies , yet emerge 40-50 years later using unthought-of technologies may be seen in a short story by Arthur C. Clark called Islands in the Sky. 16 In this story, the illustrations tell a complete story in of themselves. Everything from the crew station, the assembly process, tools, to the EVA gear and suits is described and depicted as the 1950's technology would allow these concepts to be defined. (One might wonder and speculate about how the personal data assistant and cell phone would have been drawn in those yesteryears.) However, the story is more than a metaphor to be updated to 2005 terminology. The story speaks of emerging challenges for the next generations of space faring nations to hone, to sharpen, to crystallize this vision, and most importantly, to build this vision!

How Is NASA to Achieve Operational Change?

Near-radical change or "reinvention" is needed. 8 Crew-centered control is a very worthwhile and necessary operational configuration to establish.. The crucial question is, "How can NASA go from the current operational status quo to an exploration era of conducting efficient operations?" "Operations" will take an expanded meaning and will be defined later. One cannot look backward into space history. The past has little ability to pursue better strategic methods because then "good enough" and "available now" were the order of the day during Apollo. In addition, past successes, even if applicable such as Apollo, shuttle, and the ISS, do not presage future success (Warren Bennis, Organizing for Success [10]). Well-thought out planning on multiple levels is essential for future success. The generation of plans as products should always be considered on equal footing with other programmatic decisions and endeavors. A willingness by the entire organization to make necessary changes is a required condition for success. There can be no pockets of retro-thinking. Why? Because, as Bennis puts it, *"None of us is as smart as all of us."* [10]

To understand the how of change, consider this analogy. If one wants, to change a house substantially, based on new needs, one could:

1) demolish the old house and start anew with fresh materials (clean sheet approach),
2) build additions to the house and move, as needed, walls and facilities to achieve the desired changes (incremental change),
3) disassemble the old house and use the existing components plus new needed parts to achieve the results,(full renovation),or
4) use existing structure as is, but reallocate functional space and resources to accommodate the desired changes (reallocation of function).

In these proposed approaches, the richer options, are in descending order. These methods apply equally well to a company or organization that needs change. In the case of NASA, due to the skill of its workforce and its past history of success, the demolition case should be the last choice. In the house analogy, demolition is viable only if one doesn't mind living outdoors while the replacement house is being constructed - a significant issue in cold climates. Any of the other three solutions can work. There is definable wisdom in choosing to use existing resources. The existence of the current NASA organizational infrastructure inertia that has produced the current high life cycle mission costs would tend to move the solution away from method two and towards either method three or four, with some preference for three.

The inescapable fact is that if NASA does not make substantial organizational and process changes, neither crew-centered control or overall exploration success by NASA will occur. So, which change method should be used to accomplish a crew-centered operations approach? In the spirit of acknowledging that always the devil is in the details, an answer, will be stated in terms of a "wire frame" or outline / checklist. This serves to guide change, based upon a model of the organizational performance as a definable and measurable system. The key, then, is system performance enhancement in the context of system change. The following equation is proposed here as the hypothesis for this model:

Event happening = Necessary elements + Sufficiency condition

Proper technology development is the necessary condition for on-board operations to be instituted. This technology development alone will not suffice for efficient operations to occur. The technology must also be deployed. Whether technology is deployed or not and how this done, are organizationally driven. The sufficiency condition is, then, is that the organization must be adeptly tuned into the technical environment, the program challenges, and most significantly, be aware that the critical factor is the organization itself - its strength and its purity of intention to accomplish its mission. The resources needed and time required are mandatory, and they will appear almost magically in the presence of organizations committed to the task. Therefore, using the hypothesis given above, the particular equation for success of implementing a crew-centered exploration operations system, methodology, and associated infrastructure is:

Crew-centric Operations System Establishment (COSE) = Enabling Technology Development + NASA Commitment to Change + Development Risk Acceptance + Appropriate Organizational Restructures

The center of mass of NASA's effort necessary to accomplish Crew-centric Operations System Establishment (COSE) lies in the organizational and process side of the house rather than on the technological development side . As difficult as this effort might sound, it indeed is not. There are simplistic tools available that work and work well. One tool found helpful by companies seeking to "reinvent themselves" is the use of the SWOT analysis approach: [15]

* Strengths - what does the company do well, better than the competition?
* Weaknesses - what does the company not do as well as needed to succeed?
* Opportunities - what changes in management process and structure would allow what has not been previously possible to become possible?
* Threats - what are the threats to the company and to the industry associated with making these changes?

Let's employ this SWOT analysis in three principle areas of attribution: mission, technology and culture. These broad, rather simplistic areas allow a set of metrics and goals with definitive steps to be defined that assure that NASA can accomplish its objectives in the contemporary political, international and economic environments.

Mission is the metric attribute of an organization that defines the "why" of the business. The goals and the business objectives are uniquely described by the mission. The NASA mission in the Apollo era was being on the moon first, accepting the associated risks, and accounting for costs and repercussions in areas such as work force, sustainability of mission, and infrastructure This singular focus brought great success, yet produced long-term institutional processes and structures that eventually caused unintended results mentioned in the first part of this chapter.

Technology is the metric attribute of whether or not an organization is mature, stable, cutting edge, leading edge, or bleeding edge in its employment and use of technology. If technology is of central importance, then the company will do what is must do in the technology arena in order to execute its business plan successfully. If on the other hand, technology is not at the epicenter of the company and its motivations, the company will do only what is comfortable regarding the technology. For example, in the Apollo program, NASA pioneered the use of digital computers using cutting edge Diode Transistor Logic, but for the shuttle program, however, comfortably chose to employ the AP-101, a computer historically (and currently) used in the B-52 program.

Culture is a metric denotes of how the organization sees itself as being and how it sees itself as operating in the marketplace and / or in its work environment. The culture metric reflects the organization's response to internal mission and technological stimuli, as well as to external stimuli. As with technology, culture can be an epicenter measure or a meaningless curiosity. Words like "insular", "open", "learning", and "hide bound" are often used to describe culture.

Culture is defined by the folkways, mores', and traditions as well as, more importantly, the belief systems. Belief systems, although sometimes labeled a "soft science" in human factors, cannot be ignored. As Henry Ford put it succinctly, "Whether you believe you can or you can't, you are generally right."

The reason for the introduction of these three measures of an organization and its performance is simply to measure the organization. These measurements do not replace traditional performance and product measures, but rather, add to them. In a sense, this metric prescribes and even predicts performance because it serves to measure whether the organization is "configured for success". There is no one-size-fits-all set of business metrics.

Intelligence and insight are required to establish the specifics of these metrics, conduct the measurements, and, know what is valid within the data. Meaningfully information extraction from mounds of data is anything but a science. It is an art form. And it is an art form of the highest order in situations where people and human nature are part of the equation. This is NASA's turf - highly intelligent people challenging the laws of nature and expecting to beat the odds and win often.

Table 25.3 presents a summary of NASA's importance rating for mission, technology, and culture during past and present programmatic epochs, as well as a surmise of how our nation, aware of its technology base improvements needs, would rate the factors for the exploration era.

NASA EPOCHS	MISSION	TECHNOLOGY	CULTURE	# YEARS
MECURY-GEMINI	primary	secondary	distant third	4
APOLLO	primary	secondary	distant third	5
SHUTTLE DEVELOPMENT	distant third	secondary	primary	9
SHUTTLE OPERATIONS	secondary	distant third	primary	21
ISS	secondary	distant third	primary	5
CEV / CONSTELLATION	(secondary)	(primary)	(third)	2

Table 25.3 Primary indicators of the Focus of NASA during Program Epochs

The culture rating switched from last place to first during the more recent NASA programs. This switch is unfortunate, and unfortunately it was no accident. Companies take wrong turns and have to be "righted". This is true during all stages of a program, even when resources are not an issue. In the early Apollo days, George Mueller "wreaked havoc at NASA headquarters" putting the proper emphasis of the organization on the mandatory issues. [17] This "shaking up" of the in-place management practices is quite necessary in organizing for success.

In the exploration era, it is anticipated that culture will return to last place as in Apollo. In addition, it is quite possible that careful strategic thought at the government's executive level will elevate the exploration era technology factor to primary focus as an integral part of a national program to promote excellence in technology, innovation, and education. Certainly the emphasis in Apollo placed on "systems management capability" by Gen. Sam Phillips will be just as critical in Constellation as it was in Apollo. Even with this emphasis, the Apollo cost / schedule slippage factors for the F-1 engine, the science applications, and the human space craft elements to ranged from 1.6 to 4.75 - figures absolutely not acceptable in the environment we find in 2005. [17] So we must do better!

To get to a new place, one must know from where one starts. One set of observations about NASA's past programs using the SWOT analysis is found in Table 25.4. Each "X" symbolizes that one or more activity components have been judged to have the significant programmatic impact. The table shows only one of several possible categorizations. A similar set of additional factors, see Table 25.5, could be predicted for the exploration era based on forethought and examination of historical NASA corporate information. Components such as multi-discipline integration, universal standards consolidation, increased use of automation, full-up multi-national operations, and full requirements integration would probably be in the table. For a more complete discussion, see reference [15].

CURRENT FACTORS	AFFECTS	STRENGTH	WEAKNESS	OPPORTUNITY	THREAT
Diverse Contracts	Culture		X		X
Minimal up-front integration	Culture, Mission, Technology		X		X
Independent assets and cost centers	Culture	X	X		X
Historically-based initiatives	Culture, Mission	X	X		X
Has worked well to date	Culture	X	X	X	X
First large-scale task with multiple partners and assets	Culture	X		X	
integration beyond traditional program centered integration	Culture	X		X	
new standards evolution	Technology	X	X	X	
requirements integration	Technology		X	X	X

Table 25.4 NASA Current Programmatic Component Categorizations

CURRENT FACTORS	AFFECTS	STRENGTH	WEAKNESS	OPPORTUNITY	THREAT
Diverse Contracts	Culture		X		X
Minimal up-front integration	Culture, Mission, Technology		X		X
Independent assets and cost centers	Culture	X	X		X
Historically-based initiatives	Culture, Mission	X	X		X
Has worked well to date	Culture	X	X	X	X
First large-scale task with multiple partners and assets	Culture	X		X	
integration beyond traditional program centered integration	Culture	X		X	
new standards evolution	Technology	X	X	X	
requirements integration	Technology		X	X	X

Table 25.5: Possible Additional NASA Exploration Programmatic Component Categorizations

Table 25.5 is introduced here to provide an illustrative example of how programmatic factors might relate to the organizational attributes and organizational analysis factors. This SWOT analysis needs to redone periodically due to changing issues, resultant impacts, and hence, the need to replan.

The space mission profile also affects the factors and impacts, since the technology requirements are drawn in. In an abstract way, Figure 25.4 shows the broad categories of the NASA missions past, present and future. The technology aspect is easily seen. For example, some mission scenarios—especially the human missions beyond low Earth orbit (LEO)—demand safety considerations that will require investment in technology. Against this process, Tables 25.4 and 25.5 are proposed as a first cut at a set of issues, with the understanding that the farther out the mission and time line, the less the specificity of the factors, and the more change is to be expected as the far future missions evolve, based upon successes and failures of the current missions, environment, and the rest of the factors that were introduced into the mission / technology / culture metric set.

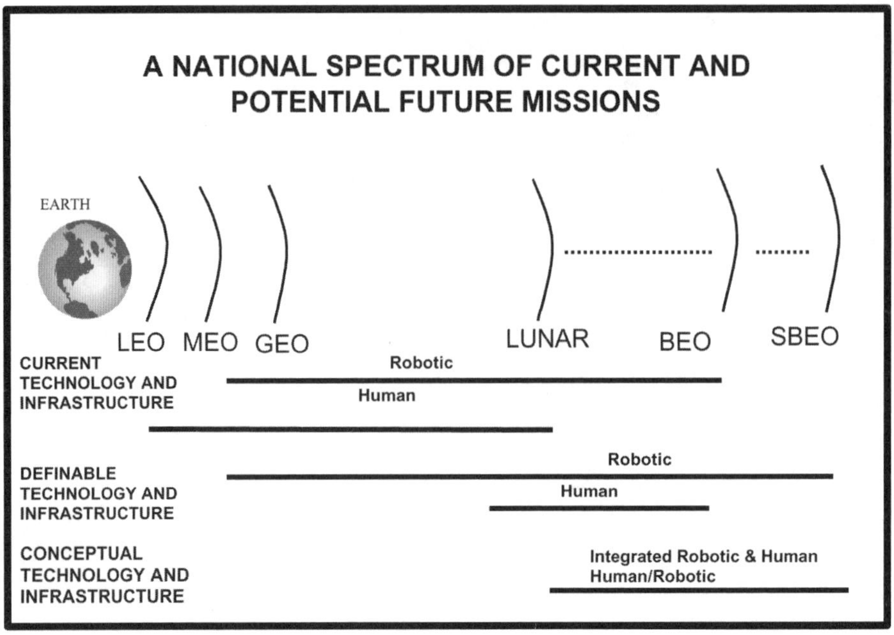

Figure 25.4 - NASA Mission Profile - Technology and Challenges
(key: L-low, E-Earth, O-orbit, M-middle, G-geosynchronous, B-beyond, S-significantly)

Once an organization is "engaged", that is, it accepts the realization that the organization itself, the environment, and the rest of the attributes all play together in the presence or absence of a plan to produce results, the sufficiency condition in the equation is supplied. The equation becomes an equality mathematically. In terms of organizational dynamics, this sufficiency means that there is an effort that is recognized to be more important than any other activity - called planning, that is, working the analyses, the issues, the challenges, the interfaces, the techniques, and the infrastructure required to accomplish the desired mission.

These engaged activities are rolled into a forward plan that is atypical of "non-engaged" organizations. Such an atypical plan is necessary to get to crew-centric operations. At first glance, the challenges to transforming the current space operations into crew-centric operations might appear to be technical. In fact the real challenges lie in the non-technical human and managerial areas, a fact that the "engaged" plan, must take into account with concrete steps to accomplish change in the mission and cultural metrics in the midst of an evolving exploration environment.

The ideas above might sound sweeping and too general to be meaningful. Let the skeptical the reader take the challenge to trace the near demise of the General Motors Corporation, that went from a corporation with the reputation of being such an economic powerhouse that, "as General Motors goes, so goes the nation" , to being an anemic corporate shell with a junk bond rating. Its market share is less than one third of what it was even 25 years ago. At a summary level, what are the issues that changed GM's situation to such a bad dream? Perhaps, it was the belief that GM was too big to be taken down, that GM was synonymous with a "job for life", where life was defined as "30 years and out with a gratuitous pension", an assumption that the forty percent plus of the market share was "theirs" to keep, that foreign competition with its low-cost, low profit vehicles could be dismissed as meaningless non-threatening competition, and perhaps lastly, that poor quality products could be foisted upon the buying public without consequences because there were no equivalent ,alternative sources of vehicles.

Could we apply some of these warnings to conventional space agencies and aerospace companies? In the constant market environment of the 1950's, GM's corporate missteps could perhaps be non-fatal, but not in the global economy of the 1980's and beyond. Corporate illness and demise can and does happen to the invincible. Tis also true in space endeavors.

An Analytic Forward Plan: Getting to an Efficient Operations Infrastructure

One pathway towards efficient operations being considered by elements within NASA [7,12] is shown in Figure 25.5. This figure models the progression of thought from concept of current operations enhancements to effective decision processes that will achieve more-efficient operations through block building and iteration converting insights self assessment into integrated action plans. This is what Peter Senge called the "Learning Organization", one that uses the performance measures of its functions in a recursive closed loop set of learning experiences thus insuring the vibrancy of the corporate entity which, thus assuring the ability of the organization to produce successful products. [8]

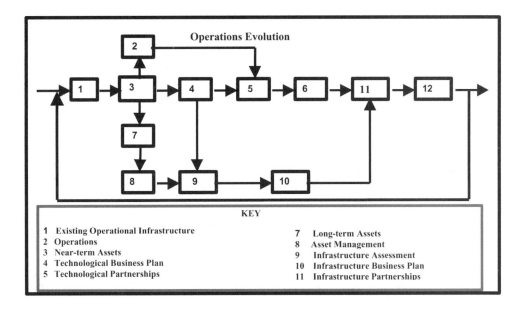

Figure 25.5 Feedback Relationship Between the Elements of the Operational Organization Evolution

Any system can and should be viewed and treated as developmental through out its entire life cycle. Definitions of the elements of Figure 25.5 are stated here in generic terms and are applicable to any sufficiently large organization involved in manufacturing, finance, commerce, and operations industries. Definitions tailored to the operations area are:

1) Existing Operational Infrastructure is the totality of the integrated systems, processes, and "hard goods" that comprise the environment necessary to execute the mission, programs, objectives and infrastructure maintenance that the organization pursues. This definition can change if the funding, life cycle definition, or programmatic direction changes, thus sometimes causing obsolescence. The concept remains, however. It is what it is at any instant in time.

2) Operations is the global task of performing activities to achieve the defined missions within the bounds of the infrastructure constraints. For NASA, the function, as defined here, does not include research on operations, although in general it should be.

3) Near-term Assets are additional, readily available systems, tools, and processes that could be inserted into the existing operational infrastructure as needed. These assets are identified separately because their adoption could enhance and add to the base of operational infrastructure.

4) Technology Business Plan is the high level plan, independent to some degree of the particular programs, wherein the decisions on whether to develop or implement particular enabling technologies are made. Enabling technologies are selected based on their estimated value to improve mission success and operational efficiency. The plan specifies how the selected promising technologies fit within the budget and

how they are to be integrated into the mainstream operational system flow. Care is taken to honor adherence to core company operating principles, goals and established partnerships. Elements of this business plan support Infrastructure Assessment.

5) Technological Partnerships are the agreements required to implement the Technology Business Plan. These partnerships will effect the Technology Business Plan in a closed- loop fashion, will help define the totality of the Near Term Assets, and may even influence Long Term Asset Management and Infrastructure Business Plans.

6) Operations Technology is the super set of technologies that provides the Operations segment with adequate "building blocks" to change existing operations in a strategic manner, utilizing both evolving technologies and existing technologies, into an ever-improving set of efficient mission operations. These employed technologies feed mission-derived information back into the planning, partnerships and long range planning processes. Once this function is fully matured, the once pinnacle of the organization - operations - becomes a building block for the Operations technology function, that in turn, becomes the main support structure for the Operations Infrastructure - the ultimate re-engineering target. (See Figures 25.6).

7) Long-term Assets are future systems, assets, tools, and processes that can be conceptualized, prototyped, and eventually inserted into the near-term asset systems. Ultimately, these assets cycle into the area defined as infrastructure. These assets can eventually imbue the operational infrastructure with synergistic efficiency. Note, as these definitions are unfolded, there is inherent recursion back on other definitional blocks.

8) Asset Management is the cost-effective stewardship of the capabilities both on-line and being brought on-line in the near term. This activity assumes the generation of and use of metrics that both support the Infrastructure Assessment function and measure the efficacy or mission performance using basic performance feedback data.

9) Infrastructure Assessment is the evaluation of overall organizational and mission performance based upon analysis of the Asset Management and Technology Business Plans. Metrics, such as the detailed SWOT analysis shown in this chapter, are used to judge the effectiveness of elements at all of the company management and mission operations infrastructures.

10) Infrastructure Business Plan is the plan that results from the integration of the partnerships, the assets, the technology plan and the current practices. This is the unifying document that defines the plan that pushes the organization to the pinnacle of Operations Infrastructure by leveraging off the Operations Technology and Infrastructure Assessment. Contained herein are the keep / cut / make / buy sort of decisions that clearly follow the incremental "lessons learned" approach.

11) Infrastructure Partnerships are the relationships that are required to acquire, maintain and operate the strategically-defined architecture and assets. These relationships differ from the technology partnerships in that the technology aspect is not the only consideration. This set of relationships is assembled to enable the long-term support of a very sophisticated, highly-integrated set of assets that ,based on broad and effective feedback, is strategically positioned to reduce cost and be sufficient robustness to continue product production or operations while accommodating internal changes and external stimuli.

12) Operations Infrastructure is the totality of the operations, assets, planning and most importantly, the responsive connection to the client community. This includes the definitions of the critical technologies, skills, external stimuli, cultural, finances, mission goals, and performance metrics that provide sufficient feedback to the eleven support functions, eventually enabling the shrinking the large pyramid back to a pinnacle that stops with Operations Technology or even Operations.

Not apparent in Figure 25.5 or in the definitions is the relationship among the building blocks. If we take the feedback diagram convention to be the "IN" arrows meaning "supported by", and the "OUT" arrow meaning "supports", the blocks may be rearranged to show the relationship that exists between the process steps. The stacking is shown in Figure 25.6, and is referred to, tongue-in-cheek,, as "building the Great Pyramid", referring to the ultimate goal of establishing the Operational Infrastructure as the highest level of integration of the organization's plan, products, and knowledge. It should be noted that a rigid, inflexible, non-learning organization will not be able to complete this reinvigoration process and will fall short at a lower level than the [12] defined block triangle, pinnacled at Operations Infrastructure.

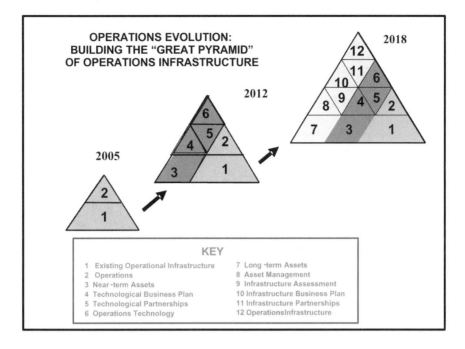

Figure 25.6: Building the Operational Infrastructure

An example would be helpful to observe the process. Referring to the partial list of potential challenges and opportunities in the area of communications, [12] we see what at first glance looks to be a technology question. There are, however, mission and culture factors sprinkled in as well which, as we have pointed out in Apollo history, are critical to success. The objective of our example exercise is to identify the necessary elements (the Issues) and the sufficiency conditions (the Affects) so that an analysis of the risks can be stated in a uniform, addressable fashion. Using SWOT analysis, the current environment (triangle elements 1-2) is reviewed and the plans for triangle elements 3-6 are formulated. At this point, NASA would have a forward plan to keep the communication system for Constellation viable for at least the duration of the program. At this point the trade studies and data gathering would barely have begun, so there is no final answer yet. However, there would have been back-of-the-envelope concepts defined that would merge assets of the Deep Space Network, NASA's current NASA Integrated Services Network, and commercial technologies.

The final steps in the process are to evaluate these options and select a strategy that could potentially allow the designed Constellation era communication system to serve as the backbone of the communications link for all of NASA as well as for the non-NASA voice, video, and data connectivity, much like the world-wide Internet web does today. These are steps 7-11, which result in a communication system that supports lunar missions, remote lunar operations, robotics, science and commercial interests. The system would then be expanded and built-out to cover communications to Mars and other destinations, hopefully without fundamental changes, (since strategic planning of the capabilities was done when the focus was lunar missions). This is block 12, Operations infrastructure.

Expensive as this analysis process might seem and indeed might be, the alternatives are less attractive. If the lunar communications system had to be modified for Mars missions to the same extent that the shuttle system had to be modified for station, the cost would be prohibitive. In addition, the risk associated with the lack of run-time experience with the product would almost insure inadequate risk abatement.

Once the twelve element plan for energizing an industry's infrastructure, the "Great Pyramid", is built, there is still no absolute guarantee of success. However, there is the guarantee that if the process is followed on a technical mission, a success path can be identified. What then is the "Great Pyramid"? It is a "System of Systems" icon for any industry seeking success to follow. In our focus here, it is NASA using the model to lead the team of nations, contractors, and entrepreneurs all of whom are seeking the safest, most reliable, most cost-effective access possible to all levels of space, from LEO to BEO. This is NASA's "value chain", as Porter refers to it in his book. [11] That System of Systems has the type of connectivity to the American people that will clearly show the benefits of space endeavor. These benefits will serve NASA well as an instrument to accomplish successfully the critically important need outlined in the President's vision in January 2004, "*engaging the American people*".

The Best Could Be Yet to Come - A Hope with Warning

With first class planning and unwavering commitment to building and maintaining the "learning organization", the application of these iterative steps can bring about a successful American space program, hosted by NASA and embraced by the American people and by people around the world. Everyone benefits from an accomplished space exploration program via spin offs of space technology being applied to everyday life and by the example provided by achievement of significant and noble accomplishments.

Be forewarned, however. Failure to institute a significantly intelligent System of Systems that embodies the "learning organization" and the accompanying analytics, metrics, and focus which enable constant, meaningful change for space projects, will inevitably lead to poor results and possible Exploration program cancellation, possibly even the demise of NASA. The competition for the federal budget (the external environment) is much fiercer than in the Apollo era. Hence there is strong need for change or reinvention.

How important is this idea of reinvention? It has been proven and documented that the probability of cancellation of a program has more to do with the number of years of execution of the program than whether or not whether the program is achieving the designated goals initially set out for the program [13]. Cancellation results from the inflexibility of the organization and the incongruence of the mission product to the societal, political, and economic environments. Harkening back to the measures of the mission / technology / culture metrics, NASA must plan well, evolve the plan and the organization, and then execute. Execution is in this System of Systems concept however, goes way beyond "build, train, and fly". It is now a recursive task, reaching back to the re-examination of the metrics, and hence the plan, on a continual basis. If the recursive process stops, so does the learning. No learning and the end game is in motion.

Summary

In summary, to accomplish NASA's planned new exploration challenges requires not only a merging of previously separated robotic and human programs, but also a radically new method of crew-centered control of operations. NASA, with its strong enthusiastic talent base, has the experience necessary to make this radical improvement in operational methods. A frame work for building the plan and executing it, based on a rich history of nearly 50 years of space flight, has been defined that creates the necessary engagement of the American people, as well as sets a charter for international participation. The sheer economics of world trade and national sovereignty, with its attendant competition, demand this change. Failure to accept and conquer this challenge will result in the shining star of space exploration excellence and all its economic and commercial benefits in the world marketplace to go elsewhere. It is ours, as a nation, to lose.

References

(1) George W. Bush, "New Vision for Space Exploration", January 14, 2004
(2) Richard E. Eckelkamp and David J. Korsmeyer, "Capability 9.1 Exploration", Human Exploration and Mobility Capability Roadmap Team, Exploration Subteam Report to the National Research Council, March 29, 2005
(3) David J. Korsmeyer, Daniel J. Clancy, James M. Crawford, and Mark E. Drummond, "Crew-Centered Operations: What HAL 9000 Should Have Been", AIAA 1st Space Exploration Conference: Continuing the Voyage of Discovery, February 1, 2005
(4) Daniel J. Clancy et al., "Automating CapCom Using Mobile Agents and Robotic Assistants", AIAA 1st Space Exploration Conference: Continuing the Voyage of Discovery, AIAA 1st Space Exploration Conference: Continuing the Voyage of Discovery, February 1, 2005
(5) Richard E. Eckelkamp and David Korsmeyer, "Efficient, Affordable Explorations Operations: Crew-centered Control", Exploration and Mobility Roadmap Team report to the National Research Council, April 4, 2005
(6) Stephen F. Zornester et al., Autonomous Systems, Robotics, and Computing Systems, Capability Roadmap, "NRC Dialogue", report to National Research Council, March 30, 2005
(7) Peter M. Senge, The Fifth Discipline, Doubleday / Currency Publishing, New York, New York, 1990, ISBN 0-385-26094-6
(8) J.S. Przemienieck, Acquisition of Defense Systems, AIAA Publishing, copyright(c) 1993, ISBN 1-56347-069-1
(9) Alvin Toffler, Creating A New Civilization, Turner Publishing, Atlanta, 1994, ISBN 1-57036-224-8Pg. 47
(10) Warren Bennis, Organizing Genius, Addison-Wesley, Reading Ma, 1997, ISBN 0-201-57051-3
(11) Michael E. Porter, Competitive Advantage, The Free Press / Simeon & Schuster, New York, New York, 1985

(12) T. E. Diegelman "Mission Operations Infrastructure", November 2000, Aerospace Technology Working Group Symposium, NASA Langley, Langley, Va.

(13) Norman R. Augustine, Augustine's Laws, Viking Press, 1983, ISBN 0-9159281-7

(14) Ferrell, O., Hartline, M., Lucas, G., Luck, D. 1998. Marketing Strategy, Orlando, FL: Dryden Press, 1998

(15) T. E. Diegelman, "Advanced Control Center Demonstration Project" (ACCDP), Mission Operations Directorate, Kick-off presentation, 1997

(16) Arthur C. Clark, "Islands in the Sky", June 1953; reprinted from book "The Exploration of Space", by the Book-of-the-Month Club, 1953.

(17) Stephen B. Johnson, The Secret of Apollo, The Johns Hopkins University Press, 2002; ISBN 0-8018-6898-X, pg. 130 -135

About the Authors

Thomas E. Diegelman is a NASA civil servant in the Mission Operations Directorate at the Johnson Space Center, Houston Texas. He is a graduate of SUNY at Buffalo, with a B.S. in Aerospace Engineering (1972) and an M.S. in Electrical Engineering, Control Systems, (1978). He as completed additional advanced degree studies at University of Houston, Clear Lake in Process Control Engineering, and Human Factors Engineering. He has been with NASA for 18 years as a project manager, system engineer, and a technology transfer manager. Mr. Diegelman was the project manager for the Shuttle Mission Trainer Upgrade, which refurbished the entire shuttle training facility, and reworked the maintenance and operations processes, yielding a substantial operations cost savings. Mr. Diegelman instituted key process changes in the Technology Transfer and Commercialization Office, and continues to participate in the Small Business Innovative Research (SBIR) program.

Richard (Rick) E. Eckelkamp received a B.A. in physics at the University of Dallas with minors in mathematics, philosophy, theology, and foreign language, and did graduate work in physics at the University of Houston. At NASA, from 1967-1987, in NASA's Mission and Planning Analysis Division, he developed extensive experience in Earth orbit, lunar, and interplanetary guidance, navigation and control, engineering analysis, and flight techniques, performed flight software verification and data production, performed Apollo through Shuttle flight control, serving as Shuttle console leads for both onboard and ground navigation, was a major designer of Shuttle onboard navigation and fault management software, as well as developed ground control center software and techniques. From 1987-1995, he performed Mars and advanced lunar program software/hardware design, performed station automated operations systems analysis and prototyping, and served as the Systems Development Manager for the Space Station Freedom command and control system. From 1995 to the present, he has worked in Advanced Life Support command and control, is serving as the Software Integration Manger for the International Space Station Robotics, and is performing systems engineering for exploration era robotics and integrated operations control.

David Korsmeyer, Ph.D. is the Chief for the Intelligent Systems Division at NASA Ames Research Center, where the research focus is on autonomous robotics and adaptive software systems for NASA Missions. He is an established Researcher with over 40 papers, and successful Project lead in distributed science systems. Dr. Korsmeyer received his B.Sc. from Penn State, his M.Sc. and Ph.D. from the University of Texas in Austin, and was a Sloan Fellow at the Stanford Graduate School of Business where he received his M.Sc. in Management.

Chapter 26

The Changing Role of Spaceports

By Derek Webber

Less Glamorous yet Enabling

When anyone speaks of rocket science, they are usually making reference to the more glamorous aspects of the space business, such as the design of spacecraft, or the provision of launch services. The terrestrial component is often overlooked. A, and yet, in many ways, the so-called "ground segment" of space activities is equally critical to the success of any mission. After all, we cannot imagine space flights that do not start from a tower of flame at a launch pad, and include telemetry, tracking and control back to base. We have become familiar with the sights of Kennedy Space Center with its line of gantries marching along the Atlantic Ocean coastline at Cape Canaveral. W, and we have seen the huge Vehicle Assembly Building and the "crawler ways" that take the assembled Shuttle stack to their point of take off for their orbital destinations.

But this may be about to change. In 2004, the only astronauts to enter space from the US departed from Mojave Spaceport. Mojave was, a site not managed by the US Air Force. It has no gantries, no high high-energy LOX and propellant bunkers, nor blockhouses. It does not even have a launch pad! Does Mojave represent the prototype of a new kind of spaceport that will take over from the earlier types as we look forward to the next stages of mankind's journey into space? If so, then however different it may be from its predecessors, we may be sure that it will still be as important and enabling to the overall success of the venture as will be the new spacecraft and launch systems.

Geopolitical Factors

The early launch sites were, established during the Cold War following World War II. They, were all military establishments, and originally they were only in the USA and the former Soviet Union. And because of their differing geographies, they happen to neatly demonstrate the two basic types of locations for a spaceport. Underneath the trajectory of a rocket heading into space is a ground track where ideally you don't want to have any people, because any such people would be at risk from debris resulting from an possible explosion caused by the typically high-energy oxidizer and propellants on board the rocket. The best zones on the Earth's surface which that meet this objective are deserts and oceans.

Another factor, resulting from of astrophysics, is that different kinds of space missions require different kinds of orbits. Polar-orbiting spacecraft can generally observe the whole surface of the Earth over a series of orbits as the Earth spins underneath. Orbits in the equatorial plane require the least energy to achieve, because advantage can be taken of the Earth's initial rotation from West to East can help to in achieveing the necessary orbital velocities, and s. Such orbits are particularly useful at geostationary altitude for providing various broadcasting and telecommunications services. Other kinds of missions may need an intermediate inclination, and the location of the a spaceport should ideally support launches into any of these types of orbit. For the desert spaceport, this is not usually a problem.

Sometimes this is hard to achieve for a coastal spaceport, and a functional split can result with equatorial launches taking place from spaceports having an Eastern coastline, and being located as near to the Equator as possible, and while Polar launches taking place from spaceports having either a Northerly or Southerly ocean overlook. For some favored countries, a fortunate single location can serve for all these needs.

During the Cold War, the Soviet Union developed a series of desert spaceports and the USA used a combination of desert and coastal spaceports. Because of the accident of a geographical accident, all of the territories of the former Soviet Union were a long way from the equator. The Soviets were (and indeed Russia remains today), therefore at a disadvantage versus the USA with regard to provision of equatorial launches. The USA can launch from much nearer to the Equator. It is somewhat ironic therefore, that because of the less favored astrophysical geography of Russia, the International Space Station (ISS) operates in an orbit placing challenges on the US Shuttle launch system in delivering supplies and components. The ISS orbit was chosen, at least in part, to allow key Russian components to be launched from the high latitude spaceport of Baikonur.

Early Launch Sites

The early coastal launch sites of the USA were at Wallops at 37.5 deg N, and Cape Canaveral (KSFC) at 28.5 deg N for equatorial launches, and Vandenberg seven years later, firing southerly for polar launches. The US also had a desert site at White Sands for early sub-orbital northerly firings. The former Soviet Union had desert sites at Kapustin Yar (48.4 deg N), at Baikonur (45.6 deg N) and ten years later at Plesetsk (62.8 deg N).

The only other active launch sites at that time were two contrasting desert sites - one in the cold Northern desert of Churchill in Canada (57.7 deg N) , used for sounding rockets, and the other being in the decidedly hot desert of Woomera, Australia (31.0 deg S), where the British tested their ICBM in northerly sub-orbital firings.

Today's Global Picture

Since that Cold War era, much has changed. There are now some 16 countries with launch sites, many with more than one spaceport, so that there are currently and a total of about 35 active spaceports in the world.

A significant development has been the move to develop spaceports nearer to the equator. This has been motivated by the need to provide launch facilities for a series of commercial geostationary communications and broadcasting satellites.

The best examples of these are the European coastal launch facilities located in French Guiana at Kourou (5.2 deg N), the Brazilian coastal facilities at Alcantara (2.2 deg S), and the international Sea Launch platform, known as Odyssey, that is positioned exactly at the Equator for its launch service. The US is also developing the capabilities of Kwajelein in the Marshall Islands, which is an atoll conveniently located at 8.0 deg N.

In developing the Ariane launch system, the Europeans abandoned the Australian Woomera site, where they had learned their early lessons trying to develop the Europa launcher, for one nearer the equator at Kourou that would enable Easterly azimuth launches. The Australians, in their turn, are currently considering developing the a Christmas Island site, on Australian territory in the Indian Ocean, near Indonesia, located at 10 deg S with clear Easterly azimuths.

A recent development at Kourou has been the decision to build a separate spaceport for the Soyuz launcher, and thus allow Russia to overcome its historical geographical disadvantage in launch economics. Simultaneously, this will provide the Europeans with their own capability to provide support a space tourism experience, if they decide to do so. Somewhat ironically, upon the demise of the former Soviet Union, Moscow lost its premier historical spaceport to the republic of Kazakhstan, and now has to lease the Baikonur facilities from that country for its Soyuz launches.

China has four active spaceports, the earliest one being Jiuquan, at 40.6 deg N, from where this country now launches its crewed missions are now launched. All of the its sites are very strictly controlled by the militarily controlled, but recent indications suggest some emphasis on developing the Hainan Island site, which, at 18.0 deg S, is the nearest spaceport in Chinese territory to the Equator, and the only Chinese site that is not a desert spaceport. Both India, from Sriharikota at 13.7 deg N, and Japan, from its spaceports at Tanegashima (30.4 deg N) and Kagoshima (31.2 deg N) are active coastal spaceports.

Predicting the Future of Launch Demand

How many orbital launches a year are provided from these 35 global spaceports? What kinds of customers and payloads are served? And how, if at all, do we see this changing over the next few decades? The historical picture, quite contrary to the trend in spaceport development, has been a rather severe decline in global launches.

From 1965 through 1985, the global launch rate for orbital missions was around about 120 launches per year. Now Today it hovers around at about 60 launches per year, and this figure includes all countries, and all military, all civilian, and all commercial launches. This at averages at only about 2 launches a year per spaceport, which cannot of course represent a good economic proposition.

And what of the future? The ASCENT Study, organized by NASA's Marshall Space Flight Center and issued in 2003, made it clear that the twenty-year forecast would remain at about the same annual level of launches for these same payload types, ie only between about 50 to 70 launches per year in total globally.

At launch rates like this, it does not make sense to design and build a reusable launch vehicle (RLV), because one such vehicle could serve to transport the entire global payload needs in a given year. The same study showed, furthermore, that the price elasticity of demand of commercial launches would not be a significant factor in increasing the annual launch demand.

However, there was also some good news. The ASCENT Study identified, (and its initial findings were incorporated in the Final Report of the President's Commission on the Future of the US Aerospace Industry), the dramatic impact that would be made by a successful space tourism industry. Even at prices around $20M, the study reported, there could be a market for 50 launches per year of paying passengers seeking an orbital experience. Note that this figure, if achieved, would result in a doubling of the launch rate projections that exclude space tourism.

And the news gets better. Unlike the traditional payload markets, it turns out that the demand for human payloads is very price sensitive. If the price per trip could be reduced to, say, $5M per orbital flight, the demand would reach about 2,000 per year. This is because while there are relatively few billionaires, there are over eight million millionaires in the world. In between, there are enough people who are rich enough to spend say 5% of their net assets on a once-in-a-lifetime orbital experience. Of course, to achieve the reduced ticket prices, it will be necessary to fly several orbital tourists at a time, so the incremental number of launches might only be, say, 500 per year. But this is still a transforming number.

Such numbers suggest a new kind of spaceflight. They lead to, the beginnings of airline-like operations, and better operability and reliability, and lower costs in general regarding the cost per pound of payload into delivered to orbit. Space tourism can thus be seen as an enabling and transforming technology and business opportunity. Yes, RLV's even make sense, but only in the context of the massive new tourism markets.

The Emergence of Space Tourism

Following the success of Burt Rutan with SpaceShipOne in winning the X-Prize in 2004, it has become a certainty that space tourism will soon be operational. Richard Branson of Virgin Galactic, out of at Mojave, and the Rocketplane team to be flying from Oklahoma, intend to provide the first space tourism offerings. Of course, these first ventures are sub-orbital only, but for this reason the prices are significantly lower. At a ticket price of around $100,000, then all 8,000,000 millionaires around the world could afford a ticket without spending more than 10% of their net worth to do it. (It' is worth mentioning that the first two tourists, Tito and Shuttleworth are reported to have spent about 10% of their assets to make their own (orbital) trips.)

What are the prospects for orbital space tourism? Well, of course the Russians already offer the service via Soyuz from Baikonur, and in a few years this it will also be possible for the Europeans from the new Soyuz pad in Kourou. The Russians are seeking funds to develop a new, larger follow-on to Soyuz, called Klipper, which could carry more tourists into orbit, and they have been trying to solicit European support in the venture. The Chinese may be assumed to eventually offer tourist rides in their Shenzhou spacecraft, once they have obtained some more operating experience. In the US, certain entrepreneurial companies have announced their intention to eventually provide an orbital space tourism experience, among them the companies being run by the billionaires Jeff Bezos and Elon Musk. And the a new company, Transformational Space Corporation (t/Space), is proposing to build a spacecraft, called the CXV, to carry US astronauts into LEO for their ongoing missions either at the ISS or as part of their journeys to the Moon and beyond, as part of the President's Vision for Space Exploration. The company will offer orbital space tourism experiences in between the trips required for the NASA crews.

How does all of this affect the spaceport scene? Recall the history making flights of Burt Rutan's civilian SpaceshipOne at Mojave, in 2004. Mojave is a totally new kind of spaceport. No "tower of flame" was seen at the launch pad. In fact no flames at all, and no launch pads at alleither. And the astronaut came back an hour later and landed right in front of the crowd. To get into space from Mojave, takeoff is from a normal aircraft runway. If all of the growth in the launch industry is going to come from the tourism sector, then attention needs to be taken to the needs of this new marketplace.

US Non-Federal Spaceport Plans

What has been the reaction within the US to these developments? Mojave became the first of a "new breed" of licensed US Non-Federal spaceports. The FAA's Office of Commercial Space Flight has previously granted licenses to six US Federal Spaceports (KSC, Edwards, Vandenberg, Wallops, White Sands and the Kwajelein site). It has also licensed five Non-Federal Spaceports (California, Kodiak, Florida Space Authority, Virginia - all based within the territories of existing Federal Spaceports - and Mojave).

Throughout the US, various many States are exploring the potential economic benefits of creating spaceports to serve the new space tourism markets, and some of them have begun the process of obtaining FAA operating licenses. These states are anticipating potential benefits in terms of employment, building contracts, tax base, and terrestrial tourism as a result.

The Rocketplane venture is doing its work based at the Oklahoma spaceport, and this is likely to be one of the next spaceports to receive a license. In Upham, New Mexico, the Southwest Regional Spaceport is being developed and will be the venue for a series of annual X-Prize Cup races into space, the first one being scheduled for 2006. Texas is proposing 4 different possible spaceport sites, and the following States propose one each: Nevada, Alabama, Washington, Wisconsin, Utah have proposed one each. In all, about a dozen potential non-Federal spaceports are currently in various stages of the process of evaluation and regulatory approval.

Can they all succeed? And if so, what will be the impact on the existing spaceports?

Spaceport Transformation

For a spaceport to succeed as a provider of space tourism opportunities, the management will need to have a focus on satisfying the needs of the new sector. This may prove to be possible, but unfortunately it would appear that the needs of the new sector can be almost directly opposite to the needs of that existing traditional spaceports that have served us in the first half century of human spaceflight. It may well prove to be the case, because of this, that the existing spaceports continue to serve their traditional markets, with their limited growth potential, while the new high growth sector goes to a series of new specially-designed spaceports of the Twenty first Century.

What kinds of transformation do the new spaceports need to represent, and which may prove impossible for the existing spaceports to offer?

The most obvious, and in a way the most emblematic, is the fact that existing Federal Spaceports are generally military establishments; among their payloads are military cargoes. Traditional rocket fuels are highly dangerous mixes of propellants and oxidants, from which the public needs to be protected.

By contrast, the SpaceshipOne flights in 2004 used a new kind of hybrid rocket engine that used combined rubber and laughing gas, a mixture of fuel and oxidizer so inherently safe that it presented no danger to the crowds who waved at the passing carrier aircraft and suspended space plane as it went by *en route* for space.

For space tourism to succeed, the public needs to be encouraged, not discouraged, from entering a spaceport to watch launches. Open access will be an important feature, even in such traditionally sensitive areas as launch control and training facilities. There will need to be new facilities for training and medical care of the space tourists. Entertainment features such as IMAX movie theaters and space theme parks will be important, and central to the experience, especially for the friends and families of tourists who are going off into space. New residential facilities and restaurants and shops will need to be built for staff, tourists, families, and terrestrial tourism customers. This will decidedly not be your father's spaceport experience!

From this I conclude that Space tourism is needed to move spaceflight from the twentieth century paradigm to the new possibilities of the Twenty-first century, and to recreate an aerospace industry with renewed growth prospects and a reinvigorated future for the younger generation. The spaceport remains an essential element of the whole space business experience, but must undergo a dramatic transformation in order to enable this new industry to develop. It remains to be seen how many of the traditional spaceports will be able to adapt to this changing role, or and how many totally new spaceports will emerge that will provide the space tourists of the future with the kind of experiences that they will demand for the price of their tickets into space.

About the Author

Derek Webber is the Washington DC Director of Spaceport Associates. He has thirty years of experience in the commercial space industry, and is a leading authority on the potential for the space tourism business. He also currently functions as Director of Business Development for the emerging new aerospace company Transformational Space Corporation (t/Space), working to introduce a viable orbital tourism business. Derek directed two landmark studies in commercial space business planning, the ASCENT study of all space markets, for NASA, and the Futron/Zogby survey of demand amongst millionaires for space tourism. He has chaired panels, given radio and TV interviews on space tourism, and provided testimony on the subject to the President's Commission on the Future of the Aerospace Industry.

Derek's career began as a launch vehicle and satellite engineer in Europe, and he has served as Head of Procurement at the Inmarsat organization (contracting for over a billion dollars worth of satellites and launch vehicles), and as Managing Director of Tachyon Europe (providing broadband/Internet access via satellite throughout that continent).

Chapter 27

Planetary and Solar Resource Management, Biospherical Security and the New Space Adventure

By Elliott and Sharon Maynard

There are three reasons why, quite apart from scientific considerations, mankind needs to travel in space. The first reason is garbage disposal; we need to transfer industrial processes into space so that the earth may remain a green and pleasant place for our grandchildren to live in. The second reason is to escape material impoverishment; the resources of this planet are finite and we shall not forego forever the abundance of solar energy and minerals and living space that are spread out all around us. The third reason is our spiritual need for an open frontier. The ultimate purpose of space travel is to bring to humanity, not only scientific discoveries and an occasional spectacular show on television, but a real expansion of our spirit.

Freeman Dyson, 1979 - Disturbing the Universe. [1]

Introduction

Over the past few decades, the development of computers, communications devices, cheap mass-transportation, satellites, and the global internet have expanded the human worldview from the few square miles surrounding a local community, to an area which encompasses the entire surface of our Planet - extending beyond, into the most distant reaches of outer space. With this historically explosive leap in technology, we humans have acquired a "toolbox" of unique new technologies, which can be applied to monitor and protect Earth's precious biospherical resources, and also to extend the limits of our awareness and intellect. These are the issues explored in this chapter. The critical need for humans to manage Earth's precious natural resources is highlighted as follows, by author Stephen Hackin in his" transformative book, Global Renaissance: "Since time immemorial the falling rains and changing seasons naturally cleansed the smoke and waste in the biosphere. But this is no longer true. Now in the global age, humans must manage the environment in ways which avoid the accumulation of materials contrary to our health, either directly or indirectly through the atmosphere, farmland and oceans which support life." [2]

A Holographic "Virtual Earth," A New Type of Global Information Resource

Through the global network of remote sensing satellites - the "Eyes" of Gaia" - we have, for the first time in history, achieved the capability to accurately monitor and diagnose Earth's vital biospherical parameters: atmospheric and water pollution, glacial and polar ice thickness, forest cover, desertification, oceanic current patterns, migrations of marine mammals, the status of global fish stocks, and the intricate dynamics of planetary weather patterns. Supercomputers connected into this satellite network can be used to create a holographic "Virtual Earth," a multidimensional model that could serve as a dynamic, real-time scientific information resource for citizens of every nation on our planet. Such a digitized Earth Program could be run either backward in time. to examine historical trends from past geological records, or forward into the Future, to simulate biospherical scenarios such as the melting of the polar ice caps, rising sea levels, or fluctuations in the Arctic and Antarctic ozone holes. Such a "Virtual Earth" would allow scientists, scholars, school children, or anyone with access to a computer to virtually "fly" over the surface of the Earth, plunge into the deep ocean depths to explore thermal vent communities, or climb to the top of Mt. Everest - all without ever physically leaving their chairs. The concept of a Central Computer, which monitors and regulates global society, was envisioned by futurists Ken Keyes, Jr. and Jacque Fresco as early as 1969, when they described a six-foot diameter sphere named Corcen, which would network and integrate computerized information, and serve as a "knowledge bank" that would regulate the lives of individuals in future global society, and coordinate what they referred to as a "humanized man-machine symbiosis". [3]

Earth as a "Natural Space Ship"

Earth has evolved in the form of a "Natural Space Ship," which orbits the Sun as part of our Solar System, and moves through Space in a spiral dance with the other celestial bodies of our galaxy. [4] Over many millennia Earth evolved a biosphere, which nurtures a rich and incredibly diverse abundance of life. To successfully sustain a sustained human presence in Space, we must first learn how to create and manage balanced, sustainable biospheres for our Orbital, Lunar, and Martian colonies. These "Oases in Space" not only provide the critical functions of carbon dioxide and waste recycling, but are also important for food production, and for providing a valuable psychological "Earth analogue" for astronauts who are confined in ships during long space voyages.

Immersion, Observation and Intuition, as Powerful Tools for Understanding Earth and In Space

In many respects, Earth's Biosphere can be compared to an incredibly complex computer program. From this viewpoint the Natural World can be perceived as a constantly shifting mosaic of dynamic interactions. One of the best ways to learn how such a complex system operates is to become totally immersed in the Natural Environment, observing, for example, the subtle intricacies of a tropical rainforest, or the colorful tapestry of interacting life forms, typical of a coral reef. Great thinkers like Leonardo da Vinci and Charles Darwin, provide us with proto-typical examples of how a keenly honed intellect, selectively focused observations, and a well-developed intuitive sense can function synergistically to decipher the complex patterns of Nature, and distill these natural coordinates into valid scientific principles. The resulting information resources can then be passed on as a legacy for future generations, and will subsequently become part of the evolving scientific social consciousness. Immersion, focused observation, and intuition are thus powerful and effective tools for understanding "the unknown" These same concepts also apply to Space exploration, especially regard to surviving in hostile environments, and creating a solid technological foundation for establishing human colonies in the "New Wilderness" of Space.

A New Paradigm for Planetary Management

A new Planetary Management Paradigm could serve as a catalyst for humans to make the shift to new and ecologically appropriate ways of thinking and action. This new paradigm would integrate the best aspects of more traditional technologies such as Natural Selection, Hybridization, and Tissue Culture, with advanced technologies such as Cloning, Genetic Engineering, and Nanotechnology. The ultimate objective of such a new paradigm would be to monitor and nurture the "health" of our planetary biosphere and to improve the basic living conditions of all its human inhabitants.

The ultimate effect of any such Super-Technology Paradigm, which synergistically combines the best of conventional, leading-edge, and alternative technologies, would be to shift our scientific thinking from simply managing natural biological systems, to improving existing systems - eventually expanding this concept to create entirely new ecosystems - "Space Oases" - which could be especially tailored for Orbital, Lunar, or Martian Colonies. This type of out-of-the-box thinking could also open up exciting possibilities for designing new, fast-growing, disease-resistant ecosystems, and intelligently managed biospheres here on Earth, such as Global Forests, or Pan-Oceanic Marine Ecosystems - evolved ecosystems, the likes of which have never before existed on our planet. Any such powerful new technologies must, of course, be very carefully tested, monitored, and managed with the best human brainpower, skills, and wisdom we can muster, as our creative scientific discretion and vision in the Present will most certainly determine the constitution, health, and value to humans of our natural planetary resources in the Future.

"Quantum Leapfrogging:" New Thinking; New Technologies

To achieve a quantum leap from operating within a traditional scientific thinking context, into more advanced modes of "super-technology thinking," we must first learn to develop clear Conceptual Models which incorporate future-focused "group thinking," bringing into play the powerful analytical, intuitive, and creative powers of the human mind and spirit. Essentially, creating conceptual models is no different from creating architectural models, except that it requires working within a constantly evolving matrix of creative thinking, and exploring new enlightened applications of collaborative synergy. We would thus shift into a new methodology of thinking, which exists in a constant in a state of flux. As is true of thinking and behavioral memes through the animal kingdom, [5] human thinking strategies will continue evolving in order to insure their own survival. As each new strategy is developed, it will, in turn, influence the leading-edge thinkers who created them, feeding back to ramp up their intellectual powers and the advanced concepts they develop to ever new and evolved levels.

New Perspectives for Earth from Space

Recent advances in space technology now allow the human consciousness to span vast interstellar distances, and to view even the most distant galaxies. This quantum expansion and evolution of the human external perspective has been reflected in corresponding shifts in the very depths and character of human consciousness itself. Similar shifts in consciousness occurred when Europeans discovered and colonized North America, and later, when inhabitants of the East Coast moved west to seek their fortunes in California. In both cases, major shifts in the social consciousness occurred when these pioneering settlers had to adjust their worldviews to function effectively in new and unfamiliar environments. Certain aspects of these shifts were subsequently reflected back to the regions where these pioneers came from. As communications and transportation improved, these shifts were correspondingly accelerated. The historical basis for shifts in human thinking from their exploratory voyages is expressed in past historical context by Indiana University School of Public Affairs Professor Emeritus, Lynton K. Caldwell, as follows: "...voyages into space had an effect similar to that of the sea voyages of preceding centuries - they added cumulatively to the true process of discovering the true nature of the Earth." [6] Yatri, in his provocative book, Unknown Man: The Mysterious Birth of a New Species, eloquently depicts the transformative effects of changes in geography and technology on the human social consciousness as follows: "With the addition of new artificial intelligences to the network and the developing richness and quality of the flowing information, Marshall McLuhan's 'Global Village' has changed into the 'Global Brain' - an entirely new kind of consciousness and a new kind of planetary species." [7]

Envisioning a New Internal Humans-in-Space Program

A new, international Humans-in-Space Program could best be developed by using an expanded global resource base, which could tap into the financial, intellectual, and manpower resources of every nation on Earth. This type of international approach would allow the human space program to function much more efficiently, by incorporating international guidelines for standardization, modularization, and manufacturing criteria. Such an approach would supersede the present bureaucratic duplication of funding and technology development, by establishing a coordinated global network of manufacturing and launch facilities, and manpower training centers. This approach would function to encourage and facilitate new levels of international cooperation (such as already exists with the global airline industry) between established space organizations such as NASA, the European Space Agency, their counterparts in Russia and China, the developing space centers in South America, Canada, and India, and in conjunction with the emerging Space Entrepreneurial Sector. Such a new international collaboration-based space program would function to effectively harness the basic human drives of corporate competition and synergy, to re-structure and re-energize the existing global space infrastructure - actively bringing it to the mainstream global media venues.

Any such international program should be structured to include a balanced group of leading-edge scientists and from Government, the Military Sectors, the Aerospace Industry, Academia, NGO's, and the emerging Space Entrepreneurial Sector. This new integrated global approach could thus effectively serve to create a far greater financial, scientific, commercial, educational, and manpower base than presently exists. Most importantly, a new Global Space Initiative would open new gateways of opportunity for people everywhere on Earth, and draw on previously untapped financial, brainpower, and scientific resources. The objective would be to maximize the "Return on Investment" from Space back to Earth, by creating new industries and commercial flows, developing new non-polluting energy technologies, and ultimately improving the lifestyles of people everywhere.

New Global Problems and a New Emerging "Collective Intelligence"

As we move forward toward developing Orbital, Lunar, and Martian colonies, and eventually move out into interstellar space, the extraordinary aspects and revelations of this pioneering experience will translate into equally profound shifts individual human consciousness, both in human society, and in the composition and health of the planetary biosphere itself. A wealth of new space-based concepts and technologies will be brought back to Earth. In one way or another, all individuals in every nation will reap the benefits, as these advanced technologies becomes integrated into people' daily lives. There is no more poignant example of this "return on investment" than has occurred during the past two decades with the social transformations driven by communications and technology, which been a direct result of space-based satellite technology. Personal computers, cell phones, and the global internet provide practical and ubiquitous examples of this transformation. What seems to be emerging from this shift is a new kind of "Collective Intelligence," perhaps best expressed by Professor Pierre Levy, Professor of Hypermedia at the University of Paris: "A new anthropological space, the knowledge space, is being formed today, which could easily take precedence over the spaces of earth, territory, and commerce that preceded it." [8]

As we cross the threshold of the Third Millennium we face a multitude of overwhelming social and environmental problems. These major global "crisis events" (i.e. international armed conflict, social violence, tsunamis, earthquakes, major storms, atmospheric and water pollution, disease epidemics, and overpopulation) have increased to the point where they have become the daily entertainment fare for the global media. Although conventional science and technology have attempted to address these problems, in most instances - especially in the case of global environmental problems - practical solutions simply do not yet exist. We need fresh approaches, new thinking, and new technological approaches to enable positive transformation at a Planetary Level.

Creating new "Hybrid Synergistic Technologies"

Entirely New Technologies can be created by combining the best of leading-edge conventional technologies with "orphaned" alternative technologies - some of which have been around since the 1940's, but were rejected by mainstream science because they did not fit within the paradigms of the times. Examples of such unique "Hybrid Synergistic Technologies" would include: Lewis Karrick's Low Temperature Carbonization Process for extracting Oil from Coal, [9] the Solar Tower, a non-polluting technology which combines solar energy and wind power, [10] and Nobel Prize Nominee, Ruggero Maria Santilli's MagnGas Process, which defies analysis using conventional scientific methodology, but is apparently capable of tapping energy from a variety of fuels at the nuclear level in a process which is analogous to the function of nuclear reactors. [11] Thus, many of the ominous eco-social problems that we now witness nearly on a regular basis can be addressed and solved with creative new applications of Space Technology.

New Evolutionary Paradigms for Third Millennium Social Transformation

The type of Evolutionary Paradigms necessary for launching the New Space Adventure would be "open-ended," with built-in flexible, evolvable guidelines. This kind of new "flexible thinking matrix," when integrated into mainstream scientific thinking would effectively enrich mental and spiritual growth, encourage intellectual diversity, and supercharge human brainpower, ultimately generating new and creative solutions for the social, economic, and environmental problems we all face as citizens of the Global Commons. As visionary author Steven Hacken puts it, "We are on the threshold of the best and brightest of times. Those entrusted with the powers of government must make decisions for the common good, or new leaders will. Given the belief that national survival is at stake, conditions considered politically or economically impossible can be accomplished in a remarkably brief time." [12]

In a sense, these new Evolutionary Paradigms might be considered to be "self aware," since they would be endowed with the innate capabilities to change, grow, and evolve in ways that would provide the most appropriate models for the social and techno-ecological of any particular timepoint. Such new paradigms already exist in their early stages, but continue to develop and mature. As these new thinking modalities continue to evolve, the boundaries Quantum Physics and Consciousness become increasingly blurred. A special new breed of creative scientific thinkers is emerging. These new thinkers might best be described as, "evolutionary explorers - pioneers who have opened up new frontiers of self-transformation. Somehow they have managed to jump across a crucial threshold to entirely different evolutionary territory." [13]

A Global Eco-Military Force

Global military forces have a strategic role to play in planetary management. A new Global Eco-Military Force could be organized and equipped as a future-focused peacekeeping organization that would combine Cultural Sensitivity with Rapid Deployment and Lightning Strike Capabilities. Functioning within contexts of "Strategic Biosphere Defense" and "Enforcement of Environmental Regulations," the basic focus of this new global military initiative would be to manage and protect natural resources on a global scale, for the present and future generations of all life on Earth.

A renewed focus should be placed on improving America's image abroad, and especially on Conflict Prevention, Resolution, and Mediation, as viable, cost-effective alternatives to War and related social conflicts, both of which negatively impact our spiritual, financial, and ecological resources at national and global levels. This New Military Ethic could be organized around Social and Cultural Issues, Environmental Protection, and Environmental Regulations, with a new emphasis on Environmental Management, Ecological Restoration, and Global Disaster Relief. Such a re-formatted Military Ethic could create positive and challenging life experiences for Military Personnel by providing Superior Educational Opportunities, and Certifiable Professional Skills, which would carry over to civilian life, and engender a heightened sense of Pride and Satisfaction as a return for service rendered.

Embedded within the new Global Eco-Military Concept is also a new perception "Home-Planet Security." In order to achieve true Homeland Security we must learn to work together to develop effective strategies for achieving this transformation. Ultimately, both "Security" and "Peace" depend directly on our ability to monitor, protect, and enforce a reasonable balance between human needs and the broader planetary ecosystem. Our future depends on it!

Strategic Initiatives for Planetary Management

A "Strategic Initiative for Planetary Management" would include the following four basic areas for development:

1) Global Ecology Management Guidelines would focus on the protection and monitoring of Oceanic and Terrestrial Biological Preserves. "Biological Species Banks" would be created for the preservation of ecological diversity, and could be integrated with Ecotourism as a mechanism for providing jobs, training programs, and financial independence for indigenous and local people in selected geographical locations.

2) Sustainable Ocean Resource Management Plan would focus on eliminating destructive fishing methods such as Bottom-Trawling, Bi-Catch Wastage, and the Overexploitation of Global Fisheries Resources. A new and socially equitable Global Fisheries Management Program could be implemented and enforced through satellite remote sensing technology. "In the future, our global oceans should be used much like land: farmed for plants, harvested for fish, mined for minerals, and the marine environment regulated for pollution." [14]

3) A Coordinated Global Satellite Sensing Network would be specifically designed to monitor Pollution, Desertification, Forest Fires, Underground fires, and Illegal Poaching of global forest and marine fishery resources. This data could be coordinated by supercomputers into a Central Information Resource, which would be available to scientists and citizens of every country in the world.

4) An International "Environmental Earth Corps" would coordinate and train Global Military Forces to develop and implement Environmental Education and Earth-Engineering Programs, enforce sustainable Ecological Regulations and Guidelines, protect Strategic Natural Resources, identify and supervise Pollution Clean-Up, and play a major role in Mega-Scale Environmental Restoration and Enhancement Projects. Military personnel would be offered a wide choice of global service opportunities, and receive technical certifications which would allow them to continue on with careers in the civilian world as productive members of the global workforce.

The New Space Renaissance

The Spiritual and Creative Aspects of a New Space Adventure would embody the seeds of a new "Space Renaissance," which could include the following elements:

Space Eco-Tourism would accommodate those adventurous individuals who desire to explore the various areas and aspects of Space as "The New Wilderness," pushing exploration to ever higher limits, and contributing to our knowledge back on Earth.

Medical Research, Rehabilitation, and Rejuvenation would accommodate advanced space medicine, and establish Space-based Facilities for Research, Medical Practice, Healing and Rejuvenation. Medical research and treatment facilities represent a major financial resource for eventually establishing what perhaps might best described as "Orbital Medical Centers."

The "Free-Flight" Aspect of Space would allow humans to free themselves from the chains of gravity, and experience Low-Gee and Zero-Gee Recreation. This free-flight environment would add an entirely new and unique dimension to the human mindset, similar to the way scuba diving opened the way for humans to discover and explore the undersea realm, or hang gliding presented new opportunities for humans to experience a technological analogue of avian flight in the atmosphere.

The Creative and Commercial Dimensions of Space would enable Artists, Athletes, Scholars, Scientists, and Spiritual Explorers to visit Space and have new experiences in orbital resorts and habitats and Lunar and Martian Colonies. Space Art, Space Entertainment, and Space Sporting Events would capitalize on already established multi-billion dollar earthbound financial industries.

In Space, a new and spirited Creative Renaissance could serve to bring profitable and exciting new dimensions to these already profitable Earth-based industries. The critical factor for implementing a New Space Renaissance is to effectively organize and activate the Creative Brainpower Synergy of our newly evolving "Planetary Consciousness Commons."

Conclusion

As we move collectively into the Future, creating Orbital, Lunar, and Martian Colonies, extending our Human Presence out into Space, this journey of exploration will undoubtedly represent the dawn of an extraordinary new experience that will be reflected in equally profound shifts in Human Attitudes, the Global Social Consciousness, and in the composition and health of our Planetary Biosphere. Although it is perhaps impossible to predict the specific effects of each new space experience, it would seem safe to assume that each new adventure and enterprise will return to Earth a set of improvements, which will add up and fractally disseminate to transform the global economy, the human social mindset, our relationships with each other, our Natural Resources - and ultimately the ways in which we humans perceive ourselves, and our place in the Universal Fabric of Time and Space. This kind of thinking is eloquently framed by authors Gary Schwartz and Linda Russek in their provocative book, The Living Energy Universe, as follows: *"Just as the energy and information of distant star systems travel in space forever, if the universal living memory process is true, our personal info-energy systems - literally composed of photons of visible and invisible 'light' - have the potential to travel, to journey and explore, and to evolve. This is the vision of The Living Energy Universe. If this vision is true, then the kind of loving intelligence that created such a process has given us both freedom and eternity. This is the kind of God that can make us all smile."* (15)

References

(1) Freeman Dyson, *Disturbing the Universe*, Basic Books/Perseus Books, LLC, New York, 1979, p. 116.

(2) Steven Hacken, *Global Renaissance: A Five-Year Plan to Save Our Planet*, Globe Press, Los Angeles, p. 40.

(3) Kenneth S. Keyes, Jr. and J. Fresco, *Looking Forward*, A. S. Barnes and Company, New York, 1969, pp. 96-98.

(4) Frank White, *The Overview Effect*, Houghton Mifflin Co., Boston, 1987 p. 20.

(5) Howard Bloom, *The Evolution of Mass Mind from the Big Bang to the 21st Century*, John Wiley and Sons, Inc., New York, 2000, pp. 57-59.

(6) Lynton K. Caldwell, "*Discovering the Biosphere*," in Paul R. Samson and D. Pitt, Eds., The Biosphere and Noosphere Reader: Global Environmental Society and Change, Routledge, London, 1999, p. 41.

(7) Yatri, *Unknown Man: The Mysterious Birth of a New Species*, Simon and Schuster, Inc., Cambridge, MA, 1988, p. 16.

(8) Pierre Levy, *Collective Intelligence: Mankind's Emerging World in Cyberspace,* Perseus Books, Cambridge, MA, 1997, p. 5.

(9) Robert Nelson, "*Oil from Coal...Free: The Karrick LTC Process*," Rex Research, Jean, NV, 1980.

(10) Michelle Nichols, "*Australia Considers Grant for Solar Power Tower*," Reuters, January 7, 2003; http://www.ennn.com/news/wire-stories.

(11) Ruggero Maria Santilli, "*Recycling Crude Oil and Liquid Wastes into MagneGas and Magne-Hydrogen*," Presented at the Hydrogen National Conference, HY 2000, Munich, Germany, September 11-15, 2000.

(12) Steven Hacken, op. cit., p. 127.

(13) Yatri, op. cit., p. 148.

(14) Steven Hacken, op. cit., p. 81.

(15) Gary E.R. Schwartz and L.G.S. Russek, *The Living Energy Universe: A Fundamental Discovery that Transforms Science and Medicine,* Hampton Roads Publishing Company, Inc., Charlottesville, VA, 1999, p. 183.

About the Authors

Elliott Maynard's education includes a Bachelor of Arts Degree at Washington and Lee University, a Masters of Science at the University of Miami, Florida, and advanced graduate studies in Zoology, Marine Sciences, Biological Oceanography, Coral Reef Ecology, and Tropical Rainforest Ecology at the University of Miami Marine Institute, Nova Oceanographic Center, and the Organization for Tropical Studies in Costa Rica. He received a Ph.D. in Consciousness Research from the University of the Trees in Boulder Creek, California, and has taught at Adelphi University and Dowling College in New York. He is also a Certified Professional Consultant (CMC). Elliott is founder and president of Arcos Cielos Research Center in Sedona, Arizona, which develops unique new paradigms in Science, Education, Fine Arts, Global Ecology, Human Potential Development, and Consciousness Research. Elliott's activities have recently focused on collaboration with key scientists and professionals around the world, working to develop futuristic research, and to introduce new paradigms in Planetary Management, and is an accomplished field biologist and underwater photographer. Elliott is a seasoned lecturer, and the author of two books and numerous scientific papers. He is a member of the World Future Society and the American Club of Budapest, and serves on the Advisory Board of the Integrity Research Institute in Washington, DC. He is a Team Leader for the Aerospace Technology Working Group Virtual Forum (Planetary Management of Universal Resources), and has collaborated with former UN Assistant Secretary General Dr. Robert Muller, to develop innovative new programs for Global Transformation, and for the creation of a New Planetary Government.

Sharon Tanemura Maynard was born in British Columbia, Canada. She earned an ARCT Degree from Toronto's Royal Conservatory of Music, and studied at Simon Frasier University in Vancouver, BC. She also earned a Bachelor of Arts in Communications from the University of the Trees Consciousness Research Institute in Boulder Creek, California. She co-founded Arcos Cielos Research Center, and has worked for the Hopi Tribe as an Education Specialist, developing a unique Bi-Cultural Educational Curriculum, which integrated Hopi Tradition with conventional educational programs. She is founder and president of two executive consulting corporations, Elshar Executive Enterprises, Inc. and Elshar International, LLC, and has worked extensively with corporate executives in the oil, music, and financial sectors. Sharon holds the titles of Certified Consultant (CC), Certified Professional Consultant (CPC), and Certified Consultant to Management (CMC). She served on the Advisory Board for the National Bureau of Certified Consultants in San Diego, California, is a Member of the World Future Society and the Club of Budapest America. Her special abilities in paraphysical perception have allowed her to create innovative new programs for success in the professional, scientific and corporate environments.

Chapter 28

The Earth Observatory

By Langdon Morris

The Problem

Today there is no regular and reliable source of information about the condition of the global environment that is available to the general public. Except for moments of extreme crisis, there is no consistent coverage of issues such as pollution, depletion of resources, human population demographics, global warming, toxic waste, or a hundred others. Nor is there systematic coverage of the extensive work that is being done worldwide in response to these vital issues, nor of the ongoing scientific research on the underlying characteristics and principles that compose the living systems of the Earth which might help us to make better choices as we confront a future that will only become more complex.

The news coverage of these issues, such as it is, is completely fragmented. Major events such as oil spills or other pollution crises, extreme climate conditions, storms, and exceptional instances of environmental degradation are reported by the media as though they were yet another sporting event, and after the immediate crisis has passed and the finals core is tabulated, they fade from the public's attention. And as we have seen with the lack of follow-through in New Orleans from the Katrina disaster, they apparently fade from the attention of politicians and bureaucrats as well.

The primary reason for this lack of effective coverage is a matter of media culture. Television, newspapers, magazines, radio, and now the internet are compelling media for conveying the immediacy of crisis, and they excel at displaying the drama of the spectacle, the emotion of conflict and confrontation, violence, risks and threats. They are also, we now know, quite adept at creating such drama, which they do through provocation, distortion, and manipulation. The conflict of a political debate is raised to peaks of screaming anger, the excitement of a championship sporting event is made into global spectacle, the heartbreak of human suffering brings out tissues by the millions. But then by tomorrow morning we forget and move on to a new disaster (and pass the popcorn, please).

However, most of the important issues concerning environmental destruction tend to be decades-long processes hidden as incremental change, and they aren't able to be conveyed in a compelling way via the event-oriented mass media. The long-term, evolutionary nature of these challenges makes them inaccessible to the "news" mentality. The notion that we are responsible for these forces that we unleash, or that we can or ought to do something about them, is rarely even whispered.

Hence, most of today's environmental news coverage is fragmented and ineffective as a source of meaningful education. Since the media do thrive on controversy, scientific and regulatory argument about the meaning of particular events or trends receives the majority of the coverage, which only results in a yet more confused public. Did the talking head for or against the issue win the sound-bite war in the 30 second news story? Ok, move on!

Even focused environmental programming is inadequate, because programs on networks like the Discovery Channel are presented without reference to any broader conceptual context. They stand alone as unconnected bits and pieces, lacking any overall framework to clarify and reflect their true meaning or potential importance, which merely reinforces the prevalence of fragmentation.

1st person: "Too bad about the rain forest."
2nd person: "Yeah, too bad. Turn the channel. I wanna see who's winning the game."
1st person: "Pass the popcorn."
2nd person: "Hey, he hit a home run!"
1st person: "Cool!"

Of course this is a cultural problem as much if not more than a media problem, because we all know only too well that the media only show us what we want to watch. Well, that's what the purveyors of mass media want us to believe, anyway. In truth, the media themselves also exert significant influence on our tastes, and at the moment it is in their financial interest to degrade our tastes to a frighteningly low level; as a nation are being significantly and rapidly degraded by an endless stream of drek, not only with respect to mass media, but in many other aspects of our lives as well. Our lives, our televisions, our refrigerators, our garages and basements are becoming filled with poor quality, shoddy, lazy, thoughtless, junk, and we fear that our minds, or the minds of our neighbors, are also becoming hollow junkyards.

In contrast, a great deal of the allure of the space movement is that it is a veritable antidote to the degradation we see all around us. Space exploration is anti-junk by its very nature. Junk science does not make rockets that reach orbit, that reach the moon, the sun, Mars, Venus, asteroids, the outer planets, and beyond, only real science does. Shoddy engineering does not make crafts that land flawlessly and operate autonomously for years on other planets, only real engineering does. Lazy thinking and successful space exploration are not compatible with one another.

Similarly, lazy thinking is no longer compatible with the survival of the Earth. And happily, millions of people, or perhaps even billions, still do admire and even revere excellent thinking, hard work, and thoughtful decision making, and they feel pride when their nation, their race, their humanity succeeds in its endeavors in space. And just as in the US, careful sociology55 has identified that there are indeed millions of people who despise the cultural decline they see around them and who desire to create a culture that carries our best qualities forward into the future, rather than our worst, and who are quite aware that insidious incremental processes are destroying precious, perhaps irreplaceable resources; and who are ready to be engaged in a systematic activity to measure, monitor, synthesize, and manage the scientific information, data, and knowledge that will lead to sound understanding and wise choices, so, then, we must engage these very people, and we must do so promptly.

The Response: Earth Observatory and EarthMedia

What's needed is to a new way to create and deliver news coverage and educational programming about environmental issues that carries with it a sense of immediacy, while at the same time delivering a larger framework of context, thereby matching the strengths of mass media with useful, contextualized information about what's really happening, so that people understand. This is the mission of the Earth Observatory.

The Earth Observatory would be a working information, research, policy, conference, and media center for environmental issues.

As an Information Center, the Earth Observatory gathers video and data on the environment from numerous partner organizations all around the globe. The purpose is not to create new data, but simply to gather as much of it as possible into one place - measurements, maps, photos, graphics, animation, time-series forecasts, news feeds, and computer-based analysis of activities, events, and trends chronicling the life of our planet, stories of environmental destruction, conservation, and preservation. Naturally, a significant proportion of the data, information, and images contained in the Earth Observatory would come from space, from satellites and other craft operated by NASA and other agencies.

As a Research Center, a professional staff of resident and visiting scholars and students are engaged in their own scientific research, as well as in helping to interpret the enormous flow of information for the media. They design, monitor, analyze, and manage research projects worldwide.

As a Conference Center, the Earth Observatory hosts researchers and policy makers in large and small, formal and informal conferences and gatherings on a wide range of environmental topics and issues.

As a Policy Advocacy Center, the Earth Observatory will support a different sort of decision making. Instead of decisions based on bureaucratic priorities, adversarial illogic, or mis-informed ideological distortions, a trusted source of sound, science-based environmental information supports participative decision making. It also provides the means to rise above prejudiced partisan politics to arrive at a reasonable discussion of issues that are surely difficult enough to handle without added ideological distortions. In the environmental field we need an organization to play the role that the Good Housekeeping Seal of Approval, or the Underwriter's Laboratories Seal, or even Consumer Reports plays in the consumer markets.

As a Media Center, the Earth Observatory is the home of EarthMedia, from which original programming and content will be disseminated worldwide on television, radio, in print, and on the web. A production set is located in the facility so that the ongoing activities of the Earth Observatory will be carried on EarthMedia, and presented by a team of hosts and moderators who will provide continuity just as a team of news anchors and reporters convey continuity in broadcast news.

By integrating these five functions, the Earth Observatory puts in place the conditions that support a comprehensive understanding of the environment and humanity's impact upon it, and the dissemination of that understanding on a global basis.

The Earth Observatory will be a vital resource for students and educators, for policy makers, activists, and media producers. It will also be a membership organization that engages citizens of all nations in the process of learning to care for our precious home.

Envisioning a Facility

To become a credible institution, the Earth Observatory needs to be a physical place where people can go, see, touch, and talk. A familiar model for it might be NASA's Mission Control, a single room where teams of scientists work round the clock to monitor and manage the activities of space missions, astronauts, satellites, and experiments. In the very nature of what they're doing there is tension and immediacy, a sense of urgency that comes from the inspiring nature of space exploration, its risks, and its complexity. New data are constantly arriving, new situations emerging, and a competent staff is on hand to collect, analyze, and explain everything that's going on.

The drawing below gives a first impression of how the Earth Observatory might look.

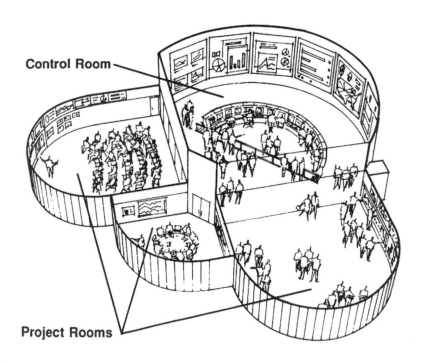

Figure 28.1 The Earth Observatory Concept Plan

The Control Room is also the EarthMedia broadcast set; the Project Rooms support various ongoing conferences and research activities. Additional facilities not shown would include offices, technical equipment, food service, and spaces accessible to the public. This drawing was prepared by Patri Merker Architects, and appeared originally in Managing the Evolving Corporation by Langdon Morris.

Not long after the first NikeTown store opened in Portland, Oregon, it also became the city's number one tourist attraction. Now maybe there aren't that many tourists in Portland, but in any case, wherever it comes to be located, the Earth Observatory should aspire to national and international renown, aspire to becoming an extraordinary place to visit in person or online, a source of amazement and insight, of stark beauty and piercing truth, of a trusted source of knowledge that is invaluable for the future of our civilization.

Conclusion

The concept of the Earth Observatory and EarthMedia as trusted sources of environmental information would build upon the success of CNN and the new paradigm for news that CNN established 3 decades ago. Prior to CNN, the news was a once-a-day event; CNN transformed it into an around-the-clock process, eliminating time lag. CNN also overcame the New York-centric orientation of the three old networks with a dynamic everywhere-to-everywhere openness, thereby surpassing the networks' intellectual dominance of news content and interpretation. In other words, CNN created a new context for globalized, 24/7news, which soon became a new context for society as a whole.

Exactly the same shift is now needed for the environment, a new paradigm that breaks the intellectual headlock of the 'event' mindset. It's time to create a new context and a new advocate for environmental awareness as only an Earth Observatory and integrated EarthMedia could. Or just as MTV created a new expert and advocate for youth culture, EarthMedia will be the expert and advocate for the environment, making it possible to coherently convey the big picture and the details of the important environmental issues that our culture must quickly learn to competently address.

Imagine, therefore, if you will ...

* Imagine a conference center dedicated to environmental issues ...
* Imagine an information center where vital environmental issues are tracked from day to day ...
* Imagine an internationally recognized center for environmental research ...
* Imagine the definitive environmental database complied from leading sources all over the world ...
* Imagine a policy advocacy group that earns the trust of all the world's citizens ...
* Imagine one place that is all these things, and a media hub dedicated to disseminating a profound understanding of the Earth to all people of all nations.
* You are imagining ... the Earth Observatory and EarthMedia.

In a short time, the Earth Observatory should become the definitive repository of trusted scientific data on the Earth's environment, an exemplar fully engaged in forging the crucial marriage of science, public policy, and public education. The unification of these three critical functions into a single, cohesive force in world politics and global policy-making could make the Earth Observatory one of the most vitally important factors in our future survival.

We have the means, now, to make it happen, and with such means also comes ... the responsibility.

About the Author

Langdon Morris - See Chapter 2.

Chapter 29

Lunar Commercial Development for Space Exploration

By Thomas L. Matula and Karen A. Loveland

Opening the Lunar Frontier

Historically, frontiers are driven by the potential for commerce. The desire for increased trade, identification of new resources or the discovery of new knowledge have driven humanity's quest to see what is beyond the next horizon. Pioneers and settlers follow the explorers; attracted by the opportunities for economic gain revealed by exploration, they put down roots to establish new communities on the frontier. It should be no surprise that the majority of the world's cities were first established to exploit unique natural resources or dominate natural trade routes; space settlement will be no different. Permanent lunar settlement will be driven not by scientific curiosity but by the potential for building economic wealth; the lunar frontier will be opened by extending the world's economy to encompass the Moon. Only by making lunar settlement economically viable will humanity extend its presence permanently to a new world and open the lunar frontier.

The frontier has long been an underlying theme of the American vision for space (McCurdy 1997 p248). President Kennedy built on America's frontier heritage in his 1962 speech at Rice University (Kennedy 1962):

"Those who came before us made certain that this country rode the first waves of the industrial revolutions, the first waves of modern invention, and the first wave of nuclear power, and this generation does not intend to founder in the backwash of the coming age of space. We mean to be a part of it—we mean to lead it."

President Kennedy also recognized the importance of commerce in the opening of a frontier when he called for America to send a man to the Moon and accelerate "the use of space satellites for world-wide communications."(Kennedy 1961). This request led to Communications Satellite Act of 1962 (Comsat Act 1962) which was the major driver in the expansion of the economic frontier of humanity to the geosynchronous orbit 24,000 miles about the Earth's center. Communication satellites have become such an integrated part of the global economy since 1962 that it is difficult to imagine how it functioned with them. It is also difficult to imagine what the global space industry would look like today without the demand for communication satellites driving the global launch market.

Perhaps no one better communicated the vision for a space frontier than President Reagan who stated in his 1984 State of the Union Address (Reagan 1984):

"A sparkling economy spurs initiatives, sunrise industries, and makes older ones more competitive.

"Nowhere is this more important than our next frontier: space. Nowhere do we so effectively demonstrate our technological leadership and ability to make life better on Earth. The Space Age is barely a quarter of a century old. But already we've pushed civilization forward with our advances in science and technology. Opportunities and jobs will multiply as we cross new thresholds of knowledge and reach deeper into the unknown."

On January 14, 2004 President Bush once again reinvigorated the vision of space as a frontier in his announcement of a Vision for Space Exploration (VSE). In his speech President provided a clear link between the opening of the space frontier and the opening of the American West (Bush 2004):

"Two centuries ago, Meriwether Lewis and William Clark left St. Louis to explore the new lands acquired in the Louisiana Purchase. They made that journey in the spirit of discovery, to learn the potential of vast new territory, and to chart a way for others to follow.

"America has ventured forth into space for the same reasons. We have undertaken space travel because the desire to explore and understand is part of our character. And that quest has brought tangible benefits that improve our lives in countless ways. "

President Bush went on to address the importance of lunar resources to opening the space frontier (Bush 2004):

"Lifting heavy spacecraft and fuel out of the Earth's gravity is expensive. Spacecraft assembled and provisioned on the moon could escape its far lower gravity using far less energy, and thus, far less cost. Also, the moon is home to abundant resources. Its soil contains raw materials that might be harvested and processed into rocket fuel or breathable air. We can use our time on the moon to develop and test new approaches and technologies and systems that will allow us to function in other, more challenging environments."

Creating a Lunar Economy

If America is going to take a lead role in opening the lunar frontier and creating a lunar economy it must change its perspective from viewing space as a

program like Project Apollo. Instead, space exploration should be viewed from the broader perspective of economic policy. Opening the lunar frontier is not about developing specific hardware to reach a specific destination like previous space programs (e.g., Project Apollo). Instead, the systematic opening of the lunar frontier to permanent settlement and economic development requires the federal government to move beyond technological innovation and develop the necessary policies to create a sustainable infrastructure for lunar economic development. Using the American frontier as an analogy is necessary to develop the policy and organizational framework to build lunar equivalent of the forts, trading posts and wagon trails that enabled the permanent settlement of the American West.

It is important to learn from the history of frontier development of America when developing policies for opening the lunar frontier. Unique organizations have been created to overcome the challenges of the frontier from the Virginia and Massachusetts Bay Companies that founded the first English settlements in America to the creation of Comsat, which pioneered the creation of a space based global telecommunications systems. That same spirit should be applied to the lunar frontier to create the unique organizations needed to open the Moon for economic development.

The Lunar Development Corporation (LDC) could be created in the spirit of other highly successful American infrastructure corporations. Key infrastructure, including the Erie Canal, the Transcontinental Railroad, the Panama Canal, the Alaska Railroad, Boulder Dam, Tennessee Valley Authority, Bonneville Power Authority and Comsat were all created using public/private partnerships. The LDC could follow in this successful path to open the lunar frontier.

The basic idea of a Lunar Development Corporation is not new:

Robert Heinlein's novel "The Moon is a Harsh Mistress" used a Lunar Authority as part of its plot line (Heinlein 1966).

Hayam Benaroya proposed a Lunar Industrialization Board to coordinate lunar economic development (Benaroya 1994).

In 1995, the United Societies in Space proposed the creation of a Lunar Economic Development Corporation (LEDA) as a private development corporation to stimulate lunar development (Harris, 1996 p295).

David Schrunk proposed a Lunar Electric Power Company to develop lunar energy resources (Schrunk 1999, p330).

All of these proposals recognized that a lunar specific organization was a necessary ingredient in any plans for lunar development. The LDC proposed here expands on these basic proposals while implementing the Aldridge Commission recommendations for developing a strategy for integrating the commercial sector into a lunar return mission (Aldridge Commission, 2004). The Aldridge Commission was created by President Bush in 2004 to develop recommendations for implementing his Vision for Space Exploration. One of its findings was that it was important to leverage the abilities of the commercial sector in reducing the costs of space exploration (Aldridge Commission, 2004).

An Act of Congress could create the LDC proposed here as a corporation in the District of Columbia. The LDC would be structured as a liaison between NASA and the emerging space commerce sector. Exemptions or modifications to federal procurement regulations could be included in the Act to streamline the procurement process and encourage new commercial firms to become vendors. Provisions also could be included for new management models to encourage joint ventures with private firms and the adoption of innovative approaches to building and operating lunar infrastructure. Most importantly, a new organization could bring a new culture of innovation and economics, one more suited to building and operating lunar infrastructure than NASA's culture of science and technology research.

Advantages of a LDC

Many advantages are associated with using the LDC to build and manage lunar infrastructure. The first advantage is that the new organization would have a new culture, one developed specifically around building and operating lunar infrastructure. One reason Project Apollo was so successful was because NASA was a new organization full of young engineers and technicians who created a culture suited to successfully implementing a scientific mission to the moon. The LDC would be like the NASA of the 1960's and attract a new breed of engineers, technicians, and administrators that bring a youthful vigor to the creation of a culture of innovation that parallels the cultures found in the new commercial space commerce firms.

Another advantage of the LDC is sustainability through multiple presidential administrations. As a government agency, NASA derives its funding from appropriations that vary due to political changes in Congress. The LDC would derive its funding from bonds and private investors (e.g., issuing common stock); thus, the LED would be able to focus on long term planning and development without the difficulties caused by funding changes. The LDC would also offer the incentive to pursue innovative revenue models to insure its independence from political budget battles for funding the development and expansion of lunar infrastructure. The LDC would be more likely to repeat the success of Comsat which was also enjoyed independence from the congressional budget cycle.

The government would still retain control over LDC's assets and any lunar infrastructure built as a government owned corporation; this advantage makes the LDC available for future national security needs. The LDC would also simplify the process of working with other government agencies like Department of Defense (DOD), Department of Transportation (DOT), Department of Energy (DOE) and the State Department on issues involving lunar policy. For example, lunar communication/navigation infrastructure built by the LDC would be readily available to the DOD for any national security needs that might arise in space or in lunar orbit.

A further advantage of the LDC is that it allows easier interaction with foreign space partners. The LDC would simply contract with foreign launch system providers for more spacelift capacity as needed. The LDC could also contract to do experiments on the moon for a foreign space agency. Complex State Department international agreements would not be needed because the LDC is a corporation and could use less complex business contracts making the bartering of services more attractive. However, international trade and arms reduction (ITAR) regulations would still govern the LDC just as NASA and private industry are governed.

The LDC relationship with NASA would be a strategic alliance/partnership. NASA would provide technical support, just as the Army Corps of Engineers provided support to the Alaska Railroad, Panama Canal, etc... and NASA provided support to Comsat. The LDC would not be limited to NASA as its only option for technical support. Comsat, Tennessee Valley Authority (TVA), and Bonneville Power Authority (BPA) contracted with private firms for needed services; the LDC would have the same options. This freedom would jump start lunar commerce among the commercial space commerce firms and help accelerate the development of a lunar economy by providing the emerging space firms with a major new customer, one with a more streamlined procurement process than NASA. Once the LDC built its infrastructure, NASA could become both a technical partner and a customer paying LDC for various services.

This business arrangement would free NASA from some of the difficulties associated with being anchored to specific infrastructure on the Moon such as those it is experiencing with the International Space Station (ISS).

Because the LDC manages the infrastructure, NASA would be free to focus on the scientific and technical aspects of space exploration. The LDC would operate the lunar infrastructure NASA requires for long range space exploration; Thus, the LDC could play a key role in the success of the VSE as a result by enabling NASA to keep its focus on the far horizon while the LDC stays focused on the Moon.

The LDC could also provide direct support for NASA in its efforts to return to the moon. Currently, under the VSE, NASA plans to use technology evolved from Project Apollo to return humans to the Moon in the 2018-2020 timeframe. Central to the NASA strategy for a lunar return will be the Crew Exploration Vehicle (CEV) which will sit atop a two-stage Crew Launch Vehicle (CLV). The CLV/CEV will be designed to place up to 25 metric tons and four humans into low Earth Orbit (LEO). In 2006, NASA is designing the new CLV and CEV based on improved Apollo-era technology and on components currently used with the Space Shuttle (Hammond 2006). Supporting the CLV/CEV will be a Cargo Launch Vehicle (CaLV) capable of launching 106 metric tons to LEO in its baseline configuration and 125 metric tons to LEO using an Earth Departure Stage (EDS). The addition of an EDS would also give the CaLV the capability of launching 55 metric tons into a lunar transfer orbit (Hammond 2006). Like the CLV, the launch vehicle for the CaLV will evolve from Apollo-era technology and Space Shuttle components. NASA plans to develop the CaLV in the 2011-2018 timeframe. As a government charted corporation (on the Comsat model), the LDC could provide NASA with additional financial support for developing the CLV/CEV and CaLV as well as serving to lower operational costs for the equipment by providing opportunities for using it for commercial payloads.

Prior to establishing a lunar outpost NASA will use unmanned orbital spacecraft to survey high priority sites on the Moon's surface. The lunar South Pole appears to be the most likely site for an initial human outpost on the Moon (King and Cook 2005). Other high priority sites on the Near Side of the Moon include the Aristarchus Plateau, Rima Bode, Mare Tranquillitatis, Mare Smythii, and Oceanus Procellarum while on the lunar Far Side NASA is considering the Orientale Basin Floor and South Pole-Aitken Basin (King and Cook 2005). One role of a LDC could be to leverage the capabilities of commercial firms for conducting the initial reconnaissance of potential lunar outposts by investing in commercial lunar missions.

NASA plans to establish the outposts themselves incrementally over multiple missions. Each lunar outpost will require a surface power system, communications/navigation systems, Rovers, habitat and laboratory facilities (King and Cook 2005). A LDC could serve as a mechanism that would allow NASA to partner with commercial firms for the construction and operation of these components, allowing NASA to supplement its lunar funding with commercial investment.

Financing The LDC

The financial structure of the LDC would be determined by Congress when the LDC is created. However, the major source of financing for LDC as a corporation focused on infrastructure development will most likely be bonds. Ideally, these Lunar Development Bonds will be issued by the LDC and fully guaranteed by the federal government. The LDC will issue the bonds through the U.S. Treasury at the prevailing market rate for federal securities. The bonds will be paid back through the revenues generated from the operation and sale of lunar infrastructure, lunar materials, intellectual property and other revenue sources identified by the LDC.

The initial funding of the LDC should be also involve the issue of public stock. Issuing a small portion of the stock to the public would allow directors from the public to be elected to join those nominated by the government.

The LDC Act should also include a "sunset" provision that requires the LDC to redeem the bonds within a specified period or face dissolution of the corporation. For example, the Act could give the LDC 30 years to generate sufficient revenue to eliminate the original bond obligation. If the LDC is unable to fulfill its obligation through revenues (and the issuance of additional public stock), the federal government would cover any balance between proceeds from the sale of LDC assets and the outstanding value of LDC bonds. This provision means LDC will have the advantage that comes from stable long-term financing and it offers it also means if the LDC fails to generate sufficient revenue it will be liquidated. Additional investigation would be required to determine the appropriate period for the sunset provision based on revenue and cost projections.

In addition, the LDC Act should provide that additional federally guaranteed bonds may be issued beyond the initial series by the LDC only after it has demonstrated the ability to generate revenue from the initial series. The amount of the new issues should be limited by revenue levels (e.g., 10 dollars for each dollar of revenue generated per year). Any amounts in excess of that limit will require Congressional approval of a business plan for future redemption.

LDC Revenues

Revenues to repay the bonds initially could come from other agencies using the infrastructure the LDC would build and operate. For example, NASA would pay LDC for lunar oxygen used for NASA spacecraft. DOD might also be a customer for lunar fuel that would be used to extend the flexibility and life of its satellites. Other revenue sources would likely include the National Science Foundation (NSF) and the DOE paying for placement and servicing of lunar experiments or for using lunar samples for research.

Benroya (2000) listed a number of activities at a lunar base that could be conducted commercially if a framework existed to do so. Blair (2000) outlined the value of mining lunar raw materials, especially oxygen for space development. Dennis Wingo's book "*Moorush*" provides a comprehensive review of the potential lunar resources including the possible presence of abundant platinum group metals (PGMs) (Wingo 2004). The presence of the LDC would greatly facilitate the commercial development of these resources by reducing the costs and risks for commercial firms.

The LDC would not be limited to revenues from U.S. agencies. It also could aggressively pursue revenues from foreign space agencies. For example, the European Space Agency (ESA) might pay the LDC to service an ESA telescope on the Moon, to send an ESA astronaut a LDC facility, or for samples of the lunar surface. The LDC could also provide support for a private lunar observatory similar to Steve Durst's proposal (Durst 2000). Other nations, private individuals, and for profit and not-for-profit organizations may pay the LDC for services. Universities also may pay LDC for experiments on the lunar surface or lunar soil for researchers to use.

Government agencies, foreign space agencies, and private firms would purchase goods and services based on a market-driven driven approach rather than costs. It is unlikely that revenues will cover 100% of the cost in the short-term. However, initial losses are common with government owned infrastructure. Mass transit systems often cover 50% of their costs or less with revenues. The lack of sufficient initial revenues is why a public corporation is necessary rather than a private one. However, the LDC will be unlike other government infrastructure systems because it will have one major incentive forcing it to search for the best deals, the requirement that the LDC pay off its debts within a specified period or be liquidated.

NASA and the LDC

NASA and its (NACA) ancestor were designed to administer technology programs, not to build and operate economic infrastructure (Centennial Commission 2003a). NACA was born in the midst of World War I when the United States had fallen so far behind in aviation technology it had no combat aircraft capable of surviving on the front lines in Europe. NACA's first mission was to close the gap and enable America to catch up with the major aviation powers of Europe. It was successful with a series of technology development programs that enabled America to dominate the skies in War World II (Centennial Commission 2003a).

NACA became NASA when Sputnik rocketed into the space in 1957, and its new mission was to develop the technology needed to allow America to overcome the Russian lead in space technology (Centennial Commission 2003b). NASA received its major input of resources, and its formative culture when its primary mission was to beat the Russians to the Moon during Project Apollo.

The culture that developed as a result of the NACA/NASA history focused on creating successful programs to rapidly advance technology. Project Apollo was such a technology program, and a very successful one. Projects Viking, Mariner, Pioneer, Voyager, Galileo, Hubble and many of NASA's other great success were also technology programs, designed to push the art of spaceflight to the cutting edge. These programs fit well into NASA's culture because each had a clear goal at the beginning that led to a clear conclusion at the end.

This contrasts with both the Space Shuttle and ISS, which were intended as infrastructure designed to enable future programs with no specific end game strategies. However, the shuttle and ISS have not been managed as infrastructure projects. NASA's culture viewed them as technology programs that required the latest, newest, most risky technology. Successfully opening the lunar frontier requires recognizing NASA does not have the culture or expertise needed for successfully implementing infrastructure projects.

NASA's key role in opening the lunar frontier will be in serving as a customer for commercial lunar services and in working with industry in developing the technology needed for lunar industries, not building and operating the lunar infrastructure. NASA should not be responsible for producing the lunar oxygen for missions, building the solar power arrays required for processing of lunar ore, nor the mining of that ore. Instead NASA's role should be the one it most suited to, namely pioneering the basic technology needed for development of lunar resources and conducting the preliminary human exploration of the Moon surface. NASA can operate then as the frontier scout it is, and move on to the next far horizon, Mars, leaving the operational work of building lunar infrastructure and mining its resources to a corporation specifically created for that purpose, the LDC.

The strategy of allowing NASA to pioneer the technology and then move on while another organization builds the operational infrastructure is not new. It has already been successfully demonstrated with the communication satellite revolution. NASA pioneered the first test of communication satellites in the early 1960's. Congress then created the Communication Satellite Corporation in 1962 to build the operational communication satellite network (Comsat 1962). NASA provided critical technical support to Comsat, and the new organization freed NASA to focus on the more demanding challenges of Project Apollo.

Summary

Although firms like Bigelow Aerospace (orbiting tourism hotels in LEO) and Virgin Galactic are developing commercial systems for Earth orbit, business models for lunar commerce are speculative at this time (Hammond 2006). The speculative nature of lunar commerce and the high investment costs needed for developing a lunar transportation system requires some form of government investment, just as the transcontinental railroad required in the 1860's. This is why NASA and/or the DoD need to provide the initial lunar transportation system. However the development of lunar commerce is critical if America's return to the Moon is to be sustainable and it is expected that post-2020 an increasing fraction of the lunar transportation system will be commercially funded and developed (Hammond 2006).

It must be recognized that opening the lunar frontier is too large a task for any single government agency. It must also be recognized that NASA's culture is that of a frontier scout, working on the ragged edge of technology. Once an organization's culture is set it requires a major effort to redirect it. Cassini, Hubble, and the Mars Rovers demonstrate that NASA is successful when the missions it is given are consistent with its culture. NASA's focus on cutting edge space technology is critical for leading the scientific exploration of space. Many have called for NASA to be converted into a space development agency instead of the space exploration agency it is today. Once such a conversion occurred, the need still would exist for an agency to explore the far frontiers of space. Why change NASA into something it will never be? Instead, those involved should recognize NASA's role and strengths and give it missions that build on these instead of trying to extend its charter to do missions not fitted to its resources. Why destroy a great agency by trying to convert it into a mission far from its core culture, expertise, and experience base?

Those building the American West recognized that frontier scouts made poor railroad builders. NASA embodies the spirit of the frontier scout, always seeking new horizons. The creators of NASA recognized this when they gave the implementation of President's Kennedy's goal of creating a satellite based communications system to the Communication Satellite Corporation and the goal of a space based weather forecasting system to the United States Weather Service (now NOAA) in the 1960s. NASA played key roles in achieving both goals by developing the space technology needed, but the systems built were owned and operated by organizations with cultures more suitable to integrate these innovations into the nation's economy. COMSAT and NOAA were successful and it is difficult to imagine how America's economy would function today without communication satellites and the accurate weather forecasts now take for granted by the average American. Frontier scouts played a critical role leading railroad survey crews through the wilderness and charting the path of the railroad. NASA engineers played a similar role in charting the technology path required for building the Comsat and weather satellite systems. However, building the railroad was left to others, and NASA left building these satellite systems to others. The same should be true for space solar power (SSP), Helium 3 (He3) mining, lunar PGM extraction and lunar oxygen (LUNOX) production. Creating the LDC to build and operate the infrastructure needed to harvest these new space resources for the future of humanity will allow NASA to once again take the point and explore the far horizons of the frontier "*for the benefit of all mankind*".

Research Questions

Question 1 - What changes in U.S. policy would be required to open the lunar frontier to economic development?

Question 2 - What are the key technologies required to develop lunar infrastructure and resources?

Question 3 - What are the specific costs and benefits of using the LDC instead of NASA to build and operate lunar infrastructure?

References

Aldridge Commission (2004) President's Commission on Implementation of United States Space Exploration Policy (2004), Report: A Journey to Inspire, Innovate and Discover," June 2004, Retrieved from http://www.nasa.gov/pdf/60736main_M2M_report_small.pdf

Benaroya, Hayam (1994). Lunar industrialization, The Journal of Practical Applications in Space, 6 (Fall), 85-94.

Benaroya, Hayam (2000). Prospects of commercial activities at a lunar base in Return to the Moon II: Proceedings of the 2000 Lunar Development Conference, July 20-21, 2000, Las Vegas, NV.

Blair, Brad (2000). An economic paradigm for commercial lunar mineral development, in Return to the Moon II: Proceedings of the 2000 Lunar Development Conference, July 20-21, 2000, Las Vegas, NV.

Bush, George (2004). President Bush Announces New Vision for Space Exploration Program, White House Press Releases, http://www.whitehouse.gov/news/releases/2004/01/20040114-3.html

Centennial of Flight Commission (2003a). "The National Advisory Committee for Aeronautics (NACA)" http://www.centennialofflight.gov/essay/Evolution_of_Technology/NACA/Tech1.htm

Centennial of Flight Commission (2003b). "The National Aeronautics and Space Administration" Retrieved from http://www.centennialofflight.gov/essay/Evolution_of_Technology/NASA/Tech2.htm

Comsat Act (1962). "Communications Satellite Act of 1962," Public Law 87-624, Proclaimed on August 31,1962, http://www.jaxa.jp/jda/library/space-law/chapter_1/1-1-2-5/index_e.html

Durst, Steve (2000). International lunar observatory/power station: From Hawaii to the Moon in Return to the Moon II: Proceedings of the 2000 Lunar Development Conference, July 20-21, 2000, Las Vegas, NV.

Hammond, Walter E. (2006), NASA Senior Engineer, Personal Communication

Harris, Philip Robert (1996). Living and working in space: Human behavior, culture and organization, Praxis Publishing: West Sussex, England.

Heinlein, Robert A. (1966). The Moon Is a Harsh Mistress, Putnam: NY, NY.

Kennedy, John F. (1961). "Special Message to the Congress on Urgent National Needs" Delivered in person before a joint session of Congress," May 25, 1961, Retrieved from http://www.jfklibrary.org/j052561.htm

Kennedy, John F., (1962). "Address at Rice University on the Nation's Space Effort" Delivered in person at Rice University, Houston, Texas, September 12, 1962, Retrieved from http://www.jfklibrary.org/j091262.htm

King, David & Steve Cook (2005), "MSFC and Exploration: Our Path Forward," Briefing at Marshall Spaceflight Center (MSFC), Huntsville, AL by David King, MSCF Director and Steve Cook, September 23, 2005.

McCurdy, Howard E,(1997). Space and the American imagination, Smithsonian Institution Press: Washington, D.C.

Reagan, Ronald (1984). Address Before a Joint Session of the Congress on the State of the Union - Delivered in person before a joint session of Congress," January 25, 1984 Retrieved from http://klabs.org/richcontent/Speeches/ReaganSpeech.htm

Schrunk, David, Burton Sharpe, Bonnie Cooper and Madhu Thangavelu (1999). The Moon: Resources, future development and colonization, Praxis Publishing: West Sussex, England.

Wingo, Dennis (2004). Moonrush: Improving Life on Earth with the Moon's Resources, Apogee Books: Burlington, Ontario, Canada

About the Authors

Thomas L. Matula, Ph.D. has a Bachelors degree from the New Mexico Institute of Mining and Technology (1983) and an MBA (1985) and Ph.D. in Business Administration (1994) from New Mexico State University. His dissertation focused on development of a model designed to identify factors that would influence public support for a commercial spaceport. He has since published numerous articles on space policy and economic development strategies for the space industry. Dr. Matula has served on the American Society of Civil Engineer's Space Engineering and Construction Committee and its Subcommittee on Space Education Initiatives. His academic career includes fifteen years of university teaching in the areas of business strategy and marketing. Dr. Matula is currently an Assistant Professor of Business Administration at the University of Houston-Victoria. He is also CEO of Edustrategy, a consulting firm specializing in workforce development.

Karen A. Loveland, Ph.D. has a Bachelor of Business Administration (1986), Master of Business Administration (1989) and Ph.D. in Business Administration (1996) from New Mexico State University. Her dissertation focused on the reliability and validity of conjoint analysis as a technique for analyzing complex consumer decisions. She has since published numerous articles on research methods and online education. She has also co-authored several papers with Dr. Matula including a recent paper on public perceptions of various space exploration goals. Her academic career includes over ten years of university teaching in the areas of marketing management and Internet marketing. Currently Dr. Loveland is an Assistant Professor of Marketing at Texas A&M University, Corpus Christi. She is also Vice-President of Marketing and Program Development at Edustrategy, a consulting firm specializing in workforce development.

Chapter 30

Managing Risks On The Space Frontier,

The Paradox of Safety, Reliability and Risk Taking

Feng Hsu, Ph.D.
NASA, USA

Romney Duffey, Ph.D.
AECL, Canada

Introduction

Is human space flight only for the highly trained and well-educated astronauts? Or is it just for a few of the wealthiest space tourists among us? Can commercial space travel be made safe yet affordable for us all? Can human space venture be successful without the possible sacrifice of lives? Attempting to answer these questions is by no means a trivial endeavor. It must certainly lead to rigorous debates, inevitably probing into the paradox of risk taking in the face of safety and uncertainty in humanity's continuous drive for the opening of the Space Frontier.

With the profound resurgence of human ingenuity in the dawn of the new millennium, technological frontiers have been created and pushed forward at such a stunning pace never before seen in the recorded history of human civilizations. In the wake of the information age, where the Internet-propelled digital packets permeates every fabric of human lives, the nano-technology and bioengineering revolutions are transforming human society permanently and irreversibly. All in all, recent advances in the space frontier has been no less than astonishing, and is quietly incubating yet another technological revolution. This certainly will change not only the fate of human spices on earth, but will perhaps also transform us from single-planet habitant into multi-planet creatures to forever survive in the vast universe! Following the President Bush's bold vision of human space explorations announced in January 2004, only a few month later comes the headline making successful flight of SpaceshipOne. Launched out of the Mojave Desert of California, the first privately financed, commercial vehicle for taking ordinary citizens to sub-orbital space and safely landed back on earth. Subsequently, SpaceshipOne's financial backers have promised to develop and promote routine space tourism in the near future. Their flight joins a list of several other flights since four years ago that took regular travelers commercially to space. A recent partnership powered by expertise from Burt Rutan and his team at Mojave, California-based Scaled Composites, and more than $25 million from software billionaire Paul Allen, called the Spaceship Company, have been formed by Rutan's team and British tycoon Richard Branson's Virgin Galactic. They plan to build a fleet of five seven-passenger "SpaceShipTwo" spacecraft using SpaceShipOne technology. Most recently, they have just announced to build a commercial spaceport in New Mexico, and are aiming to begin space tour service as soon as in late 2008.

Infused by the commercial significance of SpaceshipOne test flight, there has been a world wide heated space movement propelled at full steam by entrepreneurs, politicians, space enthusiasts and business owners alike. One year after the history-making SpaceShipOne flight, and subsequently winning the 10 million dollar X Prize, Dr. Peter Diamandis, the X Prize founder is joining forces with a venture capitalist to establish a NASCAR-like rocket-racing league (RRL) for rocket-powered aircraft. XCOR founder Jeff Greason, after successfully developed a methane-fueled rocket engine, has recently set up yet another prize for steam engine development. The 41 year old founder of Amazon.com, Jeff Bezos, a native of Texas announced several months ago the intent to use his newly acquired 165000 acres of desolate ranch land to build a world's largest space port in west Texas in order to launch space vehicles that carry ordinary space travelers. Among other space enthusiasts, entrepreneurs and privately owned aerospace pioneers, is the Las Vegas based Biglow aerospace, which has unveiled their latest effort to build an inflatable space habitat as well as space stations to help inspire and drive humanity's permanent entry and colonization into the space frontier. A recent report has revealed that in three to five years from now, a company called Space Adventures, based in Arlington, Virginia, is already planning to offer commercial trips to orbit the moon for only $100 million per passenger. In spite of the launch of the first private rocket into space, the private piloted manned space flight has just reached the tip of the ice burger. Nevertheless, it won't be very long that the civilian suborbital flights could become reality much like the way the aviation jetliners operate today that not only new jobs can be created but also a whole new industry will be raising on the horizon.

Despite SpaceShipOne's success, despite the glorious human achievements in the Apollo days, as well as the 20 plus years' Space Shuttle flight experience with recent marvelous missions of Mars exploration Landers, human space explorers, NASA and space travel entrepreneurs still have very tough challenges ahead. These challenges apparently, are not necessarily technological ones only. Do we really know how to make spaceships that can fly a couple of times a day, every day for years? How can we fly space orbital vehicles so safely and so reliably that we can fly people as a business much like the way commercial airlines do? Do we really know the economics of the emerging space flight industry and how to make it profitable enough to sustain the survival of the space travel commerce…? Key questions emerge: if we are ever going to free ourselves from the earth confined human race, how much are the societal, political and individual safety risks as well as resource sacrifices are we willing to take for opening up the new frontiers in space? Are the risks of manned space flight worth taking in the course of human exploration into the heavens? Are such risks expanding or contracting as we continue to push opening further the door to the bondless Universe? What are the balances we ought to take for benefiting human lives on earth versus societal, political, financial and human life costs in the course of continued exploration into the outer planets? How safe is safe enough for us to step ahead on the continued course for space explorations every time when tragedy and accidents struck us? When does greater efficiency approach reckless risk taking when confronted with safety and resources constraints? This chapter provides some interesting insights from various perspectives of technological, programmatic and strategic risk taking to social political issues while trying to explore the answers to these questions.

Managing the Human Elements of Risk Taking

The history of fatal accidents and tragedies in human space activities has always reminded us of an extremely high risky endeavor that requires not only highly costly social, political and financial resources, but oftentimes the sacrifice of human lives as well. The most horrific examples of risks to the US manned space programs are the fatal accidents that occurred with Apollo 1 on January 27, 1967 and with the two Space Shuttle disasters of Challenger and Columbia on January 28, 1986 and February 1st, 2003 respectively. The human reactions and societal policy responses to these tragic events have been repeatedly illustrative of exhaustive engineering reviews, detailed hazard assessments, expensive hardware tests and redesigns by system experts. These reviews have been exemplified by the search for cause and by the assignment of blame. And most notably, are the lengthy and high profiled accident investigations by Congress, Presidential Commissions that often lead to the replacement of key program managers or reorganization of institutional hierarchies and infrastructures regardless of actual effectiveness of implementing these various measures. Nonetheless, sooner or later more fatal accidents still strike us unexpectedly every time after each major accident, even after extensive, painstakingly detailed and careful investigation, exhaustive safety assessment and re-engineering, and changes in requirements and organizational structures and communications. Accidents as evident in many other industries, whether it was the Chernobyl reactor explosion, or Three Mile Island nuclear reactor melt down in 1986 and 1979, the Piper Alpha offshore oil platform fire in the Norwegian sea in 1993, or the Concorde plane crash in Paris in 2000, they all exhibit familiar interesting pattern. The primary human response to these tragic events were either to raise serious political doubts about the risks behind the economical and technological benefits of the industry in general, or to carry out extraordinarily lengthy investigations which often focused primarily on technological specifics of the accident phenomena that caused the tragedy.

Why do accidents and adverse outcomes happen all the time, in spite of most of them being preventable? As humans we tend to fix the known, what we observe to be wrong or the cause, and largely ignore the unknown since it has yet occurred. Therefore fixing not only the observed causes but more importantly, to explore the unknown in a systematic and integrated approach is the key to prevent disasters or reducing the likelihood of fatal events from reoccurrence. Even while we fear the unknown, we should still face it. While exploring the unknown harbors important implications that human exploration of the Space Frontier is inherently a high risk endeavor, only a few of us could fully recognize this very nature of our space activities. Thus, the general public perceptions of the true risk and their support or political will to continue moving forward on the space frontier have become the biggest casualty of any accident in the past. Can we imagine that without the societal and political strong support, we could have succeeded in landing humans on the moon just two years after the Apollo 1 disaster struck us; or some seven months after the near miss of Apollo 8's first manned mission to ever orbit the moon back in the Apollo era? Can we also imagine that without strong yet continued political will from their respective dynasties or regimes, the ancient Chinese could have built the Great War stretching over 6700 kilometers of mountainous terrain in hundreds of years of time span over two thousand years ago? Or the Americans could have built the Panama Canal connecting the two major oceans on earth? Or could we have built the marvelous Hoover Dam at all?

On the contrary, despite going through all sorts of lengthy safety "stand-downs" within NASA organizations and spending nearly three years on the RTF (return to flight) activities, and hundreds of millions dollars to ensure the Space Shuttle safety after the Columbia tragedy, we barely missed yet another major disaster on the subsequent flight of Discovery (STS-114). Have any of us thought about why we were able to land ourselves on the Moon some 35 years ago, just less than 9 years after President Kennedy's famous political rally on May 25th of 1961 till the successful mission of Apollo-11 in which a "giant leap for mankind" was set forth on the surface of the moon? Why after 35 years today, two years have already gone by but we still aren't quite sure if we can accomplish the first phase of the new space initiative (going to the Moon) by 2018 since President Bush's passionate announcement on January 14, 2004? Can we really say that beside the "space race" element in the 1960s (which can also be attributed to a different type of strong political will and public support motivated by cold war rather than peaceful space exploration), the Challenger and Columbia syndrome are not taking a toll deep inside our human spirit and courage, resulting in the overly conservative constellation program plan for exploring into the unknown? Clearly, perhaps the single most damage from space disasters has been "fear" or "fear of moving forward" that has seeped deep into our human minds. It is the loss of continued societal and political support, which ultimately brings about our inability to make reasonable decisions on taking risks, which of course includes exploring the unknowns in space. As the perception of risk is governed by the unknown outcomes, we must not follow such a rabbit trail repeatedly fearing failure, if we really posses the most invaluable and superb human spirit to explore the universe. The real challenges of managing the safety risks indeed are not necessarily the technological ones, but something lying inside the human spirit and willingness for continued exploration of the *unknown*. In the history of civilization, human beings have always proven ourselves able to overcome the inconceivably complex technological challenges as long as there was strong societal will and continued political support. The most severe consequence only possible to human society perhaps, of any disasters from accidents in socio-technological systems is the loss of human spirit for adventures, and for exploring the unknown, the frontier in nature and in the Universe.

Managing the Human Perceptions of Risk

While realizing the risky nature of the human space endeavor and carefully managing the risks for safety and system reliability, we must also rectify some biased risk perceptions of human space flight that have been deeply rooted in the minds of general public, and more importantly in the minds of politicians. The risk profiles and public perceptions of the space related activities have always been extraordinarily high as compared to other human activities involving complex systems. Taking the commercial airline industry for example, airplane crushes with over hundreds of human casualties occur rather regularly, at least a couple of times once every few years. However the general public seems to have developed a psychic barrier to fear that they keep on traveling by flying as if there was nothing had happened. The risk perception is that the risk is "acceptable": compared to other risks in life it is small. Thus, there has never any notable open criticism neither by the public media, nor of any political obstacles call for the demolishment or abandonment of the civil aviation industry, even after some major airline crashes that resulted in sizable human casualties. The only commercial airline crash in history that has brought an end to an invaluable and complex technology was the total dismantling of the Concorde supersonic fleet by Air France and British Airways. This became such a regrettable tragedy because it was much more of a commercial failure, or the failure of miss-calculating the risk of a strategic program decision rather than the failure of the Concorde technology itself. Can we imagine that all the commercial air travel is possible today if the politicians did not give strong backing (often times taking a risk on their political career by acting against the perhaps biased public risk perception or political views) back in the early 1930s at the infancy of the modern civil aviation industry? Similarly, the safety risk of the commercial space travel or space tourism, just like any other risks of complex technological marvels that humans have created so far, that it takes painstaking processes and time for humans to understand.

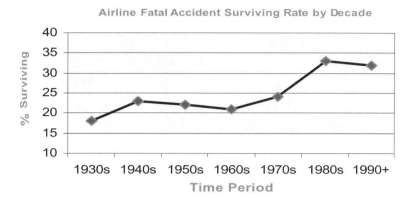

Figure 30.1. The Steady Increase of Survival Rate in Airline Safety

And more importantly, it takes the human perseverance to allow such risks be gradually reduced to a point where both aspects of technology and the human interfaces with such technology becoming highly matured. As an example Figure 30.1 shows some tangible statistics as how the survival rates or flight safety in the commercial airlines have been steadily improved over the years ever since the early days of civil aviation industry. The survivability aspect is key because we cannot reduce the rate of these (rare) events any further. Just like the automobile, we adopt mitigating measures in case of a crash.

Accepting Risk in Emerging Technologies

Every new technology is greeted with some initial fear and trepidation. It is a natural human reaction to the unknown, and an uncomfortable feeling. Examples are legion: automobile transport replacing horses; electricity replacing candles and gas lamps; genetically engineered crops replacing selected strains; bio-technology replacing traditional medicines: vaccines replacing blood letting; nuclear energy replacing coal. Obviously, if we fall short of earning public support or stop short of moving forward on the human space activity front, there could never be the future for the emerging space travel and tourism industry. With accepting some initial R&D risk of human suborbital flight today, we can then truly envisage a day for human beings in the near future such that, one could safely take-off of a Space Port somewhere in the west Texas desert shortly after his (her) breakfast, and to be safely landed at China's east coast city of Shanghai for a trade meeting, and then taking another (space) flight back to Houston for dinner on the same day!

Means of Transportation	Fatalities per million trips	Odds of being killed on a single trip
Airliner (Part 121)	0.019	52.6 million to 1
Automobile	0.130	7.6 million to 1
Commuter Airline (Part 135 scheduled)	1.72	581,395 to 1
Commuter Plane (Part 135 - Air taxi on demand)	6.10	163,934 to 1
General Aviation (Part 91)	13.3	73,187 to 1

Table 30.1. Is Driving Safer than Flying?

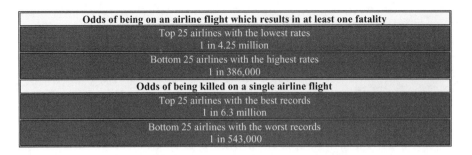

Odds of being on an airline flight which results in at least one fatality
Top 25 airlines with the lowest rates 1 in 4.25 million
Bottom 25 airlines with the highest rates 1 in 386,000
Odds of being killed on a single airline flight
Top 25 airlines with the best records 1 in 6.3 million
Bottom 25 airlines with the worst records 1 in 543,000

Table 30.2. Is Flying Really More Risky?

The news media would rarely make any more than a day's headlines or front-page reports out of a seemingly regular aviation accident. Whereas, the general public and societal responses to the Space Shuttle crash, or any nuclear plant incidents, have been profoundly overly sensitive and sensational compared to a commercial plane crash that might have killed many times more of human lives. It is human nature that the perception of risk of the unknown far exceeds the reality. The same biased risk perceptions by the general public exist in the case of automobile accidents, that no one seems to fear of driving given that the automobile risk (likelihood of being killed per single trip) is actually 1 in 7.6 million, nearly an order of magnitude higher or10 times greater than the commercial airline risk of 1 in 52.6 million chance being killed on a single trip. Table 30.1, 30.2 and Figure 30.2, 30.3, 30.4 provide more detailed risk comparison between several differing industrial and human activities. Rare events and learning curves show that airlines with more flights have lower rates, simply because of the greater number of flights before having an accident or not, which is the equivalent of being lucky or not. This rare event line is shown (Figure 30.2) as a line of slope minus one through the data, and is an *equal risk line*.

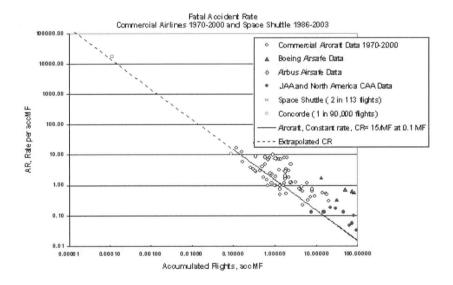

Figure 30.2. Fatal accident rate of commercial vs Space Shuttle

The paradox with a new technology like space travel is simply this. It is impossible to have a lower accident rate until many flights have occurred; meanwhile unless we have many more flights we cannot learn to reduce the number. Basically, we must have events in order to learn from them. The reason why the risk perception of Space flight in the minds of politicians or general public are so much biased is primarily due to the inherent high media and political profiles of human space activities. Society generally does not understand the differences between the matured technology, in the case of commercial airlines, and the research and developmental (R&D) technologies as in the case of human space explorations. They may tend to simply *perceive* the risk of a Space Shuttle flight from the perspectives that of a commercial airline flight, therefore wishfully expecting that the human space activities should be as safe as that of the civil aviation activities. This is clearly unattainable at an early stage of any technology, like space travel.

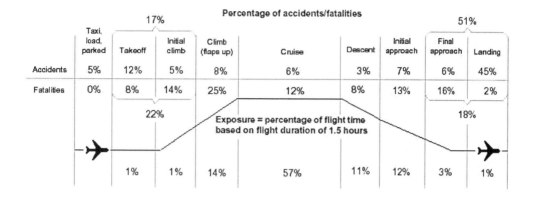

Figure 30.3. Typical percentage distribution of airline fatalities by phase of flight

Yet unscientifically founded societal and political expectations must be rectified before major endeavors can be undertaken in the course of successful human space frontier explorations. On the other hand, we must understand that the psychology of risk perceptions by the general public on various human activities are easily influenced by the visibility of the underlying events, and by the degree to which the mass media portray them. Again, as in the case of the space shuttle accidents, yes, a total of 14 brave astronauts were sacrificed in the entire SSP (space shuttle program) history of nearly a quarter of a century time span. Yet the society and space policy makers seemed so eagle to dismantling the entire program over safety concerns while blind-sighted on the overwhelming benefits in various aspects of engineering, scientific, economical and international social-political achievements. Indeed, if it was merely a safety risk concern in the minds of our space policy-makers on the total abandonment of the existing space shuttle program, then it is clearly worth the time to have a second look at the risk problem before it is too late.

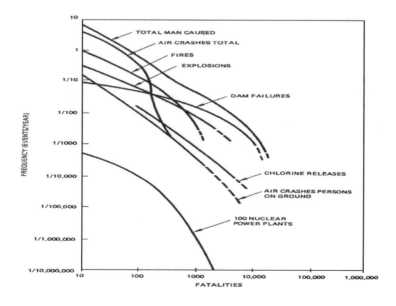

Figure 30.4. Frequency of Fatalities due to man-caused events

Dealing with Moral Issues in Risk Taking

There is also a moral issue at stake: How do we really measure the loss of individual lives as risk factor or ultimate cost to our society? Especially, how do we value the lives of 14 astronauts versus the lives of tens of thousands passengers who get killed each year either by commercial plane crashes or automobile accidents, or those who gave their lives on duties for the benefits of public good, like the test pilots, medical doctors, Army soldiers, police or fire fighters? Arguably, there are professors, scientists and well-trained technologists and important public officials among the many killed whose death to our societal loss can be equally as great as that of the astronauts sacrificed.

In pioneering any technology, risks must be taken whether they are financial, technological, operational, legal, societal, or human. These questions are profound in their ethical, social, moral and technological implications. One of the increasing obstacles facing human society, especially facing Americans today in exploring science and technology frontiers is that the skyrocketing legal risks, which undoubtedly have become key factors to the ever increased societal cost or high budgetary pressures. The human spirit of taking risks in unknown territory of yesterday is certainly worth remembering today. The inventor of the smallpox vaccine tested it on himself first but did not die; one of the discoverers of X-rays over exposed themselves and did. Likewise, is it a worthy sacrifice of astronauts in the case of an accident for the noble course of mankind's ferrying into space, in which many of us would have been dreamed of being a part of? Should the Wright brothers give up their extremely risky test fly in the face of death over 100 years ago? If American society had blindly criticized and prevented the transatlantic adventure of Charles Lindbergh's bravery back in May 20th of 1927, we would have no commercial airline industry of today, and the human lives on earth could have never been the same. Instead, he was lionized for his successful risk taking. Losing fourteen people who were very important to their families and friends in addition to being important to the whole human race in such an unexpected way is horrible and tragic. Such losses, though, are not only well worthy of the course they were engaged in, but part of life as well. As bad as the deaths of these fourteen astronauts are, sudden unexpected deaths happen much more often, all the time. It just doesn't get the notoriety that shuttle accidents do, and we accept that risk and loss.

The conclusion by the Columbia Accident Investigation Board, the CAIB report has finally offered us a much broader vision on this issue: *"Operation of the Space Shuttle, and all human spaceflight, is a developmental activity with high inherent risks."* These words are worth bearing in mind, as future spacecraft that are developed to ferry humans to the moon and Mars will be inevitably new types of spacecrafts that must satisfy even more harsh flight conditions than did Apollo or the Shuttles. So, while rigorously pursuing the scientifically based methodologies to ensure high standards of safety, we must not only rectify biased risk perceptions about human space endeavors, but also taking a more balanced psychological and moral stand to evaluate human sacrifices against all the great benefits it brings about to the advancement of mankind's searching for the space frontier. We could hardly achieve anything if we ignore to educate the public and the policy makers. If these issues are not fully recognized by the society as a whole, we will certainly achieve much less by wasting more of resources and even wasting more of human lives in the road ahead. We must also fully understand the paradox of the risks and social costs from *not* taking risks in the human space activities, or in any human development activities. Because throughout human history, the countries and nations that have led the world in exploring have been the most powerful countries on the planet, and they are looked up to for their leadership. This is exactly why we firmly believe that as the world's leading nations on earth, we need a clear vision for R&D collaborative effort in pushing the exploration into the space frontier. And inevitably, by leading the way in space commercialization, physics, biology, and all the other areas that contribute to transforming and keeping all the peace-loving nations as part of the viable and combined economical and technological forces on earth, which ultimately will move mankind outward in the solar system and beyond.

Managing the Technological Risks

For just one hundred years since the Wright brother's courageous test flight, human beings have solved the problems of how to fly and travel safely around our planet. The history of human industrialization has vividly demonstrated with certainty that science, technology and engineering advancement in the space frontier will make the next major breakout of human industrialization from Earth to space. With the sub-orbital flight of SpaceShipOne, history has once again opened up a major frontier in space science and technology development. The so-called "Space Highway" concept and subsequent achievement has brought us yet another historical opportunity in the design, produce, operate and managing the future generations of space vehicles. Meanwhile, how those technologies are developed, utilized and managed in the next few decades are of paramount important in assuring overall safety and mission success for many years ahead. Clearly, we have choices to make. Choices involving develop and apply less than matured or less experienced frontier technologies, which undoubtedly contain high risks. Not just the political, societal and human risks as discussed earlier, but involving the overall intelligent management of technological risks as well. Both the costs and the potential benefits to humans out of the space exploration endeavor will be huge. Therefore, how to manage technology risks and trying to strike a balance between these intricate, yet interrelated cost-benefit tradeoffs has become one of the most difficult challenges for human beings to overcome, especially in the new age of space movement. In probing such complex issues as technology risk assessment and management for systems and human safety in the space frontier requires us to look back and learn from successes and disasters of human space activities in the past. The safety management approach (SMS) is: (a) Prevention by design; (b) Management of risk; (c) Mitigation of consequences. Preventing accident, reducing disastrous consequences and ensure human safety under limited resources have always been in the center stage of space technologies development in the manned space flight programs. More often than not, there are echoes and strong patterns exist among all the past human tragedies from development and operation of complex technology of human-machine systems, regardless of whether they are in the space, transportation, nuclear, defense, marine or petrochemical industries. We know afterwards why we failed.

In the technology endeavor of new frontier, we observe risky outcomes all the time, and the more spectacular or costly ones are heavily reported in the media. The names of some of these are so famous that we almost do not need anything more as a descriptor to provoke our thoughts and fears. Just their utterance conjures an instant memory, an image or a pre-conditioned emotional response. Recent major events are startling, quite complex and diverse, and made headlines and caused huge inquiries. Are all the outcomes (events, disasters and accidents) in some way related? Like all other accidents, could they have been prevented beforehand instead of just understood afterwards? Are they just actually expected events, the standard outcomes in the ongoing technological dance with humans? Should we expect them not to occur in the future? How does the pervasive role of the human affect the outcomes? On the other hand, are these outcomes just the usual risk of our being around today, as depicted by Charles Perrow in his "Normal Accident" theory[1]? Could we manage such technology risk somehow? Is there really an accepted risk? Would a safety management system or organizational mechanism, as suggested by La Porte in his HRO (High Reliability Organizations) theory[2] prevent them? What is the chance of another such event? How can we track these risks so we could improve safety? What should we expect in the future?

As we carefully observe and examine these accidents and major disasters, in each case we understood the causal factors afterwards. We pointed the fingers of derision and blame, but neglected and forget the underlying reasons. They were all avoidable, if we had been smart enough to see the warning signs, better manage the situation, and to predict and reduce the risk. In addition, these accidents were in and to our most advanced technologies. To the finest designs we had produced, in major and highly visible applications, often in full view of modern cameras and news stations. Their names provide strong echoes that still sound in the industries in which they happened; of designs abandoned, staff discredited, managers removed, inquires conducted, victims mourned, reports produced, and then fading and forgotten except by those impacted or directly affected by the outcomes. Yet they were all preventable: after the fact, we knew or found out what has happened. A combination of circumstances occur, unique but unforeseen, combined with very human misunderstandings, mistakes and misperceptions that lead to the observed outcome. There are thousands of everyday auto crashes, fires, floods, ship sinking, explosions, collisions, oil spills, chemical leaks, medical malpractices, financial losses, train derailments, falls and industrial accidents. Some causing loss of life, some not. Some makes headlines for a day, some not. But none predicted beforehand. There are so many we forget them; they become part of the background noise of our technological society. They are catalogued, recorded, investigated, reported and forgotten. We assign liability, blame and damages, and the older events fade from our lives and the headlines. We move on. We hope that we have learned something from the outcome, and that it "cannot happen again". And if we are thorough or prudent, then it will not. Particularly, if we remove whatever it was from mis-operating again in the same way, but making a local change so that particular outcome cannot physically happen, or changing out the management and the procedures. Evidently, we cannot only focus our resources on the very physical cause of the Columbia accident, and we should rectify the root causes of organizational and human failures in a systematic, integrated and overall balanced approach. Otherwise, it is likely that should the accident strike us again next time, it will surprise us with a very different physical cause, a cause may not be the launch debris falling off the ET (external tank) again, but a cause may well rooted in the same human environment that harbors the making of the previous disaster!

Despite our best efforts, another similar outcome, an event or error, will occur sometime, somewhere, in a random but at the same time a systematic way. It will happen because humans and other circumstances will conspire in another unforeseen way to make it happen; with different consequences, in a different place, and at a different time. We observe the historic frequency of such events, but we really need to estimate their future probability of occurrence, the future risk. These echoes of common accident phenomena, the outcomes and their recurrences, raise major questions about our institutions, about our society and our way of handling the use of technology, and about their causes, and the responsible and best way forward. Each can or has significantly paralyzed the programs, like the case of NASA's SSP (space shuttle program), as well as the severely impacted industry like the case of commercial nuclear power, where these accidents occurred, causing inquiries, recriminations, costs, litigation, regulations and most regrettably, the devastation and down hill of the pertinent industry in general.

But do these events share a common basis? Should we have expected, or even better anticipated them? Are we truly learning from our mistakes like these and all others that just "happened"? What should or can we do to ensure these crippling losses are not in vain? How can we predict and thus prevent the next outcomes? Or should we continue to analyze them the way as usual, which has been outcome by outcome and event by event without truly understanding the systemic and deep reasons why we have these echoes of accidents? We must understand that these events (and hence most others) show, demonstrate and contain the identical causal factors of human failures and failings, and as a result, the same general and logical development path. It is quite clear that although the direct physical causes of the space shuttle Challenger disaster was different from that of the Columbia accident, however these two tragedies are intimately related to one another, in which the same type of human errors, organizational and communication failures existed throughout the history of the SSP that ultimately caused the tragic fate of the two out of five orbiters ever built. Without elaborating on the needed changes of technical approaches and methodologies for how the Space Shuttle safety risk is assessed and managed, one thing we must make ourselves absolutely clear: The underlying SoS (system of systems) architecture in the SSP, is quite an unhealthy one if not at all flawed, and we must fix it before another accident strikes us. A simple criteria to assess whether or not a complex technological system risk is well handled or miss managed in the course of human space endeavors (regardless of its failures in design, planning, engineering, operations or program management), is simply to observe the frequency of accidents and their major consequences, and to monitor accident sequence precursors (ASP), and near misses with rigor. A record of two fatal crashes out of a little over a hundred total shuttle missions is obviously unacceptable not only to our technologists, but unacceptable to our society and the general public as well. The probability of failure is now known to be about one in fifty for any and all launches. Yes, we may be able to tolerate some level of acceptable risk and taking on a considerable dose of inherent technology risks in any space exploration and developmental activities, but certainly not the risk levels as experienced by the SSP; a roughly 1.80E-2 fatal accident frequency, which is a whopping 6 orders of magnitude higher than the combined airline fatal accident risk (i.e., 1.9E-8) throughout the entire history of commercial flight industry, which in itself a frontier technology at the time barely half a century ago. Part of that risk change simply from the greater experience (numbers of flights) that have been achieved in commercial aviation: but if we must have a million space missions to attain that same experience level, we certainly cannot have 10,000 more accidents in space while we do so!

A new frontier in technical risk assessment and management in space activities is the understanding of complexity of the combination of events and their physical and phenomenological dynamics in the processes of accident occurrence. In simple words, if there is anything we know, it is that we do not really know anything at all. Our conventional understanding of reality is based on approximate physical "laws" that describe how the universe we observe behaves. Particularly, the dynamics of human elements and the complexities of human mind, when coupled with complex technological systems that have been created by that same mind, produce both outcomes we expect (results and/or products) and some that we do not (accidents and/or errors). Since one cannot expect to describe exactly all that happens, and since we only understand the causes afterwards, reactively assigning a posteriori frequencies, any ability to proactively predict the probability of outcomes a priori must be based on a testable theory that works. This is true for all the accidents that surround us, because of the overwhelming contribution of human error to accidents and events with modern technological systems. The human error events and the failures of human understanding and communications are what cause them that are the common human elements in the echoes of accidents. But when faced with an error, major or minor, humans always first deny, then blameshift, before accepting it as their very own. It is a natural survival instinct; it is part of living and our self-esteem. We do so as individuals, and seemingly also as part of our collective societies. Our mistakes are embedded or intertwined as part of a larger "system", be it a technology, a corporation, a mode of travel, a project decision, a rule or regulation, or an individual action or responsibility. They arise as and from an unforeseen combination of events that we only understood afterwards. Their consequences of events can be large or small, but all are part of a larger picture of which we humans are the vital contributor, invariably in some unforeseen way that was only obvious afterwards.

Again, the key issue in space technology safety and mission assurance lies within the thorough understanding of the dynamics of human elements in the risk and safety paradox. The first problem is the human involvement, subtle, all pervading, unrecordable and somewhat unpredictable. We cannot creep inside the human brain; the most complex technological systems ever existed of all, to see what drives it: instead we observe externally what happens internally as a result of stimuli, thinking, actions and decisions that are made inside the human technological machine as it interacts with the situation and the underlying technology. Conventionally, reliability engineering is associated with the failure rate of components, or systems, or mechanisms, not human beings in and interacting with a technological system. In other words, it is the interactive complexity between man and the technology system that we humans have yet to pay nearly enough attention to understand it and resolving it. Therefore, conventional reliability engineering based system safety methodologies fall short in a significant way to handle with issues of human dynamics, as well as the social environment within which the various human factors exist and fatal errors occur. And we now know that this conventional wisdom must change, we must redirect our focus more on the reliability, credibility and stability of human elements within the large environment of technology systems. As computer and software control techniques continue to dominant the design, fabrication of modern technological systems like the CEV (crew exploration vehicle) of NASA's next generation spacecraft, the human element becomes extremely crucial for us to attacking the increasingly complex puzzle of risk and system safety management. All events (accidents, errors and misshapes) are manifested as observed outcomes that may have multiple root causes, event sequences, initiators, system interactions, or contributory factors. It is the involvement of human that causes the outcomes we observe and which determines the intervals: rarely is it simply failure of technology itself. Figure 3.1 provide some strong evidence that the human error induced fatal accidents are dominant ones among all other causes in the global airline industry. The Federal Aviation Administration (FAA) has therefore noted:

"We need to change one of the biggest historical characteristics of safety improvements – our reactive nature. We must get in front of accidents, anticipate them and use hard data to detect problems and the disturbing trends".

This is a very clear message to us all, and to be proactive requires new theory, innovative methodology and understanding of the stochastic nature of error causation and prevention. Simply put, we need to act before the next accident strike us, and we must find accident potentials before they find us! Our fundamental work on the analysis of recorded events in major technological industries shows that when humans are involved, as they are in most activities of space flight:[3]

* a bath tub-shaped Leaning Curve (ULC) exists which we must use to track trends;
* in technological systems, accidents are primarily random, hence are hard to predict;
* accidents have a common minimum attainable and irreducible value in its frequency
* apparent disparate events in systems share the common basis of human errors.

This is demonstrated again, in the case of the two space shuttle disasters. Embarrassingly, unexpected and in full public view, the US humbled itself and its technology, first, launching on a too-cold day, and then, attempting re-entry with a damaged vehicle, a risk of debris falling during launch and causing orbiter damages, existed for years. In the Columbia inquiry, it was stated that there were "echoes" of the previous Challenger disaster present. Echoes that the key contributors were the same, the safety practices, internal management structures and decisions, institutional barriers and issues, and lack of learning from previous mistakes and events. Therefore, if without innovative safety technologies and the comprehensive risk management practices based on the in-depth understanding of human elements, if we continue the way of conducting business as usual, and identifying problems and making necessary changes only after major accidents, we are doomed to rediscover event-by-event, case-by-case, inquiry-by-inquiry, accident-by-accident, error-by-error, tragedy-by-tragedy, the same litany of causative factors, the same mistakes, the same errors.

See Figure 30.5. World-wide fatal accident causes by category in color section.

Perhaps not in the same order, or have the same importance, or the same type, but the same nevertheless. So the preventable becomes the unforeseen, the avoidable becomes inevitable, and the unthinkable, the observed, time and time again. We may call this deadly cycle of reactively and passively dealing with technology risks, the "death trap paradigm" (DTP) in the context of system safety and risk management, of which we need to do everything necessary to avoid stepping into one.

Managing Programmatic Risk in Strategic Decisions

The Space Shuttles are considered by many to be the most complex space vehicles ever designed, operated, and managed by NASA. Yet the shuttle will be retired soon, to be replaced by even more complex space vehicles and mission systems: NASA's ambitious crew exploration vehicle (CEV) will be the most complex the world has ever seen. The issue of enterprise risk management has thus become increasingly important, and the focus is now on programmatic risk management for strategic resources, and for adequate control in cost and schedule on long-term, complex, highly sensitive, or capital-intensive national or international programs. The key to enduring success of large and complex space programs, like NASA's constellation program (manned lunar mission and beyond) not only lie in the sound management of technology risks as previously discussed, but in the appropriate assessment and optimal management of programmatic risk as well. It is therefore equally important in utilizing the Programmatic Risk Assessment and Management (so called the PgRAM) techniques and tools to assess risks and their tradeoffs between all available alternatives when large program decisions of high strategic and national importance are at stake.

The history of human space programs might have taught us enough lessons in this regard, and might have reminded us to ask such a harsh question: What would the US space frontier activities have looked like now had we paid enough attention on this very issue of risk tradeoffs some 35 years ago? It may not be so hard for us to comprehend today, that, if we had adequately utilized such scientific methodology like PgRAM back in the late 60s, when serious national and strategic decisions had to be made on the fate of Apollo program in lieu of the space shuttle program, we perhaps, could have avoided a whole spectrum of serious enterprise problems.

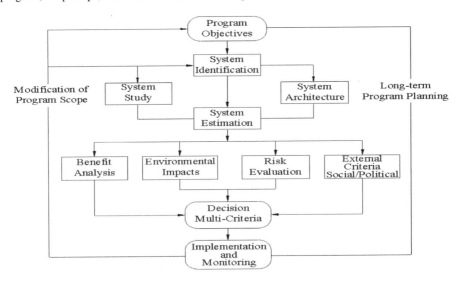

Figure 30.6. A view of risk scenario in long-term program planning

These include major setbacks later on the space shuttle program, such as the severe overrun of costs, the miss-achieved design objectives, the safety and financial disasters, the prolonged unavailability of American access to the ISS (International Space Station) and most of all, the loss of time and wasting of national resources, including the loss of most invaluable asset - the program memory, the human and corporate capital of expertise and experiences. Figure 30.6 shows a simplified view of risks in long-term program planning. As indicated in the simple process for laying out risk areas in various aspects of program assessment, benefit analysis and environmental impacts are the important elements to be considered for long-term program risks. Again, if NASA did enough serious homework on this area in the past few decades, we perhaps could have avoided several expensive investments on a number of space vehicle programs that were either short-lived or wasted, like the X33, X34 and X38 projects etc. We might have also avoided wasting money and time on switching between launch vehicle R&D efforts on the 2nd-Gen SLI (Space Launch Initiative on 2nd generation launch vehicle), CTV (crew transfer vehicle) and the OSP (orbiter space plane) etc. even though they all appeared attractive at the time.

As systems engineering and project management take on the challenge of fulfilling the project and mission requirements, areas with high programmatic risk will become the focus of program management attention to insure appropriate visibility and effectiveness in how resources are allocated. The major objective is to eliminate, as early as possible, those project uncertainties that can result in unexpected growth to cost and schedule. We must distinguish programmatic risk, which is closely associated with a project's budget, schedule, and performance requirement; it should not to be confused with the Technological risks to the safety of workers, general public and engineered systems within the confinement of the systems of the entire technology environment. As an example, take the case in the enterprise risk management of NASA's national-based Space Shuttle Program (SSP) in coupling with the International-based Space Station Program (ISS). A set of programmatic high risk scenarios could have been identified, analyzed and well anticipated with optimal and well prepared strategic responses, back in the early days when strategic program decisions, national and/or international enterprise architectures were made and determined. At the very least, if we were making more intelligent program decisions at the enterprise level with the help of such technologies as probabilistic risk assessment (PRA) and PgRAM, we could have avoided: i) the very wasteful premature shutdown of the entire space shuttle program, ii) getting ourselves into the most difficult situation for having to buy American access to the LEO (low earth orbit) in order to complete the ISS assembly, and iii) the forced cutting backs from intended scope of ISS design. These all caused loss of credibility on international partnerships and incurred unexpected waste of societal investment, in terms of curtailed ISS program, which reduces its scientific mission value. Most importantly, we could have accumulated highly desired design and operating experiences over the many years with much needed technical data on the space vehicle systems, which would have been not only pertinent to our present needs for the design and operation of long duration deep-space crew exploration vehicles. But also we could perhaps, have made the space shuttle vehicle system the technologically and economically most viable candidate to carry on forward, on the continued course in human exploration in our solar system, ferrying mankind to Mars and much beyond!

The programmatic risk approach was primarily derived from the PRA techniques used for system safety and mission assurance studies. It was particularly developed in response to failures of traditional analysis methods (like FMEA and QMS) in the planning and undertaking of complex projects and programs. In the past, poor projections of future program performance were not caused by lack of care but were the results of unrealistic analysis assumptions and ad hoc treatment of uncertainties in conventional techniques. A key contributor is that the performer, often a Program Office that may have substantial biased interests engaged solely to show program success, prepares the projections. Achievement of programmatic milestones then takes precedence over strategic program effectiveness. Therefore, it is extremely important that during the process of making strategic program decisions, an external "out of the box" independent viewpoint is an essential component of the PgRAM process. This oversight must not just be advisory: it must set directions and make effective recommendations and take efficient actions. Rigorous quantitative risk assessment and management processes at the enterprise level can truly provide an analytical framework for assessing and applying scarce resources to the management, resolution, elimination, and disposition of the highest priority program risks. It aids in identifying potential risks and quantifies the impact of each one on a program or an organization within the underlying enterprise. The process also aids programs and organizations in balancing their competing risks and optimizing the use of scarce budgetary or human capital recourses in order to eliminate much of the uncertainty from the vital decision-making process. The history of past high valued programs has repeatedly shown that performing enterprise and programmatic risk assessment and management is not only necessary for strategic decision-making. It is also an ideal vehicle for system modeling of real world problems by systematically incorporating uncertainty for each of the subtle dependencies among these areas into formal model. Programmatic risk assessment approach differs significantly from conventional deterministic analysis in its explicit inclusion of *uncertainties* in the quantification of predicted program outcomes. Decision-makers have always recognized the existence of uncertainty about the information available to them and have had to learn to deal with uncertainty intuitively in the past. However, a sound PgRAM process significantly improves the decision-makers' state of knowledge and provides a more *objective* basis for decisions. We may also deal with unknown risks (the "unknown unknowns") utilizing subjective probabilities that can bound the outcomes, and enable relative judgments on risk importance. The most important aspects of the programmatic risk management are its ability to:

* account for various kinds of enterprise or program activity constraints, such as dependencies, intra-activities and inter-activity dependencies (much like the intricate dependencies between projects across SSP and ISS programs of NASA),
* effects of external conditions (e.g., impact of government regulations, international treaties and political environment, much like the multi-national dependencies of

projects within the ISS international partnership framework), and also
* uncertainty in the input parameters, and the set of possible complications that may occur.

Such is the case that the Russians are now asking NASA to pay for the rides to allow the ISS access because there left no options for the US. So, we should have assessed the risk scenario of having no other options but to paying for access to the LEO should the continued operation of the Space Shuttle became an issue. The result might have served to the better interests of US or all other ISS partners to including another international partner in the enterprise of such joint multi-national ISS program, who might posses, the capability of manned access to the ISS at a much lower cost. In short, a sound programmatic risk management unveils the true risk of programs so that decision-makers can make informed decision about potential alternative course of action. Figure 30.7 provides a typical flowchart of risk assessment for strategic program or enterprise decision-making process. Details of such technique will be discussed in a following book.

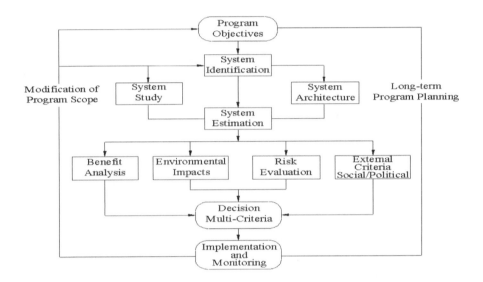

Figure 30.7. A Schematic of Programmatic Risk Assessment Approach

What It Takes to Achieve Safety and Mission Success?

With the rapid development and advancement of technologies in recent years in the system safety and assurance engineering field, as largely led by the commercial nuclear power field[4~6], mankind has achieved a great deal to assure safety and system reliability in large and complex technological systems and space program operations, despite of the setbacks and major accidents. The last few decades, in particular, have seen increasingly widespread use of PRA and management techniques (SMS) as aids in making complex technical and programmatic decisions on major space initiatives. The current "constellation program" of NASA and its rigorous assessment of CEV, CLV system architecture and design concept with respect to crew safety and vehicle reliability, culminates such effort in the right direction.

However, despite the progress that has been made in risk science and safety technology, many unresolved issues remain. There still exist numerous examples of risk-based decisions and conclusions that have caused great controversy. In particular, there is a great deal of debate in the space community surrounding the benefits PRA and of management in decision making process; the role of values and ethics and other social-political factors to safety and risk; the efficacy of quantitative versus qualitative analysis; and the role of uncertainty and incomplete information.[7~8] In addition, these debates become even more controversial around issues on how we look ahead and move forward on tough challenges in the new age of commercial space flight for safety assurance and mission success. What are the most promising solutions for system safety? Is it merely a technical issue that relies only on safety technologies, optimal design techniques, operational and organizational approaches? Or is it simply any political or organizational solutions that could help in achieving a maximum safety goal, yet still allow commercial space travel affordable to the general public? How do we regulate the imminent boom of the incoming space flight industry, the emergence of space entrepreneurships and the space technology revolution? Is there truly an acceptable risk metric for human safety in commercial space travel or in manned exploration activities? While some implications of insights to better understand of these questions are discussed in previous sections, it is the intent of the authors however, in the reminder of this chapter, to attempt summarizing and sharing with readers some intriguing thoughts, which may shade some lights on these very issues of uttermost importance in our time.

We summarize, in the following five major areas that mankind must pay great deal attention in the years ahead, in order to unveil the paradox of human safety and risk taking, and ultimately to achieve the safe and affordable undertake of human space activities, and making commercial space travel a safe and near-future reality for us all.

the compounding factors of human dynamics – a centerpiece in safety
system-based concept and systems architecture analysis – the foundation
smart decisions with Integrated Risk Management framework
design-based safety philosophy – reduce & eliminate accident by design
gaining insights by continuous R&D in accident theory development

These five aspects of high technical and programmatic importance to safety and assurance engineering activities, can be easily depicted in Figure 30.8, thus graphically defined as "the 5-block foundation" which serves as a R&D reference or principle activity guide for safety and mission assurance (SMA) programs. We now elaborate briefly on each of the five areas to further explain the main issue of concerns, which must be addressed to achieve ultimate safety and mission successes of any complex socio-technological systems in space programs, including any capital intensive engineering systems with high human safety risks in other industries. Detailed discussion on the "5-block SMA foundation" will be provided in a separate chapter.

Figure 30.8. The 5-block foundation in Safety & Mission Assurance Activities

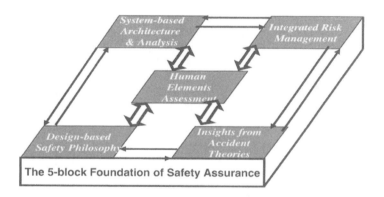

The Factors of Human Dynamics – A Centerpiece in Safety Assurance

The assessment of complex and compounding human behavior and factors are the dynamic elements in space programs, and must always serve as the centerpiece in assurance engineering activities for human safety. Although notable progress has been made in this area, more research and development (R&D) efforts are needed to meet the new challenges of the complex semi-automated homo-technological systems of today, and they must be focused on the very dynamics of the human element within the safety and risk equation. We need to fully understand, through these efforts, the profound impact of all the human factors, the socio-political and organizational influences on human safety, risk and risk perceptions within the context of human systems engineering. Adequate adaptation and implementation of safety requirements, standards and organizational policies focused on the human aspects rather than on just the mechanical systems under appropriate and enforceable regulatory framework are of uttermost important to our future success.

System-based Concept – Architecture in Systems Analysis

In the past, there have been generally little or no attentions devoted to the application of system-based view or SoS (system of systems) architecture in the process for safety and risk assessment and management activities. It is therefore crucial that we make use of systems theory and system analysis tools and methodology to safety and mission assurance assessment, rather than focusing resources only on isolated reliability requirements or hazard assessment effort focused solely on critical components and systems. Utilizing concepts from systems theory, one can approach risk assessment that actually begins with a system identification and understanding of system architecture, a step that must be carried out by the risk managers and program decision makers.

Smart Decisions by Integrated Risk Management – A "Double-T" Concept

New challenges in SMA activity require a stronger focus on utilization of innovative risk management process and practices. In the aftermath of the Challenger accident in middle 80s, and especially after the recent loss of Columbia, efforts were expanded profoundly in space agencies like NASA, on applying quantitative risk techniques such as PRA. The conduct of the present space shuttle PRA is one culmination of these efforts[9~10] although it simply highlights the present known unacceptably high levels of risk. Nonetheless, as pointed out by the CAIB report[11~12], a key critical weak-link exposed by NASA's safety & risk management practice has been the lack of a comprehensive and integrated SMA methodology. In an attempt to address such broad SMA issues from the technical risk management perspective, a SMA management framework based on integrated risk assessment process has been developed. The centerpiece of this methodology, as illustrated in Figure 30.9, is the "Triple-triplet" concept of an integrated risk management process. It largely extends the risk assessment triplet[7] to include and combine all aspects of SMA elements within a systems engineering framework. With the Triple-triplet (Double-T) integrated risk management framework, best risk tradeoff decisions can be achieved with risk insights not only from PRA but also from engineering and system safety activities all integrated into a system-based single and complete process. The Double-T conceptual framework for integrated risk management is only briefly illustrated here with the expectation that the interested readers may pursue further on technical details as cited in references[13].

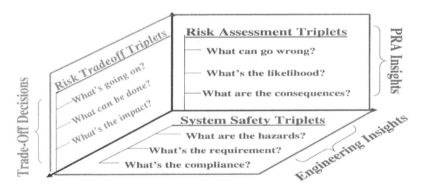

Figure 30.9. A Triple-triplet (TT) Conceptual Framework for Integrated SMA

Risk Assessment Triplet – Gaining PRA Insights:
    (1) What can go wrong?
    (2) What's the likelihood?
    (3) What are the consequences?

System Safety Triplet – Gaining Engineering Insights:
    (1) What are the hazards?
    (2) What's the requirement?
    (3) What's the compliance?

Risk Tradeoff Triplet – Making Tradeoff Decisions:
    (1) What's going on?
    (2) What can be done?
    (3) What's the impact?

Design-based Safety Philosophy – Eliminate accidents by design

The best engineering effort for system safety starts with the superior design philosophy, or the design concept selection process. Using both deterministic and probabilistic safety analysis tools for system architecture and concept studies in early design stages is critical in system safety assurance. Clearly, if a design is pregnant with a potential for disaster (a latent error, for example), the chances are accident will happen sooner or late regardless of the best engineering effort later in the program. Such as the launch debris-shedding problem rooted in the external tank design that eventually caused the loss of space shuttle Columbia. Another interesting example occurs also in the space shuttle design, where the concept of combining a cargo ship and crew vehicle into a tightly-coupled gigantic space transport truck is highly controversial with respect to safety risk over cost-benefit ratio. This likely flaw in concept selection philosophy might have been a contributing factor to the doomed fate of the entire program. Another design philosophy has to do with the consideration of variations and unknowns and to what extent they are considered. Should the design be deterministic, or probabilistic, or both? Also, criteria are fundamental to design, as well the philosophy implied within the paradox of safety and risk, or cost and reliability. One of the critical design philosophies is the essential role that failures play, thus leading us down the Universal Learning Curve. Petroski[14] has expounded this philosophy and stated that the design process must consider in depth all potential failure modes. This is true even when the new design is a moderate scaling up of a current successful system where the slightest design change can introduce effects that lead to catastrophe totally unexpected. In recent years, use of PRA in design analysis in space programs has gained widespread acceptance because of its probabilistic treatment of uncertainty is more robust than that of the deterministic based safety factor techniques, such as the use of worst case loading and 3 (or more!) sigma safety margin assignment etc., as have always been in the past. To help strike that balance between the desiderata and the actual design, the following ten most meaningful design philosophies or principles for accident avoidance and system safety are summarized below to share with our readers:

(1) Inherent simplicity and flexibility in system design often implies safety and lowest cost, and is perhaps the best design approach for safety, cost effectiveness and mission success.

(2) Prevention of accidents using defense-in-depth system design, such as use of multi-phase and multi-layered protection mechanisms and structures whenever possible.

(3) Preclusion of high consequences due to human error or reduced potential for any human induced fatal events by design, and by thorough understanding of the dynamics of human element, human factors and using sound design practices.

(4) Extensive use of "fail safe" and/or "fail operational" philosophy to guard against fatal consequence in case of total system failure.

(5) Mitigation of consequences by diversity (addressing common cause failures - CCF[15]), and avoiding tight-coupling of interface design, but use loose-coupling, or decoupled and a modularized design philosophy to isolate, contain accident propagation whenever possible.

(6) Use redundancy and always minimize the potentials of critical points of failure whenever necessary to guard against random failures or failures due to wear, fatigue or aging, but only if common cause failures is an unlikely threat.

(7) Use of passive-design safety concept in total elimination of accident or severe consequence potentials. Such as the advanced nuclear reactor design that the design precludes the possibility of core melt in case of total system failure.

(8) Use of probabilistic-based design philosophy for concept scrutiny and setting safety targets, including whenever necessary the treatment of uncertainties, as the best design practice for eliminating unknowns, and minimizing the number of variables with uncertainties as much as possible.

(9) System-based and systems engineering (as opposed to a component based) view in design concept assessment, selecting and analyzing integrated designs within a framework of system-of-systems (SoS) to avoid designs resulted from isolated view that only focused on technology side.

(10) Define the boundaries and cliff edges of the design as much as possible. And always believe in the philosophy of robust design, and choose conservatism (additional safety margins, features or systems) if it is your only option other than staying near or on the edge.

Probabilistically, we know that we may design systems to 10E-3 to 10E-6 failures over demand (reliability), by using redundancy, diversity and simplicity of inherent safety features. The dominant failure modes then always become:

a) Common mode failures – a unifying failure inherent in the design that disables many systems at once and hence can fail the whole, such as a fuel tank seal failure, power system failure, thermal tile damage or an oxygen atmosphere explosion

b) Homo-technological error – a failure of the system design that enables bypassing the safety design intent by humans avoiding, allowing, overriding, ignoring, denying, misusing, misunderstanding or misinterpreting, such as using incorrect avoidance procedures in mid air collisions, or not knowing about tile damage effects on re-entry.

Gaining Insights by Continues R&D in Accident Theories

One of the most important aspects in assuring safety and preventing catastrophic events in space or any other high risk programs is to understand and learn theoretic basis and insights about why accident occur, and how it propagates. Although there has been long debates within the safety science and technology community on issues concerning (a) why catastrophic events always happen regardless of how rigorously human beings have tried to prevent them, and (b) how much engineering effort and human resources have to be allocated to such effort. However, there still exist a great deal of confusion on this very issue among the managers and decision-makers of many high-value engineering communities. For instance, a considerable number of engineers and managers within NASA community are still skeptical about why rare event happens even with low likelihood (For the answer, see Figure 30.2). And why space shuttle orbiter safety continues to tumble whereas extraordinary amount of money and time already spent on FMEAs, hazard reviews, hardware tests, and risk and reliability analysis? Undoubtedly, only after thorough understanding of fundamental theoretic basis of the prior knowledge of how fatal accidents are increasingly happening nowadays, could we then gain critical insights from these theories, and effectively prevent and even predict future events by developing the right techniques and making right safety and risk decisions.

One of the key contributions made by Perrow in his Normal Accident Theory[1] (NAT) was the identification of the risk-contributing system characteristics, which offer significant implications on how we ought to design and operate of complex socio-technical systems. It pointed out that the combination of high complexity and tight coupling must lead to failures. However, the notion implied in this theory that we cannot assure the safety of complex technical system designs is not only a pessimistic proposition but quite disturbing to design engineers and those who operate and manage such systems. (The bathtub model and Learning Hypothesis gets one out of this conundrum). A major flaw however, in the NAT argument is that the only engineering solution he considers to improve safety is use of redundancy, despite of the paradox that redundancy introduces additional complexity and renders higher safety risk. It also ignores the human

contribution, which is unchanging. Others have argued that the safety of every engineering design is a hypothesis only to be tested during commissioning and/or operations. The issue is that not all complex systems could be fully tested; such as exist in the spacecrafts and weightlessness. The Apollo 13 near fatal accident might have been averted had they tested what would happen if two oxygen tanks would be lost. No one believed that a thermo switch failure inside the O_2 tank, induced by a pre-flight ground test was even a remote possibility for the incident. In spite of the pessimistic views contained in NAT, can we somehow organize ourselves very effectively to manage avoiding accident then? Another school of thoughts of accident theorists gave a very positive answer to this question, as represented by K. Roberts' work[2] known as the High Reliability Organization (HRO) theory, which argues that we can achieve reliability with appropriate attention to organizational management. According to HRO theory, it is not really clear that all high-risk technologies will fail. The Human Bathtub in Figure 30.10 explains this difference clearly - failure is a probability that varies with experience, and hence we can learn and reduce the rate faster by having a learning Organization that then is labeled a HRO. In contrast to NAT, HRO theory counters Perrow's hypothesis by suggesting that some interactively complex and tightly coupled systems operate with very few accidents while under highly organized "high reliability" institutions. This is consistent with the presence of a Universal Learning Curve. A most noted problem with this theory is that the systems studied by HRO researchers are neither interactively complex nor tightly coupled systems. Furthermore, the theory of HRO implies that, as far as high reliability is to be achieved in every aspect within a well-managed organization, it will no longer have accidents and therefore safety is warranted. The problem here is clearly the confusion between reliability and safety, since high reliability does not necessarily mean safety, and accident can still happen on a perfectly reliable system, and highly safe systems are not necessarily reliable. In the loss of Mars Polar Lander case of JPL in 2001, everything performed reliably as designed, but the Lander crashed into Mars during entry simply due to a design flaw in which the designer failed to account for all the complex interactions between the leg deployment and the control software. Another example could be when there is need for system operators, say airline pilot, sometimes to break the rules on predefined flight procedures in order to prevent accident, like a fatal collision during landing by averting the plane to a different runway or a nearby field. Conversely, in the case of the aircraft midair collision over Switzerland, the (human) ground controller instructed one of the planes to adopt a collision course, which was followed despite warnings to the aircrew from the collision avoidance system. Again, if the three Apollo 13 astronauts followed all the crew procedures provided to them in their training for being "reliably" performed what was documented prior to mission, they would never had any chance of surviving. We must distinguish between Reliability, Safety and Risk, which are different qualities and should not be confused. In fact, these different qualities often conflict. Increasing reliability may decrease safety or incurs more system risk and likewise, increase safety may just decrease reliability or reduces system risk. A significant challenge of engineering that deals with this issue is to find ways to increase safety without decreasing reliability. We have seen managers and lead engineers of high risk systems, who are so obsessed with focusing much of their daily effort on reliability estimates or predictions, and often forgot about "what is going on" by looking at the "big-picture" conditions of their systems - a critical insight as being one of the "triplet" element introduced within the "Double-T" concept.

The Human Bathtub

$$p(\hat{a}) = 1 - \exp\{(\ddot{e} - \ddot{e}_m)/ k - \ddot{e}(\hat{a}_0 - \hat{a})\}$$

Probability of an organizational failure

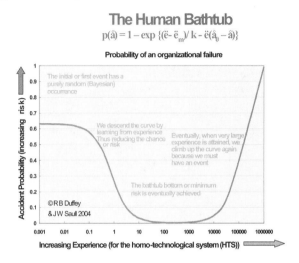

Figure 30.10. The Human Bathtub – Probability of an Organizational Failure

Obviously, both theories of NAT and HRO although provided very important insights on how we should deal with high-risk technical systems in order to achieve safety, yet they fail to offer better solutions which would complement the shortfalls existed in each of the theories. Unarguably, complex socio-technological systems not only need more sophisticated accident theories and approaches to resolve safety and reliability issues. They also demand continued R&D effort by mankind in the theoretic frontier, especially for safety challenges on the transscientific, non-random, technical, and organizational factors involved in accidents. Recently, there have been strong advocates of a system-based approach to safety[16], of which a top-down systems view is proposed in favor of the bottom-up, reliability engineering focused standard engineering approaches. In such accident theory, the traditional conception of accidents based on event sequence or chain-of-events of directly involved failures and human errors is abandoned. Chain-of-events models encourage limited notions of linear causality and cannot easily account for the none-coherent system events, such as the indirect, nonlinear and feedback interactions common for accidents in complex and dynamic systems often seen in the space industry. A systems view of understanding accident causation allows more complex relationships between events and also provides a way to look more deeply at why the events occurred. One of the weak points with the systems approach to safety, as advocated lately[17] is that it only emphasizes the top-down systems view to understanding and modeling accidents, which is mainly based on the deductive logic process that allows us a big-picture view within a broader system context. However, it ignores or underestimates the effectiveness or useful insights which can also be offered by the bottom-up inductive-based logic processes in accident modeling practice, such as the FMEA, CILs, Event tree which have long been widely used in the aerospace community. So once again a balance is required. In particular, the identification of critical (single) failure points was predominantly accomplished by these labor intensive, detailed engineering-based qualitative safety risk management infrastructure in NASA, and they are considered effective by many NASA engineers because they had served so well during the Apollo and early space shuttle era. As technologists experienced working in both fronts of theory and modeling practices in the safety risk filed, we have found that despite being strong advocates of systems-based theory and approaches, applying inductive-based modeling process can also help identify major accident scenarios, allowing us to better understand the inductive nature behind accident sequences. In light of these issues and others as pointed out in NAT and HRO accident theories, an alternative approach has been introduced, which calls for systems-based view as foundation but making use of combined risk insights from both top-down and bottom-up approaches for understanding and modeling complex accident process.[13] Such a System-based Dual-Process (SDP) approach for integrated risk modeling steams from the "Triple-triplet" (Double-T) concept, as discussed briefly in previous section, and the details of its theoretic basis is beyond the scope of this chapter. Interested readers should pursue further details in other references.[10,13] The significant benefits of the SDP approach to safety however, result from the combined insights gained from other accident theories like NAT, HRO and general systems approach. Several key elements of the SDP approach to safety and accident modeling, which can be summarized in these four aspects:

* Adopting a Systems-based view to accident that focuses on the socio- homo- technological system as a whole, and understanding and managing the interactive complexity by modeling the relationships between technical, organizational, human and social-political aspects.

* Understand and modeling accidents with "dual-process" logical thinking, from both Top-down deductive logic for overall system architecture and performance, combined with the complimentary Bottom-up inductive logic thinking, which is scenario-based focusing on accident scenario identification and reliability engineering aspects.

* Emphasis on the dynamics of human aspects for managing uncertainties in safety and reliability organizational infrastructure by taking full insights and advantages as offered by the HRO theory. Fully recognize that HRO does provide insights for improving human, and system reliability

* Applying System-based Integrated Risk Management Framework based on the Triple-triplet concept, and managing total safety risk by making decisions based on integrating all phases of a system life cycle: from design and concept selection using NAT risk insights to engineering, operation and program management.

Looking Ahead

History has brought mankind to the brink of an unprecedented era of opportunities for science and technology advancement in the space frontier. With the continued escalation of mankind's quest for unknowns outward, it is inevitable, as headed by ambitious US space endeavors that a whole new array of worldwide space sciences, commercial space flight industries and space commerce are emerging on the horizon. There are, of course, full of uncertainty and risks on the road ahead: uncertainties that might well beyond human comprehensions, and risks that almost certainly will cost enormous resources, and perhaps human lives as well. Nevertheless, we cannot be afraid of moving forward in the face of risk. Safety risk can never be zero: mankind will never migrate beyond earth if we fail to take risks on the space frontier – paying the probable price for what we gain. This is simply Darwin's law of nature, and wishing to benefit in space without human sacrifice is simply against the law of nature. We cannot imaging an answer otherwise! Human civilization as we know it today would not exist if we were afraid of risk taking. Taking risk is part of human spirit, a key attribute of mankind built into our genetic propensity ever since our very existence. To answer the question, can commercial space travel made safe yet affordable for us all? If we move one million people each year to another planet, or between continents in LEO, and the desire is for no accidents, the risk goal is much less than 10E-6 per year , and with, say, 100 passengers on each flight, it is a goal of less than 10E-4 per flight. This target has been achieved with commercial flights today, but not yet with present launch rockets and shuttles, which are still at the ~ 10E-2 level per launch. We have two orders of magnitude reduction in failure rates that are required to achieve the same comparative risk level. Can we make space exploration safe? Yes, absolutely but this doesn't mean accident will never happen again. Taking risk always come with high returns, yet always implies adverse consequences. We still need a technological advance or breakthrough, which itself carries risk. The benefit of tomorrow from human space endeavors today will never be fully understood, or anticipated, it is even beyond the wildest imagination of mankind capable at the current stage of civilizations.

The key is how we manage risk and dealing with uncertainty. We must realize managing and taking risk is a major part of our daily activities working in the space frontier. The right question is how we can take smart risks and taking it most intelligently? The most important issue here is uncertainty: technical, organizational and social. It is uncertainty that makes engineering difficult and challenging and sometimes unsuccessful. Deciding which outstanding problem ought be giving priority is a difficult problem in itself. Also, because many high-tech systems do use new technologies, understanding of the physical and social phenomena that may cause problems is often limited. Space and most high-tech systems have unresolved technical uncertainty. If it were necessary to resolve all uncertainty before use or operation of any spacecraft systems as politically perceived by many of us within the executive circle, most of high-risk space systems would need to be shut down and important functions provided in our society would come to a halt. Obviously, mankind cannot wait for complete understanding before endeavoring ahead to launch technically complex space systems.

The past few decades have seen increasingly widespread use of risk assessment and management techniques in making complex technical and strategic decisions. As safety technologists working in the risk management front, we must fulfill our vital responsibility to the human society. There is no doubt if we continue to advocate use of scientific-based methodology and gaining insights through continued research and development in the theoretical front, mankind will ultimately succeed in living beyond earth and ferrying into the cosmos with safety!

References

1. Perrow, C. Normal Accidents; Living with High Risk Technologies. New York: Basic.1984.
2. La Porte, Todd R. and Consolini, Paula. Working in Practice But Not in Theory: Theoretical Challenges of High Reliability Organizations. Journal of Public Administration Research and Theory, 1, 1991.
3. Duffey, R. B. and Saull, J. W., Manage the Risk: The human element, Butterworth-Heinemann, New York, 2002.
4. Apostolakis, G. E., Bickel, J. H., Kaplan, S., Editorial: Probabilistic Risk Assessment in the Nuclear Power Utility Industry, Reliability Engineering and System Safety, Vol. 24, No. 2, 1989.
5. W. E. Vesely, Editor, "Special Issues on Developments in Risk-Informed Decision Making for Nuclear Power Plants", Reliability Engineering and System Safety, Vol.63, No.3, 1999.
6. Frank, M. V., Probabilistic Risk Assessment in Aerospace: Evolution from the Nuclear Industry. Proc. PSAM5, Osaka Japan, Nov.2000.
7. Kaplan, S. & Garrick, B. J. On the Quantitative Definition of Risk. Risk Analysis Vol.1, No.1, 1981.
8. Azarm, M. A., Hsu, F., Role of Risk Assessment and Reliability Assurance Program for Space Vehicles, Brookhaven National Laboratory Technical Report for NASA Code Q, 1992.
9. Vesely, W. E., Hsu. F. et al., Performance of A Probabilistic Risk Assessment for the Space Shuttle, NASA Johnson Space Center Report, Jan., 2001.
10 Hsu, F., Railsback, J., The Space Shuttle Probabilistic Risk Assessment Framework – A Structured Multi-layer Multiphase Modeling Approach for Large and Complex systems, Proc., PSAM7, Vol.3, Berlin Germany, 2004.
11. CAIB, The Columbia Accident Investigation Board, report Vol-I, August, 2003.
12. Fragola, J. R., Space Shuttle Program Risk Management, Reliability Availability Maintainability Symposium (RAMS)96, Las Vegas, NV, January 1996.
13. Hsu, F., An Integrated Risk Management Framework – The Triple-Triplet Concept for Risk-informed SMA Management. NASA RMC, October, 2004.
14. Petroski, H. Design Paradigms, Case Histories of Error and Judgment in Engineering. Cambridge University Press, 1994.
15. Mosleh, A., Rasmusen, D. M. Guidelines on Modeling Common Cause Failures in Probabilistic Risk Assessment, NUREG/CR-5485, Nov. 98.
16. Hatfield, A. J. and Hipel, K. W., Risk and Systems Theory. International Journal, Risk Analysis, Vol. 22, No. 6, 2002.
17. Rasmussen, J. and Svedung, I. Proactive Risk Management in a Dynamic Society. Report of Swedish Rescue Services Agency, 2000.

About the Authors

Dr. Feng Hsu, Ph.D. (Email: Feng.Hsu@NASA.GOV, Tel: 301-286-3416) a US expert with over 20 years of experiences in the field of Risk Analysis, Safety and Mission Assurance (SMA) assessment for complex engineering systems. Formerly a staff research engineer at world renowned Brookhaven National Laboratory (BNL), Dr. Hsu has worked extensively on reliability, probabilistic risk assessment (PRA) and management theory and methodology research for nuclear and space launch systems since the 1980s. He then became Sr. staff engineer/scientist and joined NASA's SAIC team in the Space Shuttle and Exploration Analysis Department at Johnson Space Center in Houston since 2000. Dr. Hsu has been a lead engineering analyst and project manager working as technical expert in the space center on NASA' key program areas, such as PRA, SMA for the Space Shuttle, International Space Station as well as the Risk-informed design assessment for the new generation space launch & crew exploration vehicle (CEV/CLV) systems. Dr. Hsu is now a leading NASA engineer working on frontier space missions at GSFC, and he has over 60 publications, including Journal articles, NUREG/CRs, BNL and NASA technical reports. Besides being referees for several International Journals, Dr. Hsu is a member of technical committee of IEEE SMC, and has chaired technical sessions in various professional conferences. He has won numerous research and service awards from BNL, SAIC and NASA.

Romney Duffey Ph D., is the Principal Scientist with AECL (Canada). He is a leading expert in commercial nuclear reactor studies, is active in global environmental and energy studies and in advanced system design, and is currently leading work on advanced energy concepts. He has an extensive technology background, including energy, environment waste, safety, risk, simulation, physical modeling and uncertainty analysis. Romney is the author of the original text about errors in technology ("Know the Risk", Butterworth-Heinemann, 2002), and of more than 150 published technical papers and reports. He is the past Chair of the ASME's (American Society of Mechanical Engineers) Nuclear Engineering Division, former Chair of Nuclear Energy Department at Brookhaven National Laboratory (BNL), an active Member of the American and Canadian Nuclear Societies, and a past Chair of the American Nuclear Society's (ANS) Thermal Hydraulics Division. Recently elected a Fellow of ASME for his exceptional engineering achievements and contributions to the engineering profession.

Chapter 31

A Magnificent Challenge

By Richard E. Eckelkamp

We as a nation and as a world have been presented by the U.S President a magnificent new vision - a challenge in space exploration to return to the moon and continue to the beyond. [1] Positive visions and challenges are inherently good for a people, just as for individuals. "Without a vision, the people perish." [2] Without challenges, individuals and mankind stagnate, vegetate, and turn inward to pursue little of value.

This new space visionary challenge is quite significant and significantly difficult. What at first might seem almost blasé and not worth one's while, quickly becomes a massive and ambitious endeavor when the hood is opened for detailed observation. The challenge, however, is not so much financial as it is human and technical.

On the human front, our traditional space industry, both governmental and corporate, crawls inefficiently through present and toward future projects with a highly-viscous, heavy-organizational inertial, an inertia inherited from the "early days" projects like War World II aircraft, early missile programs, the atomic bomb project, and Apollo. In these projects the emphasis had been on quick decisive results with cost not being a primary driver. Humans were desperate for these items. "Decent" human existence was judged to be at stake.

Strangely enough, this time around with the President's challenge, "decent" human existence is again at stake. The enemy this time is our own internal failures to be flexible, to be adaptive to needed changes and innovations, and to employ common sense design and execution of cost-efficient and meaningful programs. As with any vision, the "meaningfulness" part is supra important and necessary. There must be a significant reason "why?" for a project in order for that project to take hold in the heart and mind of both individuals and society.

For the earlier U.S. projects mentioned, the "whys?" were grounded in fear. For this present exploration challenge there is no discernable, immanent fear. Worries over future meteoric collisions, solar burnout, environmental disaster, or some possible interplanetary alien invasion either generate only brief mental acknowledgement or, in the last case, minimal credence. Considerations that society's lack of clear purpose or direction towards substantive endeavors usually results in decay seldom enter the human consciousness, much less rise to be a driving force.

If not fear, then, faith and or hope seem logical alternate motivators. Subtracting out the fear factor from Apollo reveals that hope was also present in the hearts of mankind. We all wanted to make some positive progress that would be oriented towards improving the future of all on this planet, rather than bashing each other to bits in new, "exciting" ways. The spiritual side of us felt good during Apollo.

What about now? What can stir our hearts to support and work these new space endeavors? The answers so far are dribbling in a few pieces at a time, not yet enough to be able to spin one or more inspiring tapestries. Our special opportunity is to search for and provide some of the pieces. I, for one, judge that a strong component is the meaningful and noble work that space quests can provide. For example, we have enough unprocessed data from previous space projects to provide interesting work for whole nations of people of all ages: vintaged seniors, middle-aged adults, students, even those of elementary age. There also exist forward tasks of many magnitudes that could be farmed out to all who are willing to participate in Earth's space work. These and other candidates will eventually be assayed for inclusion in a solid vision for the future. Without such a vision our technical efforts, impressive as they might be, will falter.

On the technical front, the challenges to accomplish the exploration vision are just as significant as on the aforementioned human front! Substantial, but achievable, advances are needed in robotics. This chapter will focus mostly on robotic needs. Also treated in summary fashion are command and control, communications, navigation, advanced life support, radiation protection, and space suits. Though not treated in this chapter, advances are also required in planetary surface and in-space construction, planetary mobility, thermal control, field medical equipment and techniques, science data and specimen gathering and processing, equipment maintenance, off-Earth societal and residential infrastructures, and education, cultural, and religious support. There are assured more than on this list.

Robotics

One of the greater exploration challenge areas is in the field of robotics. The President in announcing the space vision placed strong emphasis on the use of robots. The ratio of robotic tasks to human tasks should probably be 1.0 or greater. This will require many advances in robotics and in human-robotic cooperative operations. To date, use of robots and automated machines in space has consisted mostly only of mobile data gathering in free space or on planetary surfaces or the remote control of single mechanical arms to perform simple tasks. Translated, this means that we have a long way to go before robots can play a major role in exploration activities.

Though the robotic challenge is great, there is great expectation of success. Terrestrial use of robots has and is flourishing. Beginning with imaginative anthropomorphic constructs in movies and other entertainment centers, proceeding to tele-operated single and multiple arms and hands extending man's reach into environments too dangerous for humans to tread, then to programmed arm/hands performing repetitive assembly tasks, and now to a myriad of robots and automated machines performing a like myriad of scientific, military, and personal assistance tasks. We employ flying drones to perform reconnaissance and soon active military operations. In a Saturn car factory six robots assemble a car in three hours. There are robot dogs for companionship and entertainment. There are automated home vacuum cleaners and lawn mowers. There are smart prosthetic human limbs. And the laboratories across the planet are rich with developments, a number of which will come to the market place to meet eagerly-waiting organizational and personal consumers.

With all this progress in earthy robotics, why is exploration robotics such a challenge? The answer consists of several hard facts: the environment, the human-less factor, and the human factor. First, extraterrestrial efforts whether in free space or on heavenly body surfaces are conducted in harsh environments. Outside the comforting cocoon of Earth, survival, whether it be for carbon-based living entities or silicon and metallic-based mechanisms, is far from certain. High and low temperatures are extreme. Lighting conditions are glaring or dimmed. Atmospheric pressure is absent, near-absent, or too great. Gravitational and similar force effects are much different than those on Earth, where our bodies grew up and where our machinery was built.

We are making some strides in building robots and automated machines that can work in space. Successful photo and scientific data gatherers have been sent to our moon and all but one of our known planets save Pluto, other moons, and recently to asteroids. Table 31.1 lists successful mission suites from multiple countries.3 Added to this list should be a host of unspecified military reconnaissance craft.

Body	Spacecraft	Countries
Sun	Pioneer, Explorer, Helios, Yohkoh, Ulysses, SOHO, Genesis	USA, Germany, Japan, ESA
Mercury	Mariner, Messenger,	USA
Venus	Venera, Mariner, Zond, Pioneer, Venus, Vega, Galileo, Magellan	USSR, USA, ESA
Moon	Pioneer, Luna, Ranger, Surveyor, Langley Lunar Orbiter, Explorer, Zond, Muses, Galileo, Clementine, Lunar Prospector. SMART, Lunar-A, Selene	USA, USSR, ESA, Japan
Mars	Mars series, Mariner, Zond, Kosmos, Viking, Phobos, Mars Global Surveyor, Pathfinder, Nozomi, Mars Odyssey, Mars Exploration Rovers, Mars Express	USSR, USA, Russia, Japan
Jupiter	Pioneer, Voyager, Ulysses, Galileo,	USA, ESA,
Saturn	Pioneer, Voyager	USA, ESA
Uranus	Voyager	USA
Neptune	Voyager	USA
Asteroids	Deep Space, NEAR, Galileo	USA, ESA

Table 31.1 Successful Robotic Planetary Mission Suites

Besides the successful missions listed in Table 31-1, there exist a larger number of attempts that failed. For example, of all the spacecraft launched towards Mars from 1960 through 1999, over half failed before achieving their objectives. Of thirty-two attempts, six failed during launch and seventeen after launch. Nine achieved their objectives.[4] Please realize that this record should not be a source of discouragement. Such is the nature of space endeavors.

The successes do attest to our ability to operate complex machines in the harsh space environment. In other instances, however, the environment has won. In 1967, The USSR Venera 4 spacecraft entering Venus became inoperable before it reached the surface due to the severe high-pressured and acidic atmosphere. The USSR Mars 4 landing craft hardware and software were no match for the Martian dust storms in 1971. Contact with the US Mars Surveyor Polar Lander was lost as it entered the Martian atmosphere in 1999. We have more than enough lessons on the challenge of building automated machines for such environments.

The human-less or human missing factor of difficulty is quite interesting. Our best results in building things and in accomplishing activities come when we humans can be present to modify, to adjust, to compensate, to replan, to rethink. Many of our automated failures have been accompanied by the "Oh, no!" moans as we watched our projects go awry from afar. We humans could not in these instances be present close enough to the craft to prevent or fix the problems. Had we been there, our ingenuity might have saved the day. Thus the human-less challenge factor of fielding space machines becomes evident. Many of these robotic devices must in large part be operated without humans in the loop. Despite intense and talented design efforts, conditions, unforeseen or discounted, sometimes bite us hard as our machines fail.

Theoretically these latter shortcomings can be minimized radically. We have plenty of experience in real space life. We have developed useful techniques such as contingency analysis, component redundancy, fault identification software, and automated reconfiguration. Still to be developed are methods to repair robots in the field. Each problem associated with building and operating space robots is solvable. Success lies in applying significant effort to the problems, one at a time.

Since the challenges associated with the off-earth environment and the remoteness of human intervention are solvable, what else remains to hamper a flourishing robotic exploration program? A minor problem is a latent fear of robots, concerns that they might run amuck or "take over", and a worry that they might rob man of his livelihood. These issues do need to be addressed and are solvable with proper designs.

The main challenge, however, is the human factor of getting organized to define the specific robots needed and to develop them in an integrated cooperative program spread across multiple companies and space centers while minimizing development and operation costs. This human or organization factor is achievable with concentrated and unified effort. Historical examples of the kind of organization and drive required include the Manhattan World War Two atomic bomb project and the Apollo lunar landing program. During these two projects substantial new managerial efforts were invented to get the end results. The leaders of these programs built a carefully choreographed set of development assignments, tests, contingency analyses, and new paradigms to achieve gigantic results.

Our task now is to formulate and execute a similar choreograph. We must weld a considerable number of our individual university, space center, and private company robotic experiments and programs into an integrated effort, add pieces that are yet to be done, and build an efficient system of exploration robots.

Figure 31.1 [5] illustrates a logical pathway to developing such a robotic exploration system. One derives objectives based on the vision. The objectives are used to design missions. The missions are used to determine what types of robots are needed. In some instances, candidate robotic concepts must be tested either in earth-based or space-based facilities. Parallel to this thread of activity, two others can be performed. We can formulate design principles for space robots consistent with the exploration environment, with sound engineering practices, and with the objectives of flexible and cost-time efficient operations. We can also glean design and operational information from our past successes and failures with space and terrestrial robots.

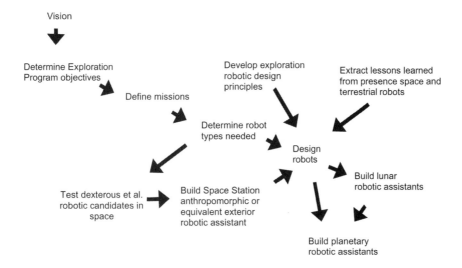

Figure 31.1 Pathway to an Integrated Exploration Robotic System

Where are we now in the pathway to integrated robotics illustrated in Figure 31.1? Work is currently progressing well in defining program objectives and missions. [6] [7] [8]. There are, of course, plenty of questions to be answered. For example, associated with lunar objectives as a class:

1) Do we build lunar base(s)
A) to serve only as a temporary proving ground for interplanetary efforts,
B) or also to accomplish permanent moon-based endeavors?

2) Which scenario do we use for human-robotic time task partitioning:

Scenario A
(1) Robots go first to build the infrastructure
(2) Humans come later and
  a) work alone
  b) work with robotic help, or

Scenario B
(1) Humans go first with robotic assistants
(2) Human go alone and work forever without robotic assistants
(3) Humans go first alone and acquire robotic assistance later?

Questions at lower levels must also be answered before robots can be designed. For example in the mission category a few questions are:

* permanent base versus mobile base versus wandering "pack mule" prospectors?
* prospecting for ice?, valuable ores? in highlands or mares?
* roads versus hopping versus. flying vs. all terrain vehicles?
* power transmission via wire or emf?
* navigation via beacon or/and satellites?

Eventually decisions will have to made on how to use robots on the moon and beyond in:

* surveying and reconnaissance
* power generation, transmission, and storage
* fuel and water generation, transmission, and storage

* environmental control
* communications and navigation infrastructure buildup
* dirt manipulation
* human and materials transportation
* mining
* building materials production
* construction of buildings, roads, landing port, and science/industrial facilities, and maintenance including that of robots

One task that can be done now is to develop robotic design principles, since these are independent of most aspects of specific mission definitions. Some principles developed so far include:

1) **redundancy** - use either one more-massive and expensive large-capacity robot plus a similar backup to do the job or use several smaller less expensive and robots who together could do the job in the same time or the same job in a longer time in the presence of failures? (The answer may vary with each specific task.)
2) **interoperability** - use components of robotics in multiple machines, both robotic & non-robotic, e.g., power supplies, computers, limbs, end effectors, wiring harnesses, ...
3) **market choice** - use commercial-off-the-shelf hardware, software, and standards - e.g., web-based packet communications, standard interfaces (e.g., USB, Firewire, data and mechanical connection), and commonly-available components and subunits
4) **standardization** - develop and use new international robotic standards
5) **generalized workers** - build robots that can perform a variety of tasks
6) **alterability** - to be able to re-program and or physically reconfigure robots to perform new tasks or recover from unplanned events or failures
7) **controllability** - provide autonomous operations with override capability and provide for various levels of tele-robotic control

Some of the more important specific challenging work items to be performed to produce successful exploration robots include:

1) Build robots that can work in space environment - Most developed robots work only in the Earth's environment. Space robots built to date include the Shuttle SRMS and Station SSRMS arms and many successful interplanetary orbiter, probes, and landers.
2) Minimize the time latency effects inherent in controlling space robots from Earth control. - Round trip signal path and processing times vary from a few seconds for the moon to many minutes for Mars and beyond. Potential compensation methods include use of automated scripts, on board autonomy, and robot movement anticipatory algorithms.
3) Be able to conduct surface robotic operations at lunar noon - The high temperature and albedo within several earth-length days of lunar noon cause heat rejection problems and preclude many types of optical navigation and some types of optical perception techniques.
4) Be able to conduct surface robotic operations during Martian blowing dust times - How much of this can be achieved is unknown.
5) Make further advances in robotic perception and "cognition" -
6) Develop secure and redundant command paths to robots - It is essential that our exploration robots not be high jacked or "control-jacked" by enemies or fools.
7) Repair robots in the "space field" - It is impracticable to bring robots back to earth for most repairs since the lunar one-way trip is days and is months or more from Mars and beyond.
8) Human-Robot coordination - How does one coordinate operationally remote multiple robots and humans working together in a confined area on the lunar/planetary surface or outside in-transit vehicles?

There are often many developmental subtasks contained in each of aforementioned work areas. For example, subtasks under perception and "cognition" advances include:

1) Translating the input of visual, auditory, and tactile sensory inputs into rapid "recognition" of objects by robotic software
2) Physical path planning in presence of many multiple types of constraints
3) Logical layering of hierarchical control mechanisms, i.e., building simple behaviors and task skills into

complex behaviors and specific tasks involving multiple skills [single joint movement ...basic skill (grasp) ... tasks with multiple skills (go fetch panel)]

4) Transformation of high-level commands to a sequence of skills including constraints checking and intelligent reaction to these constraints

5) Inter -robot information and skill transfer

6) Operating in the environment of non-cooperative agents (e.g., robots avoiding interference with the actions of other robots and with carbon-based units)

7) Self-starting actions based on perceived needs (within constraints)

There also exist some non-technical challenges associated with building and using exploration robots. These latter challenges can be as strong a barrier to overcome as are the technical ones. On the human front, on both the individual and on the organizational level, there exists some fear of robots and of automation in general. The more common of these fears are associated with potential job loss and potential physical harm from robots. These fears can and should be answered. For example, automation and the increasing use of robots produce about as many jobs as are displaced. These new jobs are in the design and construction of robots and automated tools and in new areas of accomplishment that have not been able to be worked due to excess manpower has been required to because there has been little or no automation.

Today, many of our more advanced robots are, with some exceptions, being designed and built in austere, low-organizational-overhead environments of small companies. This seems a good trend. Innovative robotic research projects and products often come from this type of environment. The type of robots coming out are myriad, as are the target markets for these robots. It is a similar trend for robotic work being performed at universities and space centers.

What is needed to have successful, time- and cost-efficient robots systems in space exploration is to built and execute an integrated, thoughtful plan of attack. To accomplish this great work we must evolve our current sets of robotic activities scattered among individual space centers, private companies, universities, and, in other countries into an integrated endeavor to propel robotics into the Exploration Era. For the sake of success we must do this.

An integrated Exploration Robotics System is our goal. This Exploration Robotics System can be built in stages as mission needs unfold. In fact the system, much like a living person, will continue to grow and to develop, and to refine new capabilities. Stasis will not be achieved in this robotic system, anymore than stasis will be achieved in exploration endeavors in general. Exploration and change are parts of our future. The possible roads and choices are endless.

Command and Control

Ranking as one of the two most significant exploration technical and managerial challenges is a major overhaul of the command and control infrastructure and its use to conduct space operations. For a detailed analysis of the problems and potential solutions to solve the problems, see chapter 25 of this book. In summary, our challenge is to build an efficient, cost effective exploration infrastructure, to coordinate exploration robots and human crews from multiple earth sites to accomplish science and exploration objectives, and to maximize the self-sufficiency of the lunar/planetary exploration teams while preserving exploration and science activity participation by Earth-based personnel when needed or logical to do so.

The command and control area includes the computer timing and execution processes, both manual and automatic, of exploration activities, the communication, procedural, software, and control process methods employed, and the partitioning of control functionality between astronauts and equipment in the space field and Earth-based controllers and control centers. Much detail on requirements, measures of success, and methods to achieve efficient command and control is given in Chapter 25.

Because of the long distances involved, efficiency and safety concerns dictate heavy use of automation and robots working in cooperation with humans to accomplish exploration objectives, a significant number of technological advances are needed. Required are:

1. Order of magnitude faster computers,
2. Large distributed data bases,
3. 100 megabit per second space communication links,

4. Web-based or similar space network,

5. Inter-control center voice, data, and video connectivity,

6. Exploration Vehicle autonomous navigation,

7. Advances in ground-based robotic control, and

8. Advances in autonomous fault management for equipment.

Many interesting new operational techniques need to be developed. For example, how does one coordinate operationally remote multiple robots and humans working together in a confined area on the lunar/planetary surface or outside in-transit space vehicles?

As stated strongly in Chapter 25, much hard work is required to establish a new paradigm for operations for the Exploration Era. Those agencies, countries, and, companies who fail to do so will find themselves spectators watching the innovators explore and profit from space.

Communications

Space communications for the Exploration Era presents marvelous opportunities to establish unprecedented richness in possible space missions as well as to improve mankind's earthly existence through better and faster information flow and global capabilities development. This is a continuation of mutual benefits between celestial and planetary efforts that has been occurring since the late 1950s. We have gone from local low-power crystal radios and underwater limited cable communications traffic to continuously available intercontinental color television and trading information over the internet while one is bicycling though cascades or walking on the surface of Antarctica. As we have been impressed continually by communication advances and we will be more so during the Exploration Era.

In the Exploration Era the types, number, and of communications links will need to be multiplied by at least two orders of magnitude. Modern operational concepts require significant bandwidths and multipoint communication capabilities. The exploration communications architecture has requirements for an adaptable, high-rate communication backbone infrastructure with access links to space and ground networks, with inter-spacecraft communication links, and with close range wireless proximity links. The human and robotic endeavors will require a communication infrastructure that supports bi-directional, multiple video, voice, and Internet-like data transfers and will enable simultaneous communications among local work site personnel and equipment, planetary bases, orbiting facilities and Earth-based control centers. When feasible, planetary orbital satellites will be deployed to aid in both communications and navigation. [9] [10]

Communications must occur among multiple spacecraft, multiple planets and moons, multiple exploration-faring humans and robot. Arenas of communication need include:

1. Local In-Space Communications - provides communications for EVA crew personnel and equipment outside of a traversing spacecraft.

    a. Provides two-way voice, video and data communications while on EVA outside a traversing spacecraft:
        i. among EVA personnel,
        ii. between EVA personnel and the spacecraft,
        iii. between EVA personnel and detached or docking equipment, and
        iv. between detached or docking equipment and the spacecraft.
    b. Communications with surface bases with Earth are relayed through the spacecraft,
    c. Provides situational awareness for EVA personnel
    d. Enables EVA crews to execute procedures using up-to-date textual and graphical data.
    e. Enables crew to monitor and control detached or docking equipment
    f. Enables automatic docking
    g. Allows spacecraft and ground/planetary personnel to monitor EVA procedures and operational status.

The bandwidths required to achieve the needed communications [11] among EVA crewman, the spacecraft and robots are one megabit per second (MBS) in data plus compressed, scalable high-density television. For automatic docking equipment a 0.2 MBS data link is needed.

Necessary technological developments can begin with personal surveillance devices and prototype three-dimensional high definition prototypes. Low mass cameras, antennas, and transmitter/receiver devices need to be built. The surveillance market seems to be heading in the required direction. Current space communication methods need to be upgraded to use multi-point techniques and to provide expanded bandwidth needed for operational efficiencies.

2. Lunar and Planetary Surface Communications - provides communications among elements at local surface worksites, planetary bases, and home planet facilities. Provided are voice, video, and data communications among personnel and equipment at a worksite on surface of the moon (or Mars/other similar body) including EVA personnel, robots, rovers, fixed equipment, local transports, surface habitats, and home planet (Earth) facilities. Types include:

a. two-way voice and live high-resolution, compressible video from each EVA crew person and each robot to the planetary surface base and to Earth (can be relayed through surface base),
b. two-way voice and live video among EVA personnel, and
c. two-way data/command transfer among EVA personnel and local robots.

The bandwidth provided needs to be of sufficient width to send instructional photographs, data/command, and videos from the surface base to each EVA person. The local work site area is defined as a range that extends approximately 1 kilometer from the center of the work site or permanent base.

The bandwidths required to achieve the needed communications are in Table 31.2. The first number in each cell is the bandwidth in MBS and the second number is the video resolution required

From / To	EVA Person A	Site Base & Remote Base	Robot X	Fixed Equipment	Surface Vehicles	Earth
EVA Person B	1, scalable	10, CHDTV[12]	1, CHDTV	0.2, scalable	1, CHDTV	n/a
Site Base & Remote Base	10, CHDTV	10, CHDTV	10, CHDTV	10, CHDTV	10, CHDTV	100, CHDTV
Robot Y	0.2, n/a	1, n/a	2, n/a	n/a, n/a	0.2, n/a	10, n/a
Fixed Equipment	0.2, n/a	10, n/a	n/a, n/a	n/a, n/a	0.2, n/a	0.2, n/a
Surface Vehicles	1, scalable	1, CHDTV	1, CHDTV	1, scalable	1, CHDTV	1, CHDTV
Earth	n/a	100, CHDTV	1, CHDTV	n/a	1, CHDTV	n/a, n/a

Table 31.2 Lunar and Planetary Surface Communication Requirements

Using these capabilities EVA crew will able to execute procedures using up-to-date textual and visual information sent by elements external to the site. EVA crews can perform coordinated tasks via information exchange with each other and can monitor and control robots, rovers, and equipment platforms.

3. Surface and near-surface mobile communications - provides voice, video and data communications among vehicles moving along the surface, vehicles in suborbital transport or reconnaissance, surface elements, and home planet facilities. This capability allows traversing vehicles and crew to:

a. receive procedures, maps, systems' data from surface bases and home planet facilities,
b. perform coordinated tasks via exchange information exchange with each other, and
c. have full command and control of robotic rovers, platforms, and stationary equipment.

The bandwidths required to achieve the needed communications are in Table 31.3.

Significant communication capabilities between the moon and the earth must be replaced. The equipment used in during Apollo is old, doesn't use current protocols, isn't not web-compatible, and to a significant extent inoperative13. In the 40 plus years since Apollo, communications technologies have improved dramatically. The ability to transfer megabits of information on Earth is near trivial. In space for the Exploration Era 100 megabit-per-second links are needed. Our current capability, however, lies at one megabit per second [11]. Technological developments required also include some ability to provide communications over the horizon perhaps by surface beacon or satellites, either of which will need to support navigation requirements.

From ⟍ To	Surface Vehicles	Sub-orbital Transports	Mobile Robots	Site Base, Remote Base, & Earth	Reconnaissance Vehicles	Fixed Equipment
Surface Vehicles	1, CHDTV	1, CHDTV[12]	1, CHDTV	1, CHDTV	1, CHDTV	1, scalable
Sub-orbital Transports	0.2, low	0.2, n/a	1, CHDTV	1, CHDTV	n/a, n/a	0.2, low
Mobile Robots	0.2, n/a	n/a, n/a	0.2, n/a	0.2, n/a	n/a, n/a	n/a, n/a
Site Base, Remote Base, & Earth	1, CHDTV	1, CHDTV	1, CHDTV	1, CHDTV	10, CHDTV	10, CHDTV
Reconnaissance Vehicles	0.2, n/a	n/a, n/a	n/a, n/a	0.2, n/a	n/a, n/a	n/a, n/a
Fixed Equipment	0.2, n/a	0.2, n/a	n/a, n/a	0.2, n/a	n/a, n/a	n/a, n/a

Table 31-3 Surface and Near-surface Mobile Communication Requirements

It is judged that all that with focused efforts and funding, all the aforementioned exploration communications equipment and infrastructure can be built using existing earth-based developing and marketed technologies. The challenge lies in combining all the individual required kinds of communication links into a comprehensive system of integrated capabilities that are compatible with existing earth-based communications, uses common hardware with them and is both expandable and upgraded by nature.

Navigation

Whereas nautical navigation often uses the stars to determine ship and airplane locations on the surface or in the air of the Earth, celestial navigation uses the stars and other equipment to determine where and how fast Earth's spaceships have progressed towards the stars themselves. From the inception of space flight, considerable amounts of under-the-hood effort have gone into the endeavors, part art and part science, labeled as space or celestial navigation. Its task is to determine vehicular position and velocity, thus enabling automated guidance schemes to perform thrusting maneuvers to propel the spacecraft to the desired location. Throughout the history of space accomplishments, navigation techniques using specially developed navigation sensors and software algorithms have been used to take our spacecrafts to are near all of the planets in our solar system save Pluto, to fly by, orbit, or land on asteroids and copious moons, even to achieve pinpoint landings on our own moon, to rendezvous spacecraft, to propel spacecraft out of the solar system, and to traverse the lunar and Martian surfaces in unmanned and, in the Apollo program manned, surface vehicles.

In the Exploration Era, navigation will need to expand its breadth and techniques to provide needed services to many types of space and surface vehicles going to many destinations. [10] Local in-space navigation will provide relative navigation information for EVA crew personnel and equipment outside of a traversing spacecraft. Information provided includes attitude, relative position and velocity, relative range and range rate and any other needed relative navigation parameters for:

1. EVA personnel relative to the spacecraft,
2. Detached and docking equipment relative to the spacecraft,
3. EVA personnel relative to detached equipment, and
4. EVA personnel relative to each other.

The information provided will allow EVA personnel to estimate traverse times, provides the needed elements to enable capture or rescue of crew or items that have drifted away accidentally, and enable the automatic docking of co-orbiting equipment. Examples of equipment needing navigation are miniature flying camera systems, automated tool carts, equipment carriers, and robotic crew assistants. The required one sigma relative accuracies for EVA crew or non docking equipment are 2 meters in position, 0.2 meters/sec in velocity, and 1.5 degrees in attitude. [10] For docking equipment docking the accuracies need to be 2 centimeters, 1 centimeter/sec, and 1 degree.

Though space relative navigation systems have not been developed for multiple users, the algorithms and navigation hardware changes are achievable by building on previous efforts by the U.S. and Russia. Encouraging the development of items such as optical LED devices, optical shape and color algorithms, and multi-element local traffic control algorithms is advisable.

Lunar and planetary surface and near-surface navigation consists of navigation tasks performed at local surface worksites, on vehicles moving along the surface, on suborbital transports, and on overhead reconnaissance vehicles. Local worksite surface navigation provides position, velocity, bearing, and other navigational parameters for personnel and equipment on the surface of the moon (or Mars/other similar body) relative to a local rectangular or similar site grid. The local work site area defined as a range of approximately 1 kilometer from the center of the work site or permanent base. Work site personnel and equipment include, but are not limited to, EVA personnel, robots, rovers, surveyed navigational beacons or landmarks, fixed equipment, specimen locations, and local transports. This capability enables situational awareness for the personnel and autonomous equipment and provides the location of the other equipment or personnel with respect to each other and to the work site or the permanent base. One application, for example, is the ability of EVA personnel or robots to return to locations where previous specimens have been gathered

The required one sigma position accuracies, relative to the local grid for fixed equipment, such as recharging stations is 10 meters for mobile equipment and 20 meters for personnel, For specimen locations, excavation sites and navigation beacons the requirement is 5 centimeters. In addition, the local site grid must be matched to local overhead photography within 3 meters (1 sigma) and the local site grid must be tied to an inertial coordinate system within an accuracy of 100 meters. Though some of these requirements are stringent, the techniques developed during Apollo, Martian, and more recent lunar programs along with current advances in terrestrial hardware and software are a good starting point for the required moderate development of surface and near-surface exploration navigation systems that can achieve these accuracies. . Surface beacons or equivalent and, probably orbital navigation satellites need to be developed, as well as small, low power navigation sets that can be worn by personnel and robots and sensor reflectors and transponders that can be affixed to mobile and stationery equipment.

Surface and near-surface mobile navigation provides both relative and inertial navigational information for personnel and equipment traversing the moon or Mars/other similar body. The range of operation extends outward from the main surface base to distances defined by remote work sites, eventually on the order of several thousand kilometers. Personnel and vehicles include but are not limited to pressurized crew transports, unpressurized equipment movers, suborbital transports, aerial reconnaissance vehicles, mobile rovers and robots, and EVA personnel. Provided are both relative position, velocity, bearing, and other navigation parameters with respect to a planet-wide surface grid, as well as absolute position, velocity, and other navigation parameters in an inertial coordinate system frame.

Traversing relative navigation enables personnel and surface mobile equipment in transit to find their way to remote work sites and to return to a permanent base using some of the same equipment and techniques used for local worksite navigation. Both the traversing inertial navigational system and the relative navigational systems enable suborbital or other transports to land close enough to designated sites to accomplish mission objectives safely. The inertial navigation capability allows over-flying or orbiting vehicles to perform required overhead surveys of remote worksites

The required one sigma position accuracies of in-transit surface moving vehicles and personnel are:
100 meters with respect to the planetary surface grid (relative nav) and 350 meters with respect to an inertial coordinate frame (inertial nav). For sub-orbital transports landing at a surface site and for overhead reconnaissance vehicles, the accuracies one sigma must be 100 meters in position with respect to the site surface grid (relative nav) and 350 meters in position and .35 meters/sec in velocity inertially.

Realtime lunar inertial navigation is quite challenging due to the anomalous gravity field. There is a strong need due to efficiency of operations and autonomy considerations that dictate the need to develop inertial navigation sensors not dependent upon earth-based equipment and personnel. Satellites, beacons, and optical sensors are good technical candidates.

Finally, a more efficient navigation system architecture needs to be developed for spacecraft traveling to and from planetary surfaces, for Earth and planetary orbits, and for deep space destinations. Our current system is quite manual and quite old. New trackers and other sensors and new navigation techniques and algorithms are needed that will enable smooth traffic control of many vehicles without the use of a large workforce.

Advanced Life Support

When one goes on an extended backpacking trip in the wilderness, one must take along all that is needed for survival. This includes shelter, clothing, food, water, and other necessary equipment. The amount one takes depends on what is available in that wilderness. If there is no or little chance of food and water replenished, then one must take a supply adequate for the journey duration. The longer the trip, the more one must take. Thus it is in space. Up until now most of our space adventures have lasted several weeks or less. We have taken all the food, water, and air needed for the duration. In the case of our space stations, we have relied on periodic resupply from our mother planet.

For our planned future planetary outposts, remote space platforms, and longer space flights, however, this complete reliance on resupply is not tenable. Excepting the resources perhaps discovered on distant worlds, e.g., water on the moon or Mars, most, if not all, life-sustaining consumables must be recycled in a closed-loop fashion. This applies to air, water, and food. Almost from the inception of manned spaceflight, both the U.S and Russia have advanced, tested, and fielded to space some technologies to this end. But these efforts have not been enough, nor have they been designed to work in an integral fashion. A working, comprehensive, and integrated advanced life support system needs to be developed if we are to become truly a space-faring society. As with any similar endeavor, the technologies developed will provide quite useful terrestrial applications.

The required Exploration Era environment system would:

1. close the loop entirely for water
   a. Water in all forms would be recycled - that contained in the human transport craft, habitats and spacesuits as well as in hydrous materials, waste water, and other waste products (urine, etc.),
   b. The amount of water would be a constant. No extra would be needed form Earth.
2. close the loop entirely for air,
   a. Oxygen would be produced from plants and manmade devices.
   b. Carbon dioxide would be removed by plants and manmade devices.
   c. Contaminants and pathogens would be removed by systems that employ no non-replenishable consumables.
3. Close the loop to a high degree for food,
   a. Grow or raise as much food as practicable.
   b. Supplement with thermo stabilized-irradiated and freeze-dried foods.
4. Close the loop to a high degree on solid and gaseous wastes,
   a. Recycle plant and animal wastes to compost and fuel.
   b. Recover fuel gases.
   c. Incinerate solids for energy production, if healthy to do so.
   d. Compact or convert remaining solids for building materials.
   e. Have a super-small ("thimble-sized") trash container for the remainder.

All of the elements of such a system are possible using current and approaching technologies [15,16]. All are elements of this system are quite necessary to enable long and distant space missions, to provide a safe, good quality lifestyle for humans traveling and working in space, and to prevent contamination of solar system heavenly bodies, including Earth.

There has been considerable progress made in some of the above areas. [17] Water recycling technologies such as vapor compression and rotating distillation processes, organic and inorganic filtration, adsorption / ion exchange beds, catalytic oxidation, ultraviolet sterilization, and the use of microbial check valves has brought water purification to a state near ready for deployment. Air cleaning technologies are also fairly well developed. Devices such as the International Space Station's Solid Polymer Electrolyte Oxygen Generation Assembly and the regenerative 4-bed molecular sieve process (that also provides for water recovery), solid amine absorbent beds (under study) plus the demonstrated use of plants for carbon dioxide absorption (doubles as a food source) holds great promise that we are moderately near to closing the loop on air.

The growing of plants for fresh food, as well as air will certainly contribute to crew health and performance.[15] Likewise the raising of fish such as the space-efficient tilapia would be cheered by spacefarers. How much fresh food can be provided depends on the nature of the vehicles, missions, and crew quarters. The amount possible is expected to increase as the exploration program progresses. Further development needs to occur on in-space plant production, when desired crop densities are twenty times that in open air home planet fields. This development includes the reduction of input energy required (mostly lighting), the increase in the varieties of crops used, and increasing plant resistance to diseases, as well as recovery from disease infestations.

Closing the solid loop will take more effort than will the others loops. Compact composting and energy extraction on the small scales that will be required for exploration need to be developed. The overall philosophy of throwaway or disposable objects, just as in our modern terrestrial life, needs to be transformed to one of conversion and second use, third use, fourth use, etc.

Finally, much advancement needs to be made to produce an integrated advanced life support system that is controlled for the most part automatically without human intervention. Today the various closed loop components are not integrated, are controlled manually, and require a lot of crew intervention. For example, one earlier envisioned system, designed by logical extensions from Russian and U.S. flight and experimental experiences would require 1.75 persons working one eight hour shift daily just to maintain and perform routine tasks associated with running the system. Not included are repairs.

All of this said, the developments needed to bring an integrated closed loop life support system to deployment status are quite achievable with moderate effort. If such a system is not developed, we will have a very limited and unfulfilling exploration program.

Radiation Protection

The most serious of Exploration Era challenges is the need to protect people from radiation once they leave the safety of Earth's environment, including its Van Allen radiation belt. Potentially lethal barrages of event radiation can and have occurred. Thank the Lord that no such events occurred while the astronauts of Apollo were out on lunar surface EVAs 19,20. Even the continual normal background radiation has its cumulative harmful effects on the human body. It is imperative that both exploration vehicles and habitats, as well as associated crew timelines, be designed to minimize human susceptibility to radiation.

The goal of radiation protection is to enable astronauts to live and work safely in the space environment, one that does include significant radiation. To the extent possible the safety will be achieved by stopping the radiation via shielding. Only secondarily, should the safety be achieved by limiting the astronaut career lifetime exposure to the danger.

Space faring humans must deal with four types of radiation: that radiation trapped in Earth's Van Allen belt formed by the interaction of our planet's magnetic field with continuous solar wind particles, galactic high energy cosmic radiation, occasional and often deadly special energetic solar particle events, and secondary radiation formed by the interaction between incoming radiation and shielding [17, 21]. The second type of radiation is peculiar to spacecraft in low Earth orbit. The remaining types dominate above low Earth orbit, even out to the planets and beyond. All types of radiation exposure need to be limited because they can produce damage to the central nervous system, degenerative tissue disease (cataracts, heart disease, etc.), acute radiation sickness any of which can easily lead to a shorter life.

Astronauts have been classified as radiation workers and, monitoring their radiation exposure has been a key requirement for spaceflight since Project Mercury 19. The minimum radiation protection required is stated as the lowest acceptable maximum astronaut career time in the space environment. One current candidate set of exposure limits for career space time is three 180-day missions in low Earth orbit, six 90-day lunar missions, or one 1000-day mission to Mars. These mission limits are not valid if a large special energetic solar particle event would occur. Such an event of sufficient magnitude could severely decrease or end an astronaut's in-space career. Therefore, it is very important that all possible feasible steps be taken to lessen the danger. Astronauts on the moon or other planetary surface will have to live in shielded quarters, probably underground. Traversing spacecraft whether making the several day back-an- forth trips to the moon or making the many-months trips associated with Mars will have to be shielded. Astronauts will need to minimize their exposure in outdoor areas and must wear shielded spacesuits.

There is hope. Lighter, more effective shielding materials are being developed (e.g., polyethemides et al) that produce less secondary radioactive particles [21]. Smaller, lighter, and closer-to-full-spectrum radiation dosimeters are being manufactured. Advances are being made in the prediction of energetic solar particle events. Experimental genetic DNA repair of radiation damage are being sought in laboratories. Judicious adherence to the principle of minimizing human exposure to radiation via shielding, the use of surrogate robots to perform outside tasks, and the use of faster spaceships give the best protection. With vigilance, the careful design of operational scenarios, and a continual focused development of specific pertinent technological items, men and women should be able to flourish in space.

Space Suits

Working, even existing in space requires humans to have extensive bodily protection. For many tasks the humans can work in relatively large constructed cocoons capable of keeping the body within the necessary narrow limits of temperature, pressure, atmospheric, and luminosity needed for survival. The cocoons can be space ships, enclosed planetary rovers, or surface habitats. Within these, humans can exist without extra epidural apparatus.

However, when man ventures out of the cocoon into the harsh environment of free space or onto extraterrestrial surfaces, he must wear a spacesuit. Space suits exist today. They were derived initially from suits worn by high altitude aviators and were evolved by both the Russian and U.S. space programs for use for a few hours during extravehicular excursions in earth orbit , lunar transit, and on the lunar surface. These suites are used for short work sessions, need replenishing before reuse, expend consumables, and work within narrow temperature ranges. For the Exploration Era activities suits that overcome some of these limitations need to be built. More that one suit will probably need to be developed. For example, the environment on the surface of the moon is quite different from the surface of Mars. And there are other environments in which we will wish to function, such as the deep cold of asteroid work.

The greatest challenge for the lunar suit is cooling. [18] First there is no known way for humans to work on the lunar surface within several "earth-days" of lunar noon. No space suit built or currently conceivable can overcome the intense heat of lunar midday. Apollo surface EVAs were accomplished at low sun angles. This is a serious and mission limiting problem. There is no known solution. For now, man's outside activities on the lunar surface will need to be restricted to locations near the shadows of topographical features or be restricted to lunar early mornings and late afternoons when the sun angle is 50 degrees or lower. This same temperature mission limit probably applies to robots.

How to cool the lunar suit is also a challenge. The Apollo suit sublimated water to free space. It was an effective technique, but such a technique cannot be used for extended lunar stays. Water is much too scarce to use in this way. Techniques being considered include the use of closed-loop phase change substances or using cryogenic oxygen as an expendable.

Additional suit challenges to be worked include: [15] [18]

1. extending suit use time - the limit for EVA time is 5 to 8 hours for the lunar surface and 3 to 4 hours for the Martian surface(due to stronger gravity), based of carrying weight,
2. CO2 control - the LiOH canisters of current suits must give way to consumable-less closed-looped systems such as solid amine absorbent beds,
3. dust - a major problem on both the lunar and Martian surfaces. It was estimated that dust alone might limit suit use to three outings based on Apollo crew experience reports.
4. lower pressure suits to decrease pre-breathing time, and
5. reduced-thickness gloves - necessary for better work dexterity. This item will need continual development as better materials are introduced.

All of the above challenges can be met. The toughest, the mission-limiting challenge of lunar cooling, needs intense work.

Summary

This chapter has listed the more important technological challenges in seven major systems that must be worked and solved before much space exploration can be done. These challenges will not go away. They must be faced before mankind can become truly space faring. Those organizations that delay the start of meeting these challenges delay their own trek into space. These work items, along with additional ones in other systems, may be grouped together and be labeled as part of our magnificent challenge. Come join us in working this great endeavor.

References

(1) George W. Bush, "New Vision for Space Exploration", January 14, 2004

(2) Proverbs 29:18

(3) Calvin J. Hamilton, Visions of the Solar System, 1996-2005

(4) Jim Plaxco, "Mars Space Missions", Astronomical Society of Long Beach website

(5) Richard E. Eckelkamp, "Space Robot Laborers", September 13, 2004, Aerospace and Technology Description
of Human and Robotic Surface Activities", NASA/TP-2003-212053, July, 2003Working Group conference

(6) Michael B. Duke, Stephen J. Hoffman, Kelly J. Snook," Lunar Surface Reference Missions: A

(7) Owen K. Garriott, Michael D. Griffin et al., " Extending Human Presence into the Solar System, An
Independent Study for the Planetary Society on the Strategy for the Proposed U.S. Space Exploration Policy, July, 2004.

(8) NASA's Exploration Architecture, September 19, 2005, NASA website nasa.gov/pdf/1133654main_ESAS_charts.pdf

(9) Robert Spearing and Michael Regan, "Space Communications Capability Roadmap Interim Review",
presented to the National Science Foundation, March 24, 2005

(10) June Zakrasjek and Richard E. Eckelkamp, "Human Exploration Systems and Mobility Capability Roadmap
Progress Review: Capability 9.2 Mobility", presented to the National Science Foundation, March 29, 2005

(11) Conversations with Orin L. Schmidt, Communications and Tracking, Johnson Space Center, January through April, 2005

(12) Compressed High Density Television

(13) Conversations with Thomas E Ohnesorge, Advanced Operations Development, Johnson Space Center,
March through October, 2005

(14) Conversations with Ellen M. Braden and Emil R. Schiesser, Aeroscience and Flight Mechanics, Johnson
Space Center, January through October, 2005

(15) Al Boehm and Dennis Grounds, "Human Health & Support Systems Capability Roadmap
Progress Review", presented to The National Science Foundation, March 17, 2005

(16) Crew and Thermal Systems, JSC, "Advanced Life Support Requirements Document", JSC-38571C, February, 2003

(17) Bruce E. Duffield, "Advanced Life Support Technologies List, Lockheed Martin memo HDID-2G42-1130 to
Crew and Thermal Systems, JSC, June 24, 1999

(18) Conversations with Michael N. Rouen, Crew and Thermal Systems, Johnson Space Center, February through December, 2005

(19) "Experiment Description for: Radiation Protection and Instrumentation (AP003)", Life Sciences Data
Archive @ Johnson Space Center, Houston, Texas, Spaceflight Radiation Health Program at JSC

(20) E. V. Benton, Editor, "Space Radiation", Nuclear Tracks and Radiation Measurements, Vol. 20 No. 1,
Pergamon Press New York, January, 1992

(21) Nassar Barghouty, "Space Radiation: Understanding & Meeting the Challenge for Exploration", September
13, 2004 meeting of the Aerospace and Technology Working Group

About the Author

Richard (Rick) E. Eckelkamp received a B.A. in physics at the University of Dallas with minors in mathematics, philosophy, theology, and foreign language, and did graduate work in physics at the University of Houston. At NASA, from 1967-1987, in NASA's Mission and Planning Analysis Division, he developed extensive experience in Earth orbit, lunar, and interplanetary guidance, navigation and control, engineering analysis, and flight techniques, performed flight software verification and data production, performed Apollo through Shuttle flight control, serving as Shuttle console leads for both onboard and ground navigation, was a major designer of Shuttle onboard navigation and fault management software, as well as developed ground control center software and techniques. From 1987-1995, he performed Mars and advanced lunar program software/hardware design, performed station automated operations systems analysis and prototyping, and served as the Systems Development Manager for the Space Station Freedom command and control system. From 1995 to the present, he has worked in Advanced Life Support command and control, is serving as the Software Integration Manger for the International Space Station Robotics, and is performing systems engineering for exploration era robotics and integrated operations control.

Chapter 32

Planning the Oasis in Space

By Richard Kirby and Ed Kiker

Theory

A model is a theory, and it may be represented as a picture. When we create a model of a space oasis, we are creating a theory or hypothesis of what it may be like or should be like, at least in part.

It is said that the first model was created by Anaximander (c.610-546 BCE), who was "the author of the first surviving lines of Western philosophy. He speculated about 'the Boundless' as the origin of all that is. He also worked in the fields we now call geography and biology. Moreover, Anaximander was the first speculative astronomer, originating the world-picture of the open universe, which replaced the closed universe of the celestial vault." [1]

When we build a model we are also engaging our will and our intelligence to create either a larger scale version (or smaller scale version) of the subject we are modeling. We may also be communicating our model as a theory of that which we are trying to understand. The first sense of model is technological, preparing to build a different scale version of the model. The second sense is scientific, our attempt to understand something such as the inner workings of the solar system or an atom.

Aesthetics offers us another kind of model, a transcendent dimension that seeks the essence beauty. Aesthetics - sometimes called the general philosophy of art, or more precisely the general philosophy of the beautiful and the sublime - has also been called a search to reproduce from the Real, the Ideal. For example, the great painter J.M.W. Turner [1775 - 1851] shows an idealized version of a ship in a skyscape in his work 'The Fighting Temeraire.'

But when we ask what we mean by 'idealized,' we also introduce a different realm: the emotions. Impressionists depict how people feel or might feel about the meaning of something that they see, such as a rainbow, a landscape, or a bolt of lightening. Hence, many portrait painters do not attempt to create a "painted photograph," but rather to bring out the deeper, hidden meaning of their subjects.

Thus, a model builder may be a scientist seeking understanding, a technologist seeking the means to operate successfully upon the environment, or an artist looking to probe and discover the ideal beauty that is beneath the surface of the subject being modeled.

A fourth perspective is the spiritual dimension, where models integrate different viewpoints and address the realm of action or deeds. The moral contours of the human mind are, in a current idiom, 'hard-wired' to admire what is done more readily than that which is merely described. Accordingly, the spiritual perspective on modeling oases in space seeks to integrate the scientific, the technological, the artistic, and the spiritual components of model building.

***

Experience shows that what is most personal is also what is most universal. The truth that is beauty, which is discovered by the poet, the artist, or the model builder, has the peculiar quality of being of more or less of universal appeal. The study of aesthetics leads not only to the statement of John Keats that "truth is beauty, beauty is truth this is all you know," but also to some ancillary truth, namely that truth, beauty, and love are all the same thing too.

The International Space Station (ISS) (above)
Rick Guidice 1980 NASA Research Painting (left)
US Astronaut Michael Fincke outside the
ISS in a Russian spacesuit

PAST

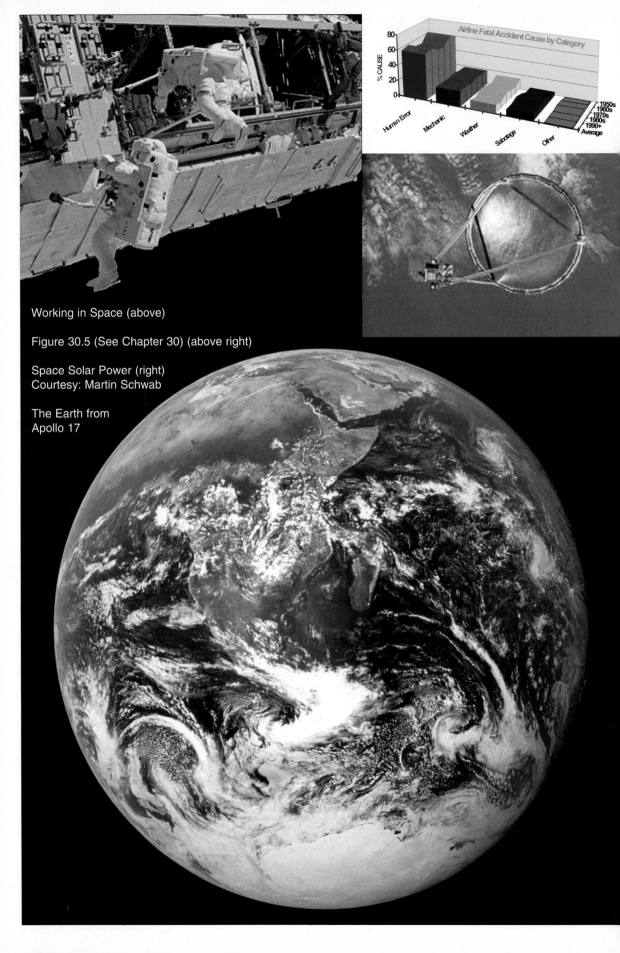

Working in Space (above)

Figure 30.5 (See Chapter 30) (above right)

Space Solar Power (right)
Courtesy: Martin Schwab

The Earth from
Apollo 17

Airline Fatal Accident Cause by Category

% CAUSE

Human Error   Mechanic   Weather   Sabotage   Other

1950s
1960s
1970s
1980s
1990+
Average

PRESENT

Author Derek Webber with SpaceShipOne.

Hubble Space Telescope image
"Heart of the Whirlpool Galaxy"

Bacterial Engineering (below)
See Chapter 13

Cell Colony Organization (below)
See Chapter 13

Searching the cosmos (above)

Astronaut Space Walk (right)

"Space" by
Candice Schoenherr,
Age 12, Iowa (below)

Sir Richard Branson and Burt Rutan speak to the press after Rutan's X-Prize winning flight

The Crew of Return to Flight Space Shuttle Discovery (above)

The US Space Shuttle (above) and the first private spacecraft, SpaceShipOne which won the X-Prize. (below)

Career Astronaut's Wings

Andy Schlebecker, Age 12, Maine. (above)

Kids in space (below)

FUTURE

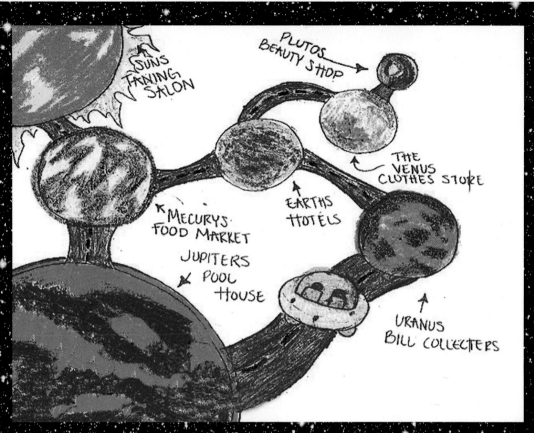

John Albert, Age 14, Mississippi, "Living in Space" (above).
Chad Ingulsrud, Age 14, North Dakota, "Me on Mars" (below)

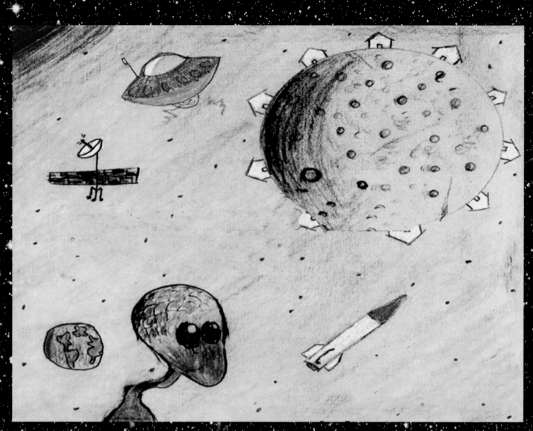

Sean Fay, age 11, and Cole Christine, age 15, Maine, "Aliens" (above)
Space Habitat, Alberto Zamoro, Mexican Artist (below)

When we create a model of a space oasis, we see that they need not be out in space itself: the time has come for us to create Earthly models of space oases. And if we ask ourselves what an Earthbound space oasis would be, we begin to ask, "What kind of Retreat Center would be suitable for Earth and space citizens?" Thus, the orbiting oasis should be modeled and lived in on Earth; we don't have to wait to escape terrestrial gravity in a literal sense.

What about the relationship between astronomical, or space, or cosmic culture, and spiritual civilization? What if then our subject suddenly becomes, under this gaze, a different one; and that is the question of turning all existing congregational temples into space oases.

Here we have the vision of a new norm for churches. Is it so strange to imagine, for example, that every steeple would sprout fins, and become a rocket? Is it too much to ask that we should have an orbiting satellite, at least in pictorial form, in every Narthex? Why not have an astronomical telescope in the courtyard of every church to signify that the church has healed the terrible rift between Galileo and the Church, between science and spirituality, thus giving us a new spiritual civilization for the Space Age.

Nicholas Eftimiades has proposed that space cities be formed from existing ones via the training of youngsters in a federation of galactic explorers. But the larger truth here is that the space perspective on the values and cultural norms concerning the city transform the whole theory of the city beyond just the question of how to educate the young. All these reflections follow from the simple question of modeling a space oasis.

Accordingly, the task before professionals in the field of space science - and even more so those in the field of space culture - is to ask the question, What will it mean to have as an oasis, the ideal recreational center for space citizens?

The question of the oasis is about far more than designing a place to replenish with water, fuel, or even palm trees. It has to do with touching and renewing the core of the soul's values through inspiration in the "natural" environment, by the activities, sacraments, liturgies or sacred rituals of love and communion, of community, and wealth creation, and beauty creation. We begin to inquire into the nature of the ideally balanced environment for work and play in the space age. Our inquiry then becomes a more general inquiry into the theory of the intelligent building, in the broad sense of social as well as spiritual intelligence.

What would be the experimental architectural theory best suited to the design of a space oasis? To this end, of course, we have to ask rather more foundational questions, such as What is a human being? For this purpose we would answer, simply, that a human being is one who lives in a community of love, among the ethos of a solar system, the galaxies, and the heavens.

Thus we begin to model our ideal space oasis, and we now move on to consider the physical and technical aspects of the fully realized oasis in space.

Practice

Oases in space are space stations, stepping-stone support communities from one space exploration region to another. We envision a Lunar Oasis, perhaps eventually several. There will be another on Mars, and several more in the Main Asteroid Belt. There will certainly be a Hohman Transfer Oasis in the Hohman transfer orbit crossing the orbits of Earth and Mars, where personnel in transit between Earth and Mars may live safely on the several-month journey.

Oases need not be only in outer space, but also under the oceans, in Antarctica, and other areas on Earth not served by existing infrastructure.

The essence of Oases is community. Each provides not only the necessary bare-bones consumables and communications tools for the explorers it serves, and provides the social support necessary for social and mental recreation. Humans are communal creatures - we require communal settings in order to function at our best, and so oases change space exploration from personal journeys into collective journeys. Oases in space are modeled on small towns: they include not only complete facilities for mission support, but all the amenities for rest, recreation, and permanent gracious living as well.

Oases are designed for large-scale expansion from Oasis to Town to City. They are designed to support families, and from these families come the space explorers, miners, settlers, artists, administrators, restaurant operators, barbers, store operators, plumbers, mechanics, teachers, children, farmers and gardeners, builders, writers, publishers, actors and all the many other occupations of a balanced community in a psychologically healthy environment.

From each Oasis, human-crewed and robotic ships reach out to explore the local region. Because of the constant danger of solar flares, meteorites, and high background radiation, many missions will be robotic, and humans will follow to places where human attributes and capabilities are needed.

Space provides a unique industrial environment with which we have little experience. It allows limitless energy from the sun, huge spaces, micro-gravity, and asteroids with massive amounts of metals, ices and minerals. We would do well to have orbital facilities where researchers can examine these attributes just to see what new things can be done. There will be surprises.

Rotating space stations can provide variable artificial gravity, from almost nil to many times Earth gravity. The space environment has the potential of tremendous wealth creation in intellectual property and industrial products.

Oases are as self-sufficient as possible. They obtain their material needs from in situ resources processed by hydrothermal and solar energy from orbital solar power satellites, or power beamed from the surface of the Moon or Mars, or from orbital nuclear stations in orbits that will not degrade and pose any hazard to the Earth. They recycle wastes, import as little as possible, and export as much as possible.

Medical facilities in space will allow researchers to work with very dangerous biotechnologies without any possibility of endangering the Earth. Microgravity stations will also allow burn victims to heal without touching their bedding.

The dangerous aspects of the space environment predicate that law enforcement personnel will control many aspects of behavior and possessions. Projectile weapons and explosives will not be allowed, as they endanger the integrity of a space station. Personal behaviors will be monitored to avoid dangers to the station.

Creative financing for very expensive aspects of space exploration will be developed, perhaps accommodating many transnational corporations and national governments in public-private enterprises. World-wide shareholding in single enterprises will become common. The principles of Oases can be studied on Earth in examples such as the Biosphere Project in Arizona, naval submarines and capital ships, and remote military bases and civil exploration and work camps. An excellent recent example is the chain of Pipeline Camps across Alaska supporting the construction of the Trans-Alaska Pipeline.

Another "enclosed city," already well-studied, is the Naval aircraft carrier, which may house 5000 personnel of both sexes. Modern aircraft carriers have nuclear power plants and are capable of operations at sea for several years. They not only have machine shops and fuel stores, weapons caches and aircraft, but also shopping stores, theaters, libraries, snack shops, post offices, glee clubs, choirs, barber shops, arts and crafts facilities, full-function gymnasiums, and all kinds of clubs for sports, games, private scientific interests, model building, bowling, and many more. They even carry selective absorbents to restrict and clean fuel spills, and processes seawater into fresh water for drinking and bathing requirements. Throughout a carrier cleanliness is a fetish.

An Oasis in orbital space would be much like an aircraft carrier, although it will have no seawater to process. Virtually all stores must be brought to it at great cost, so recycling and conservation will be of even greater importance.

On the Moon and Mars, however, more in situ resources are available. Minerals and ice can be mined and processed to provide water, hydrogen, oxygen, methane, helium, helium-three, carbon, carbon-dioxide, carbon monoxide, glass, iron, steel, fiberglass, carbon fullerenes, gems, building stone, sinter block and cinder block, concrete, and many other materials. Solar energy can be obtained either on the surface or beamed from orbital solar power satellites, and on the Moon helium-three will be available for fusion energy plants.

Most of the material requirements can thus be obtained locally on the Moon and Mars. However, whether on the Moon, Mars, or orbital space, more than just material requirements must be met to create a good psychological setting for the people living and working there. There must be room to move about, so that people don't feel buried and claustrophobic. There must be spaces that are open, wide and high, and flooded with natural sunlight perhaps brought in by light pipes from the surface. There must be growing green plants in abundance for freshening the air in a natural manner, providing natural variety in shape and color for visual stimulus, and providing the smells of Earth which we will surely miss if they are not present.

Even towns and cities that are not remote may organize themselves towards self-sufficiency along the Oasis concept lines, and may be awarded the title of Oasis if they meet certain criteria. We already see this trend in large cities, as local community centers coalesce around neighborhoods of a few thousand homes, with their own local supermarkets, theaters, shopping malls, bowling alleys, auto repair shops, warehouses, grade schools, community colleges and other local interest groups. The trend towards local community activity and identity is innate in humanity, and it happens naturally without central planning.

As in Earth-bound towns and cities, there must be the possibility to alter the environment from time to time - to change the functions of buildings, to re-arrange the open spaces, to add and remove waterfalls, small forests, planted fields, or parks. There must also be silent places, where there is not the constant hum of machinery, which astronauts complain about so frequently.

Art must be readily available and commonplace throughout an Oasis. Statues, paintings and sculptures should be ubiquitous. There must be associations of artists, and plays and operas performed by local people. The more people can be brought into artistic endeavors, even if only as painters and prop-makers, the healthier the community will be.

There must be modern libraries with books and videos, and classrooms with teachers in many subjects. Many of these teachers will be workers during their workdays, teaching on their own time in whatever interests that may attract students. On-line learning must be available in practically every subject known.

There should be game arcades, swimming pools, parks with trees, bowling alleys, saloons, craft shops and craft shows, flea markets, and most all of the recreations normally found on Earth. There should be so much available for occupying one's spare time that no one would ever get to take advantage of it all.

In orbital space, on the Moon and on Mars the opportunity should be made as often as is practical and reasonably safe, depending upon meteorite and radiation hazards, to go out on the surface for a complete change of scenery and the opportunity to do some genuine exploring. No matter how large an underground facility can be made, they are still underground, and being outside under the sky and the stars will be psychologically necessary. Open panoramas of the outdoors should also be available from indoors, as underground facilities shielded by overhead cover should have large windows for viewing the Lunar and Martian landscapes. If this is not practical, then panoramas underground can simulate the surface with IMAX-style projections from video cameras on the surface.

We have also proposed to build a restaurant overlooking the Sea of Dreams on the Lunar surface. Individuals could purchase clear or colored lights, solar-powered, beaming up into the sky. Some could even be sent from the Earth and automatically landed and actuated before humans return to the Moon. Others could be purchased at the Lunar Sea of Dreams restaurant, or on Earth, and Lunar workers would emplace them. Each light would be dedicated to a dream, an inspiration, or a person such as a wife, husband, or child, or an inspirational person from history, or even to celebrate a concept such as love or passion. The Sea of Dreams would eventually be covered with thousands or even millions of white and colored lights, creating a vibrant display symbolizing our dreams, and realizing them at the same time.

Research Questions

1. What are the best technologies to use in raising food and feeding balanced meals to all the personnel of a space Oasis colony instead of importing it? Will it be necessary to grow entire plants, or can just the edible parts of plants be grown in culture facilities? How much space will that require per person?
2. Are there any substances that are biologically necessary for human health that can only be obtained from meat, fish or poultry, rather than from plants or synthesized at the Oasis?
3. What volumes of gases are required per person per day for breathing, and what recycling technologies are available and at what efficiencies?

4. What are the most efficient air cleaners available? What are their drawbacks? What molecules cannot be adequately removed from air with current technologies?

5. What is the best governance system for different-sized Oases? How is law enforcement to operate?

6. What is the optimum design for protection against solar flares, cosmic radiation, and meteorites at a deep-space Oasis? What thicknesses of rock, water, plastic or other materials will be required?

7. For an Oasis on the Moon, how can materials for the construction and operation of mineral and gas mining machines be made from in-situ resources, involving creation of mining equipment, refining equipment, machine tools, and assembly tools? What is the bare minimum that absolutely must be brought from the Earth?

8. How would you design an early, robotic Lunar gas-mining machine that can cook gases from the Lunar surface without processing large amounts of materials? This may be a robot that simply cooks the surface with solar or nuclear power and collects the gases, or perhaps drills rods into the surface a few meters to heat a deeper fraction. The robot should separate helium-three from the collected gases and store it in a tank that can be delivered to a robotic Earth-return spacecraft.

9. What are the five Earth-orbit crossing asteroids larger than 100 meters diameter easiest to approach and robotically mine and/or deflect to another orbit as proof-of-principle missions?

10. Do we have any indications of lava tubes on the Moon?

11. How might lava tubes or mined galleries be sealed to keep atmosphere in?

12. How to Design a robotic Lunar Light lander for the Sea of Dreams?

Reference

(1) http://www.iep.utm.edu/a/anaximan.htm The Internet Encyclopedia of Philosophy

About the Authors

Edward B. Kiker is the General Engineer for the Office of the Chief Scientist, Operational Support Office for the US Army Space and Missile Defense Command/Army Forces Strategic Command in Colorado Springs, Colorado. He is responsible for searching out new technologies that can be applied to space to keep the United States the pre-imminent space power.

He attended Harvard University in the Army ROTC program during 1966-70, with major in Lunar Geology. He served 1971-75 on active duty as an Engineer Officer in Alaska, Korea, and Fort Belvoir, Virginia. He served 1975-79 as an Armored Cavalry Arctic Scouts Officer with the Alaska Army National Guard. His civilian service began in 1975 as Natural Resources Specialist at Ft. Greely, AK, with LANDSAT satellite imagery work and environmental management. Between 1981-84 he served as Alaska State Director, Project High Frontier, developing a notional National Ballistic Missile Defense System and Commercial Space Activity.

In 1987 Ed moved to the Army Space Institute, Ft. Leavenworth, KS, where he wrote the first Required Operational Capabilities Plan and the Organizational and Operational Plan for the functioning of the current Ground-Based Missile Defense System operated by the US Army Space and Missile Defense Command. He was also the Army Training and Doctrine Command Point of Contact for the Space Exploration Initiative. He represented the Army Corps of Engineers at the Second International Fusion Energy Conference, and at International Lunar Base Design Conferences here and in Beatenberg, Switzerland.

Richard Stephen Kirby is a Cosmic Theologian and Astronomical Chaplain. His most important publications to date are "The Person in Psychology" and "Individual Diffferences," both written with John Radford, "The Mission of Mysticisms," and "Christians in the World of Computers" with Parker Rossman. Dr. Kirby also co-wrote "Temples of Tomorrow" with Earl Brewer, "The Leadership of Civilization Building" with Richard J. Spady, and "Nurturing Civilization Building" with Barbara Gilles. Kirby's Ph.D. thesis at Kings College London in 1992 was on the subject of Theological Definition of cosmic Order. Dr. Kirby attended the General Theological Seminary of the Episcopal Church in the ordination program from 1982 to 1985, and worked in various church ministries on both sides of the Atlantic for the next 10 years. While finishing his Ph.D. in Christian doctrine and history he worked developing a theological think tank which became known as the Stuart C. Dodd Institute for Social Innovation. In 1988 Dr. Kirby worked to develop the World Network of Religious Futurists. He became the chairman in 1993 and served in this capacity for 12 years when he began to serve as the organization's Chaplain. In 2001 he formed the chaplaincy program for the World Future Society. In 2005 he became the first President and Chaplain of Kepler Academy, an astronomical and theological college.

http://www.newgenius.com, http://www.wnrf.org, http://www.stuartcdoddinstitute.org

PART IV.

STRATEGY & SYNTHESIS

We have covered a tremendous range of issues and topics in Parts I, II, and III. Now in Part IV we bring it together with four chapters that synthesize the key strategic issues for the future of humans in space, including a view of the critical role of advocacy, and alternative strategic considerations and approaches. The final chapter examines theory and action; theory, since correct theory is essential to achieving the right results, and action, since useful action is the necessary complement to a theoretical formulation.

Chapter 33

The Genes of Space

By Hylan B. Lyon, Jr., Becky Cross & Bruce Pittman

Introduction

We all feel wonder and awe at the majesty of the universe, and our ancestors, as far back in antiquity as we can see, have shared this wonder at the movements of the stars and planets. With the extension of mankind's "large steps" onto the moon during the last century, human space activities provided even more captivating and compelling experiences for people worldwide. Add to this the exploration of our solar system and the profusion of scientific knowledge from the Hubble space telescope and other instruments in Earth orbit and we have profoundly changed our understanding of the Universe and our place in it.

The entertainment industry has tapped into this underlying resonance across the world's population. They have created productions that have dominated the media, with blockbuster movie series such as Star Wars and Star Trek, and the earlier science fiction stories of authors such as Arthur C. Clarke and Robert Heinlein that all echo the heroic "archetypes" of human drama and outward expansion. The vastly improved communications media of the last several decades have brought compelling images of space, both real and computer animated, from the deepest reaches of space to theaters, living rooms, and computer screens everywhere.

The title of this chapter, "The Genes of Space" reflects the fact that while each of us has a different set of genes that determine our individual potential, we are learning from the results of the Human Genome Experiment that it is the stimuli that express (or turn on) these genes that determine our capabilities. Thus, we base this chapter on the notion that our shared wonder about the majesty of the universe can be thought of as a "cultural gene," and that these words and images about human exploration of space will evoke in each reader a positive or negative response. To ensure that this cultural gene potential is activated we must provide the proper stimulus that will enable or even compel its expression. In this chapter we will discuss what can and must be done to create the required environmental stimuli.

At present these latent "cultural" ties to space have not manifested themselves into a political force that is potent enough or consistent enough to mobilize the needed increases in federal funds or private sector resources. This to a large extent is due to the fact that these space initiatives are presented to the public in a way that is not adequately engaging nor compelling. The people are interested in these projects but do not see how space projects can make any kind of significant contribution to their lives or their fortunes. It is often said that public support for space is a mile wide and an inch deep.

Thus, although the underlying instincts are universal and profound, they have not yet mobilized large-scale financial support from the free market. Reflecting this perspective at a recent NASA-sponsored conference on the Vision for Space Exploration, James Cameron, award winning director of Titanic, The Terminator, Aliens and The Abyss, offered NASA two important pieces of advice: "You need to tell your story better," and "You need a better story to tell."

Hence the hypothesis that is the focus of this chapter: "There are powerful forces latent within the broad reaches of human society that are sufficient to empower the emergence of a true spacefaring civilization on a scale that is appropriate and necessary for mankind. But they have not yet been activated."

The challenges then, are to position compelling arguments within the fabric of public discourse to activate these latent forces, and to identify individuals who command the credibility to empower visions and ability to communicate the vision. Only thus will we simultaneously enable: changes in administration of our governments, the emergence of commercial interests and shifts in the mood of society.

Ambivalence, Resistance, and Confusion

The willingness of professional astronauts and private enterprises around the world to risk their lives and financial resources to open the space frontiers is both stunning and troublesome at the same time. The public appears to be proud that individuals of such talent and dedication are willing to put their lives at risk for us, but at the same time many wonder if the risks they take are really worth the benefits that result. They also doubt the value of mankind's progress in space and the vast expense it entails.

From a cynical perspective, NASA's post-Apollo efforts can be seen as bureaucratically self serving, and both the Shuttle and the International Space Station have been labeled as "mistakes" by current NASA Administrator Mike Griffin. Broken promises, cost overruns, and slipped schedules have depleted the confidence of the public, Congress, and the White House not just in NASA, but the entire space program.

And at the same time the public takes for granted space based systems that affect their life everyday; weather satellites, GPS, and satellite radio have become unquestioned parts of the fabric of modern society,

Furthermore, the public doesn't understand why it's so difficult and costs so much to go to space, and while they take pride in past space accomplishments, they question the cost and extensive bureaucracy that the space program has fostered. In almost all other technical fields, computers, telecommunications, televisions, etc., as the technology matures the capabilities increase and the costs come down. But not space. The reliability of rockets today remains about where they were 30 years ago and the cost of putting a pound of payload in space has remained stubbornly high despite massive amounts of money that have been spent on projects to reduce it.

The values of the scientific discoveries that have come from the space program are not universally accepted, and the relationship between science and public opinion has been and will continue to be a complex issue. Over the past 30 years the National Science Foundation (NSF) has published Science and Engineering Indicators on several interesting aspects of public opinion on science. Yet we hear that public opinion related to science is driven as much by ideological and religious beliefs as it is by thoughtful reasoning. A recent article by Bill Moyers (New York Review of Books, March 24, 2005) presented an agonizing appraisal of the impact of religious beliefs on government policy toward science.

Thus, the future vision for space must be compelling not just to NASA and the big aerospace companies, but also to the general public, and particularly to young students who must decide if there really is a payoff in taking challenging science and math classes. This future vision needs to provide a new paradigm, not just of armchair exploration in which we watch astronauts on TV do things that the rest of us can never hope to participate in, but one that is genuinely participative. This future should be laid out so every 6th grader knows that if they really want to go to space, or participate in space projects, they have at least an even chance to do so if they are willing to invest their own efforts in developing the required knowledge and skills.

Key to our case is that these differentiations within our public, in a sense reside within us all, we all have: the sense of wonder, the challenge of preparation, the sense of separation, and the questions about the costs/and benefits of space. In many ways our receptivity to space challenges at any one time is determined by what we have recently been exposed to, and how that is reinforced by other ideas embedded in public discourse.

The Key Questions

Against this backdrop, we propose that we can indeed overcome the lack of active support for manned space activities through the proper use of the words and ideas that are contained in this book, and in doing so we must recognize that what we do with these words and ideas will determine in part the following outcomes:

* Will space activities be seen as simply another discretionary expenditure within an increasingly complex set of demands on our discretionary resources?
* Or will space activities compel and propel the imagination and desires of mankind?
* Will the rare probability of space catastrophes (i.e., impact of a giant meteor) and the perception that we are incapable of reacting in timely fashion discourage the public?

* Or will the awareness that we could respond to some catastrophic events shift the expenditure from discretionary to mandatory?

* Will space activities constantly be imbedded in bureaucratic programs that are constrained by hidden agendas and micromanaged to less than compelling ends?
* Or will space activities compel and explore domains of thought and human presence many of which do not exist in even today's most creative of minds?

* Will space be a venue limited to government-sponsored astronauts performing science and research tasks?
* Or will space be a place of commerce and community where people can work, live, build businesses and raise families; can it truly be the "next frontier"

We are conflicted as a society as to how to proceed, and the exploration of this dilemma is the underlying debate that needs to happen, but isn't. What we do in the space advocacy community to help guide and communicate the vision will impact whether NASA programs inspire and motivate the latent demand or evoke a "been there, done that" attitude.

Engaging the Children
The Texas Essential Knowledge and Skills statement is used to guide one of the largest public school curricula in the US, and as a participant in drafting the Science and Mathematics portion the lead author found consistently strong interest in science and space topics.

Like adults, children have a deep-seated awe of the cosmos, and they feel excitement about anything having to so with space. Through K-12 enrichment activities sponsored by The IGNITE Foundation, an education enrichment non-profit organization, it has been noted that children of all ages have an interest in space and space exploration, but most do not feel that it is something that they could ever be a part of. [See Chapter 19, Sowing Inspiration for Generations of "Space Adventurers" by Becky Cross.] Even young children seem removed from the topic despite their fascination with it, and it's clear that K-12 children have little or no exposure to the potential career opportunities that future space initiatives hold, while older students do not see the possibility of a space-related career path.

Academic standards on both the national and state levels essentially remove space science from the curriculum, and the drive to participate is nearly non-existent unless a child happens to have a parent, mentor, or teacher who is knowledgeable on the subject.

Stimulating children with information on current space activities is thus essential to generating a future workforce for space initiatives. As their interest is already there, it simply needs to be cultivated and encouraged. As an added bonus, educating children also educates the adults around them.

If this challenge is to be met, we must understand how to influence public opinion about the importance of space activities, shifting the focus from the high expense to the value and importance of the investment in:

* The future of global society
* Potential new commerce opportunities
* Human's quality of life in an increasingly constrained global village
* The survival of human activity in the context of the solar system and its challenges, and
* The increasingly important role space will play commercially as more nations and companies develop space capabilities.

In undertaking these educational efforts we must understand how to utilize the appealing synergy between the manned and unmanned systems. The question is not one of "man or machine" but instead one of "man and machine," and how can they work best together. As the cost of getting people into space falls and the market for supplying such access increases, then a more realistic balance can be identified between man and machine. Just as on earth, robots have removed much of the drudgery from manufacturing, robots in space can enhance human productivity and safety.

The Role of the Private Sector and Commerce.
The original intent of the NASA charter provided a role for NASA in stimulating space commerce, but this

role has been narrowly defined due to the politics of bureaucratic survival in a hostile environment. Without a repositioning of roles and missions, NASA's future will be a predictable disappointment. A government agency is prohibited from "selling itself or its programs," which in NASA's case inhibits progress, but promoting space is a role that the private sector will soon be ready, willing, and able to help perform.

Until recently, the investment costs and risks of space commerce have been too high, and the potential payoffs too low for anyone other than the government to participate, but this is starting to change. Entrepreneurs such as Bob Bigelow, Elon Musk, Jeff Bezos, Paul Allen and Richard Branson are investing hundreds of millions of their own dollars to open up the space frontier. Other than Bigelow, who wants to provide inflatable space hotels in orbit, they are all focused on supplying launch and flight services, but as yet the demand for cheap, reliable access to space to produce goods and services is not as well developed as it needs to be.

The most plausible market opportunities apparent today are on the moon or in Earth orbit, and they include:

* Mining the lunar surface for Helium 3 to fuel fusion energy stations on Earth
* Biotech research in microgravity (see Chapter 14: Biotech: A Near Future Revolution from Space)
* Flying tourists into space for out-of-this-world vacations
* Placing large solar cell arrays in space and beaming power back to earth

While many of these concepts seem far out and unrealistic it is good to remember that it was less than 66 years between the Wright Brothers first flight and the first flight of the Boeing 747. Little would the Wright Brothers have guessed that one of the main markets for their invention was carrying cut flowers and perishable produce between continents.

Can anyone bring closure to these business cases with or without government assistance? Not yet. But further progress in any of these potential markets could stimulate a broader consensus on Earth. It is important that the link be made between government space programs and their ability to assist in jump starting these commercial ventures. NASA is making all of the right noises about this but there is a potential problem with the planned NASA Crew Exploration Vehicle and its potential competition with private companies that want to provide many of the same capabilities.

In support of a commercial space vision, NASA's roles should be defined as:

• Leading in building the infrastructure
• Leading in space exploration/science
• Following wherever private capital worldwide is willing to accept the risk, and
• Getting out of the way when commerce can take over.

In addition to supporting commercial activities, and its exploration role, NASA may also find an appropriate role to play in the development of energy sources, other than nuclear fission, to replace liquid petroleum; lunar helium 3 is one possibility.

NASA should also play a role in developing a global surveillance system that promotes global, international, national, and domestic security, a function that can only be adequately performed from the surface of the moon.

New Concepts of International Participation and Global Collaboration

International participation in these projects should be much more comprehensive than joint scientific collaboration projects or cost sharing and collaboration on mega-engineering projects.

It can and must be a catalyst to stimulate the hopes, dreams, and aspirations of the societies of the world in many dimensions. It can provide a framework for true international cooperation and community that gives us new insights in how to work and live together in peace and prosperity both on earth and in space.

Conclusion

Every successive US government administration, every presentation before a corporate boards of directors in the aerospace and biomedical industries, every educational non-profit organization, and every planning session for congressional committee staffs and members of Congress will be confronted with the dilemmas described here.

If the force toward space that lies latent within society is powerful enough to inspire us, then why do we have to work so hard to make it compelling? What generates the latency? In our view, latency is due to the gap between the lure of space, and the widespread perception that most individuals cannot participate, nor be affected in any real positive way, by the outcomes of today's space programs.

Those who bridge this gap, such as young "astronauts to be" who in their early years become obsessed with doing whatever it takes to fly in space, do not need a push from us, but for the rest of society a compelling argument must be made.

Our challenge is thus to define and empower the space ethic in enough people to carry the day. We seek to activate a significant number of individuals who as yet may not perceive that their lives will be affected in any meaningful way by initiatives in space, but when they hear the right story told well, they will. Many avenues must be pursued, some will work and some will not. We must listen, learn, refine and move onward.

You, the reader, may be an author of a white paper, a proposal, a planning document, or a budget analysis, and you have a critical role to play in persuading an indifferent society which nevertheless has the potential imbedded in their genes to become enthusiastic supporters.

The piece that you write is a bridge, an invitation to reconsider, to reassess, to reawaken the almost primordial instincts and to take a step toward space. By your prose and choice of words you too can activate the genes of space that reside in all of us.

We leave it to the authors of the other chapters of this book to provide you with the necessary fodder for your argument, and we know from experience that you, as author of your own political document, will have to recast these concepts in fresh terms suited to your audience. But if you accept the idea that your prose contributes to unlocking support for something that is appropriate for all humanity, then this chapter did its job. The originality that is critical to compelling prose supporting these goals is based upon the sincerity and authenticity of the emotions of the author.

Why is this important? It's important because you, and only you, are in position to chip away at the "competing arguments." Your commitment to the overarching necessity of venturing into space in order to properly conduct our shared stewardship of humanity and the Earth is the key. Your prose becomes part of the fabric of public discourse that activates latent demand, and your "voice" contained within that prose will command the credibility to empower the vision.

About the Authors

Hylan B. Lyon, Jr. is President and COO of Gamma Design, Inc. In the early 1970s he was a member of the President's Science Advisor's staff for space and aviation, and since then he has held numerous business and policy assignments in high technology and aerospace in both government and industry.

Becky Cross has written 2 children's books, including I Am A Space Shuttle. I LOVE TO FLY!. She is also founder of The IGNITE Foundation (www.ignitefoundation.org), an educational enrichment organization operating in New England. Through IGNITE programs she has interacted with more than 15,000 students from kindergarten through high school on the topic of space. Ms. Cross has spoken at large on the topics of future space and the benefits of inspiring future generations of space adventurers. She presented at the 2005 International Space Development Conference on the topic of the Benefits of Space.

Bruce Pittman has been involved in high technology product development, project management and system engineering for over 25 years. He spent 11 years working for NASA managing projects in the areas of planetary exploration and infrared astronomy. From 1985-1998 he was a consultant and educator for both industry and government in project management, system engineering, concurrent engineering, risk management and planning, both in the US and abroad. From 1999-2002 Bruce was a Director for two high technology firms, one in telecom hardware manufacturing and the other in enterprise software for new product development. Now a consultant, coach and trainer, Bruce has worked with a wide variety of both Fortune 1000 companies and smaller companies, including Align Technology, Varian, Westinghouse, Lockheed Martin, Northrop Grumman, TRW, Cepheid, Behring Diagnostic, and government agencies such as NASA, the Dept. of Defense and the Dept. of Energy. Bruce also teaches as an adjunct professor in the graduate engineering management program at Santa Clara University. Bruce has a BS in Mechanical Engineering from UC Davis and an MS in Engineering Management from Santa Clara University. He is a Vice President and member of the Board of Directors of the Society of Concurrent Product Development. He is also an Associate Fellow of the American Institute of Aeronautics and Astronautics.

Chapter 34

Alternative Strategic Approaches to Space

By Martin Schwab

Introduction

The theme of this chapter is *redirection* of military capabilities by the major world powers into global space missions. The following overview of the major divisions of strategic thinking in the West and the scenario sketches serve as reminders for decision makers that there is a wide range of philosophies as to *why* humans should explore outer space, and that all of these philosophies need to be considered when deciding *how* we can best explore outer space. In order to promote interactive dialogue in our meeting rooms, boardrooms and classrooms, scenario sketches are given for intervals of 25 years out from 2006 (2031, 2056 and 2081). The last section argues that a variety of clear and present threats to humanity exist now and are natural rallying cries for policy makers to justify working globally, collaboratively and urgently in outer space.

| Weaponized anarchy | Realism | Economic Liberalism | Idealism | Dynamic co-exploration |

Figure 34.1 Continuum of Relationships in Outer Space

The strategic approaches to space that are introduced in this chapter are located at various points on the above *continuum of relationships in outer space.* It will be shown that the directions of these approaches have a common pull toward co-exploration and that it is only the starting points on the continuum where assumptions differ. If all of us on this continuum are to better arrange human institutions to bring about co-exploration *beyond Earth*, we need to briefly review the classic theories of international relations that divide us, and not just in the West: Realism (balance of power), Liberalism (economic liberty) and Idealism, often referred to as constructivism or radicalism of which Marxism is one of many varieties.

All three theories have intellectual blind spots. The perspective of pure balance of power realism or "peace through strength" does not always account for naturally existing progress and genuine good relations among nations, nor does it recognize that political legitimacy can be a source of power equal to military power. Pure economic liberalism or "peace through trade" often fails to recognize that democracies survive only if they safeguard military power, and that transitions to democracy are sometimes violent. Pure idealism-constructivism, or "peace through ideas," often does not explain which power structures or social conditions are able to consistently yield new realities through the persuasive communication of values. (1)

The constructivist bias expressed in this chapter favors an international negotiation framework at the presidential level that would allow *small, annual, incremental and reciprocal transfers* in terms of percentages of military budgets into large scale and integrated efforts to explore, understand, preserve and enjoy our solar system. These transfers might conceivably be deposited into an international budget or a private fund, with appropriate safeguards.

Peace through Strength in Space

"Peace through strength" or balance of power realism is a strategic approach that dates back far into our ancestry. This approach assumes that *human nature is fundamentally bad.* Sub-groups of humans that our political scientists refer to as nation-states require "sticks" to maintain "order." Under this approach, human nature and individual self-interest must be safeguarded against in order to secure the greatest security for the greatest number of nation-states. The peace through strength approach in the United States (U.S.) contains two sub-divisions, illustrated below by scenario sketches. At times, these two camps appear to be diametrically opposed, although in reality they share the assumption of realism on our continuum, that militaries are the primary components of world order, to be responsibly influenced through U.S. leadership.

Global Stability through U.S. Dominance of Space

Scenario sketch: U.S. military dominance in low Earth orbit either exists or is perceived. All other militaries and para-militaries on Earth choose to invest in areas other than military-space technology, resulting from sober cost benefit analyses of challenging the supremacy of U.S. military-space power. As with British naval dominance of the high seas during the 18th and 19th centuries, U.S. dominance of outer space guarantees the security and predictability of global commerce from 2006 until 2081.

Professor Everett Dolman of the U.S. Air Force School of Advanced Air and Space Studies at Maxwell Air Force Base has introduced the concept of "astrogeography," or geographic positioning in space as it pertains to the evolution of current and future military space advantage. While Dolman does not advocate any one "astropolitical" future for the exploration of space, he does analyze what the future might be like, given the continuing and his view likely nationalistic pursuit of outer space. Dolman formulates a neoclassical astropolitical dictum to guide our way: "Who controls low-Earth orbit controls near Earth space. Who controls near-Earth space dominates terra. Who dominates terra determines the destiny of humankind." Dolman argues that the agreements against the weaponizing of space forged during the Cold War were achieved because neither the U.S. nor the Soviet Union thought that they could achieve dominance in space over the other. [2]

Now the U.S. alone enjoys the position of dominance, and can choose to either build upon or squander this advantage as other nations will inevitably catch up if the U.S. does nothing. Dolman asserts that because U.S. citizens are not comfortable with the exercise of global dominance, this makes the U.S. the best candidate for monopolizing space weapons - to keep the peace on Earth so that humanity can begin her first *global golden age*. [3]

Global Stability through U.S. Negotiation of Space

Scenario sketch: The U.S. acts as a major contributor to a global consensus on robotic and human exploration of our solar system until 2031. A contentious treaty-making process outlawing weapons in space results in a secret military pact between China, Russia and Europe for "control of the high ground." Once the secret pact becomes obvious, the U.S. with the aid of Japan, India, Malaysia, Australia and a reluctant Canada rapidly mobilize a space weapons program. These allies in outer space reassert dominance of low Earth orbit by 2042, 100 years after the Battle of Midway Island in World War II. By 2056, the secret collaborators in China, Russia and Europe are ousted within their own governments. The U.S. is granted moral authority and political capital on Earth by all nations at the Conference on Disarmament in Geneva because it did not start a space weapons race in the 2010 timeframe. The U.S. leads the peaceful development of the solar system through the rule of international law, in keeping with the spirit of the Outer Space Treaty of 1967. This landmark treaty emerges as the defacto "Constitution of the Solar System" until 2081.

Below is an example of the type of strategic thought that is already taking place toward dynamic co-exploration on our *continuum of relationships in outer space,* taken from an address given by Theresa Hitchens, Director of the Center for Defense Information (an independent oversight group) at an international conference on the prevention of an arms race in outer space (PAROS) in the Council Chamber of the Palais des Nations in Geneva:

...While the United States may be unwilling to work toward a ban on space weapons, it remains a major - and for the most part responsible - player in space. Isolating the United States because of its position on space weaponization is simply a waste of time; or worse, attempts to do so may well backfire by promoting the views of those in the United States who see unilateral approaches to security as the only approaches. Meanwhile, other spacefaring nations need to be discouraged from treading down similar destructive paths. This brings even more urgency to undertaking initiatives that promote cooperation amongst the spacefaring powers in areas where they have mutual interests. There is little time to waste. [4]

Peace through Trade in Space

"Peace through trade," also known as the classic theory of economic liberalism assumes that *human nature is fundamentally good.* In outer space as well as on Earth the curious, creative and restless human spirit, the individual self-interest must be given priority over the interests of nation-states. The poor cannot be helped by hurting the rich through tyrannical economic redistribution of wealth by Earth-bound governments, but only by government tax relief, "carrots," provided to wealthy individuals to employ the masses. *Each individual* within the mass of workers is not a cog in a space wheel but rather a potential future employer or manager of an enterprise in outer space.

Scenario sketch: Global free enterprise, not governments, dominate the initial development of our solar system until 2056, followed by another generation of limited government that supports and does not hinder the individual in space until 2081. Rapid advances in the space elevator make space tourism throughout the Earth-moon system easily affordable by the upper middle class in modernized nations from 2031 to 2056. Asteroids and Earth's moon are mined by many small private firms operating in perfect competition from 2031-2056. At least one small subterranean private settlement project to be constructed at Mars has been fully invested in by multiple private banks as of 2031. Private constructions of space wheel settlements, still only affordable by Earth's economic elite are established at L5 from 2056 to 2081.

"Space is a place not a program." This has been one of the rallying cries of the Space Frontier Foundation since its founding in 1988. This influential membership advocacy group believes that "the permanent human habitation of space can only be accomplished by unleashing true free enterprise." [5] If global mobilization to explore *our solar system and beyond* becomes driven primarily by a free and fair global market, the human space endeavor in reality might look something like the list of hypothetical public-private space initiatives during the 2031-2081 timeframe below. Note also the capitalization of the astrogeographical regions, potential units of interplanetary representative government (or private sector governance) for the post 2081 timeframe:

* Hypothetical public-private space initiatives from 2031-2081
* Eurasian Space Consortium colonizing Northern Latitudes of Mars.
* Pan-American Space Coalition colonizing Southern Latitudes of Mars.
* African Space Agency mining the Sea of Tranquility of Earth's moon for helium 3.
* Oceanic Space Launch Authority sending scientific rovers to the far side of Earth's moon.

Peace through International Endeavor in Space

"Peace through international endeavor in space" is not so much a national strategic approach as it is a transglobal design, grounded in the classic theory of constructivism. This transglobal design assumes that *human beings have an equal capacity for good and evil.* In this thought model, the choices that individual human beings make in their lives are primarily influenced by those external conditions that have shaped us all in our local communities and through various philosophies of child development. In this model of idealism, nurture, more than nature determines who we are as individuals, and how we interact with the rest of our world.

Medieval literature scholar Joseph Campbell argues that the form of nurture that our world desperately needs today is a modern myth that has the vitality to move the human soul to offer its temporal life for others in a great cause. Our current myth that success is measured economically is a recipe for societal discontent, resulting in various forms of neuroticism and schizophrenia, including through drug use. Campbell argues that the economic model of history fails to explain why the pyramids of Egypt and the cathedrals of Europe were built over many lifetimes. Campbell cites July 20, 1969, the night Neil Armstrong walked on our moon, as the great event of the modern era - a public demonstration of the standard to which all human beings in our global tribe can now aspire. [6]

In the 21st century, with planning, dedication and most importantly, a better sense of *global team spirit,* it could be possible for any human being, who so desires to participate in public or private sector space endeavors, each promising direct and indirect benefits to all of Earth over multiple generations.

Scenario sketch: Massive and sustained redirection of the militaries of our planet into a supreme migration of human and robotic missions to Mars and all celestial bodies in our solar system. These joint human "expeditionary missions" are negotiated, planned and executed by all state and non-state actors on a near continuous basis from 2006 to 2081.

This scenario echoes the precedent of massive international action that was set by President Dwight D.

Eisenhower when he was Supreme Commander of the Allied Expeditionary Force in Operation Overlord. History teaches us that this global invasion of Nazi occupied France was critical in bringing World War II to an end so that Europe could integrate itself back into our present day world system. In 1961, nine months after transferring the title of Commander-in Chief to President John F. Kennedy, General Eisenhower addressed the faculty and students of the Naval War College on the topic of how to help President Kennedy and future U.S. presidents win the Cold War. Below are a few telling excerpts, from his address, which echo Campbell's thoughts above, indicate his views on the military conquest of space in 1961, and suggest just how far behind schedule Eisenhower might view the human endeavor in outer space today:

...If there is one difference between Communism and representative government, it is that we believe in a Supreme Being of some sort, while they say we are cattle...we are dedicated to the defense of something that is even more precious than life itself. We must recall from time to time the majesty of Patrick Henry's statement in the Virginia Convention, "Give me liberty or give me death."...I believe that someday humans are going to circle the moon, take some pictures of it, and maybe even get to a planet and back if there's time - I don't know - but I believe those things ought to come about as a by-product of all the research we are doing today in missiles and in bigger engines and so on. I think to make the so-called race to the moon a major element in our struggle to show that we are superior to the Russians, is getting our eyes off the right target. I really believe that we don't have that many enemies on the moon... if we will put our eyes on the values that bind us together [referring to the transatlantic alliance], then I think we may at least control our side of this material destructiveness, and this may be the strongest element we have in making the other fellow [referring to the superpower of the Soviet Union] be very careful himself. This is the best defense we have both against the other fellow and against ourselves... [7]

To be fair, it must be recognized that it was the Eisenhower administration that institutionalized the legacy of government secrecy in U.S. outer space activities with the advent of the Corona reconnaissance satellite and the subsequent National Reconnaissance Office, a name that was classified until 1992. However, President Eisenhower cannot be blamed for using secrecy in space to protect his citizens from possible Soviet aggression, as it was the Soviet Union that rejected his offer of an Open Skies initiative in 1955. [8] This initiative would have provided mutual aerial observation to promote openness and transparency of military forces and activities, a provision that is now in force as of 2002 under the Open Skies Treaty. The Open Skies Treaty is currently signed by 34 nations including the U.S. and the Russian Federation, due to the initiative of President George H.W. Bush in 1989. [9]

Peace through international endeavor in space is another idea that has been lingering around policy circles for decades, now ready to come to fruition. During the Ronald Reagan administration, the National Commission on Space issued its report entitled *Pioneering the Space Frontier,* which captures the fundamental idea that human settlement of the solar system in unified effort would be the ultimate fulfillment of the American Revolution. The report contains a "Declaration for Space," a clear reference to the U.S. Declaration of Independence in 1776. The Declaration for Space is reproduced in its entirety as a postscript to chapter 35, "Theory and Action for the Future of Humans in Space." Below are condensed excerpts from the Declaration for Space, including a "Rationale for Exploring and Settling the Solar System."

The Solar System is our extended home...space technology has freed humankind to move outward as a species destined to expand to other worlds...We must stimulate individual initiative and free enterprise in space...Now America can create new wealth on the space frontier to benefit the entire human community by combining the energy of the Sun with materials left in space during the formation of the Solar System...American investments on the space frontier should be sustained at a small but steady fraction of our national budget...America must work with other nations in a manner consistent with our Constitution, national security, and international agreements... [10]

If we agree that Joseph Campbell is right, that walking on our moon was the greatest contribution to the human psyche by the U.S. then it follows that the U.S. decision regarding Hiroshima and Nagasaki in August 1945 was perhaps the least great contribution in this regard. These two stress points; colossal expectation as well as enormous apprehension of the U.S. by the rest of the world, born in the years 1969 and 1945 respectively will still guide world events into the near future. Pioneering the Space Frontier reiterates the words inscribed on the plaque that the Apollo astronauts placed on our moon nearly 24 years after Hiroshima: "We came in peace for all mankind," yet humankind is still not at peace with itself. Why? One way to grapple with this question may be to ask another question: How might current global political sensitivities change if citizens worldwide were together included in the planning, deliberating and dieing that will be necessary for the co-exploration of Mars and then stayed there, represented by an international crew? Or, perform in unified action similar feats of common valor on other fronts *beyond Earth?*

Clear and Present Threats to Humanity

The following paragraphs describe some of the major threats that humanity faces today, and will continue to face in the coming decades, offering an outline for possible countermeasures by redirecting the world's military capabilities into co-exploration of our solar system.

* **Asteroids and comets of our solar system.** There is extensive evidence on Earth, Earth's moon, Mercury and Mars of major collisions in the past, and many recent observations of minor impacts and near impacts between Earth's orbit and the orbits of asteroids and comets of various sizes, velocities, and masses. In 1994, we even witnessed large multiple comet impacts into Jupiter, each of which would have obliterated Earth. [11] Everyday, humanity lives under the threat of local, regional, and potentially global environmental catastrophe because of the substantial risk that asteroids or comets on Earth-crossing orbits go undiscovered. [12] Our world system has the inherent capability to confront this peril of nature but we need to establish an appropriate global chain of command and control of *detection, inspection and redirection* of these near Earth objects. It would be irresponsible of the leaders of our world system to not work together to create and maintain a more than adequate common defense of our home planet and her citizens.

* **Natural and human influenced change to the Earth system.** Citizens and their representatives need to know that we are *now* winning or losing the battles against multiple threats to human existence, in the wider war for our progeny. We are *now* experiencing the effects of climate change around Earth. We are *now* experiencing potential pandemics of disease around Earth. We are *now* experiencing fresh water scarcity around Earth. We are *now* experiencing biodiversity decline around Earth. These global threats can be overcome by an expanded human presence in our solar system, if for no other reason than micro-evacuation followed by back-population of Earth, in a worst case scenario. Closer to home, continued medical experimentation aboard the International Space Station (ISS) could potentially yield breakthrough defenses against SARS, the Ebola virus and AIDS, each of which potentially threatens global civilization, as we know it.

* **Human to human violence.** Surveillance satellites, in addition to monitoring Earth's natural sub-systems can aid human intelligence efforts around our world in preventing nuclear attack. This form of violence is designed to spawn terror among the global civilian population, serving the interests of various parochial political objectives. Sustained genocide and other forms of local intimidation are other recent tools of these objectives around Earth, of which surveillance satellites are able to provide detailed evidence to a vigilant global community, willing to take necessary action at a minute's notice.

* **Gamma-ray bursts (GRBs) of radiation from hypernovas within our galaxy.** GRBs belong to a threat category entirely of their own. They are not only a threat to our world, as they threaten possible life in entire solar systems that happen to lie within the narrowly beamed cones along the rotational axis of a progenitor star that collapses into a black hole, within a given galaxy. [13] GRBs involve the extragalactic release of energy equivalent to a billion trillion of our suns (the most powerful type of explosion in the known universe), and no one is exactly sure what causes them or even from where they originate. [14] On December 3, 2003, the European Space Agency (ESA) Integral space-based observatory detected a troublesome GRB. Months of Earth-based observations of this event, called GRB 031203, concluded that this was the closest cosmic GRB on record, but also the faintest. This suggests that an entire population of GRBs has been missed. [15] As with global asteroid and comet defense, space-based detection of GRBs is optimal, along with the *simultaneous* conception and preparation of responsible protections or countermeasures by our world's militaries, working together.

* **Three critical applications for space solar power (SSP).** 1) Solar power satellites can harvest solar power in space for transmission to Earth which can help fight climate change by serving as one of many distributed renewable energy sources to go online globally, once global civilization decides to move beyond fossil fuels. 2) SSP can be used to power energy intensive seawater desalination, necessary in the fight against global fresh water scarcity. 3) A more controversial area in which SSP could play a role is serving as a power source for interplanetary probes, which currently require small nuclear power sources for optimal performance. These three applications add a great deal of justification for global collaborative investment in this uniquely synergistic technology that can aid us in confronting multiple global threats.

Conclusion

The mandate of human history and the mandate evident to us as we look to the human future demand that we as citizens, policy makers, and policy implementers *now* do our duty to prepare for the inevitable. It is possible to address the assorted threats to our planetary system and our solar system in unison. During the Cold War, the U.S. and her allies followed a grand strategy of containment of Soviet and Chinese communism. In the post Cold War strategic environment, the most compelling replacement of the grand strategy of containment may be that of Thomas P.M. Barnett's grand strategy of economic and political connectivity. [16] The concept of redirection that has been presented in this chapter is complimentary to Barnett's connectivity. Without connectivity, it will be hard to redirect the backward slide of our military capabilities into the darkness of weaponized anarchy in space and on Earth. Six billion together cannot fail.

Research Questions

What missions in outer space (besides planetary defense) are best suited for the military systems of our world, which by achieving create political accord among the major world powers?

Could a negotiation framework at the presidential level be initiated by the U.S. to allow *small, annual, incremental, and reciprocal transfers* in terms of percentages from military budgets to an international civil space pioneering and defense budget or private fund?

References

(1) Snyder, Jack. (2004, November-December). One world, rival theories. Foreign Policy, 53-62.

(2) Dolman, Everett Carl. (2001). Astropolitik: Classical geopolitics in the space age. London: Routledge (Taylor and Francis Group), Frank Cass Publications. Abstract retrieved February 14, 2003, from
http://search.barnesandnoble.com/booksearch/isbnInquiry.asp?userid=6WLS.

(3) Dolman, Everett C. (2005, September 14). U.S. military transformation and weapons in space. Paper presented at the e-parliament conference on Space Security, Rayburn House Office Building, Washington, D.C. Retrieved November 19, 2005, from http://www.e-parl.net/pages/space_hearing_images/ConfPaper%20Dolman%20US%20Military%20Transform %20%26%20Space.pdf.

(4) Hitchens, Theresa. (2005, March 21-22). Engaging the reluctant superpower: Practical measures for ensuring space security. Paper presented at the conference on Safeguarding Space Security: Prevention of an Arms Race in Outer Space, Council Chamber, Palais des Nations, Geneva. Sponsored by the People's Republic of China, the Russian Federation, the Simons Center for Disarmament and Non-Proliferation Research (Canada) and the United Nations Institute for Disarmament Research. Retrieved April 23, 2005, from http://www.cdi.org/friendlyversion/printversion.cfm?documentID=2946&from_page=../program/document.cfm.

(5) Space Frontier Foundation. (2005). Space is a place, not a program: Help us to open the space frontier [web page]. Retrieved November 19, 2005, from, http://www.space-frontier.org/OtherVoices/19580608spaceisaplace.html.

(6) Campbell, Joseph. (Lecturer). (2002). Joseph Campbell Audio Collection, Mythology and the individual (Cassette recording No. 5, The vitality of myth). Minneapolis, MN: HighBridge Audio.

(7) U.S. Naval Institute Proceedings. (1971, June). Eisenhower at the Naval War College. Annapolis, MD. Forward by Col. Robert M. Krone (USAF).

(8) Taubman, Philip. (2003). Secret empire: Eisenhower, the CIA and the hidden story of America's space espionage. New York: Simon and Schuster.

(9) U.S. Department of State. (2005, June 13). Open skies treaty (Bureau of Arms Control Fact Sheet). Retrieved January 19, 2006, from, http://www.state.gov/t/ac/rls/fs/47801.htm.

(10) National Commission on Space. (1986). Pioneering the space frontier. New York: Bantam Books.

(11) Meteor Crater Enterprises (Director and Producer). (2001). Collisions and impacts: The role of meteors and craters in our solar system [videotape]. (Available from Meteor Crater Enterprises, P.O. Box 70, Flagstaff, AZ 86002-0070.)

(12) NASA. (2003, August 22). Study to determine the feasibility of extending the search for near-Earth objects to smaller limiting diameters. Report of the Near-Earth Object Science Definition Team. Prepared at the request of NASA Office of Space Science, Solar System Exploration Division. Retrieved November 19, 2005, from http://neo.jpl.nasa.gov/neo/neoreport030825.pdf.

(13) Dar, A. & A. De Rujula. (2002). The threat to life from Eta Carinae and gamma ray bursts. In A. Morselli and P. Picozza (Eds.), Frascati physics series: Vol. 24. Astrophysics and gamma ray physics in space (pp. 513-523). Retrieved November 20, 2005, from, http://www.citebase.org/cgi-bin/fulltext?format=application/pdf&identifier=oai:arXiv.org:astro-ph/0110162.

(14) NASA Goddard Space Flight Center. (2001, November 7). NASA's HETE spots rare gamma-ray burst afterglow (Top Story). Retrieved November 20, 2005, from, http://www.gsfc.nasa.gov/topstory/20011025heteburst.html.

(15) European Space Agency. (2004, August 6). ESA's integral detects closest cosmic gamma ray burst (News Release). Retrieved November 20, 2005, from, http://sci.esa.int/science-e/www/object/index.cfm?fobjectid=35670.

(16) Barnett, Thomas P.M. (2004). The Pentagon's new map: war and peace in the twenty-first century. New York: G.P. Putnam's Sons.

About the Author

Martin Schwab, see Chapter 3

Chapter 35

The Past is Not Sufficient as a Prolog

By Hylan B. Lyon, Jr.

The necessary conditions for a successful enterprise or policy are often self evident, intuitively obvious, and compelling to a number of advocates. But the sufficient conditions that determine if success will be achieved are often virtually invisible due to blinders of attitude. By "necessary" I mean the absolutely essential elements that must be in place for success; by "sufficient" I mean that all elements are in place in the appropriate context, amount, and proportion. Necessity therefore addresses conditions from a minimalist perspective, and is regularly subject to distortion by policy makers who have an agenda to push, while sufficiency is pragmatic and realistic.

This is important because many of the more intractable problems facing society have sufficiency conditions that are far beyond the conception of those concerned with necessity, so let's reserve a name for those that seem to resist solution as Major Classical Problems, and let's us see what we already know about them and see if that knowledge can be converted to a new understanding. Why? Because the initiatives called for in this book will require solutions of the same character as many of these major classical problems, solutions that address sufficiency as well as necessity.

One of the characteristics of our culture is that the effort required to understand sufficiency conditions is suppressed as a consequence of "paralysis by analysis," "Pareto analysis," and "focus, focus, focus." These methods assume that sufficiency conditions will be confronted and solved as a solution is worked out. But major classical problems have proven that they are of a different class; unmet sufficiency conditions often stifle solutions or result in counterintuitive and counterproductive outcomes.

We don't lack for statements of the major classical problems the world has to solve. Lists abound in the literature of analysis, such as this one from 1976:

Arms negotiations
Joint space ventures
Law of the sea
Global health care
Nuclear safeguards
Regional development programs
Materials buffer stock network
Global food allocation and production enhancement
Brain drain
Regulation of Multi Nationals

Compiled by Franklyn P. Huddle, Congressional Research Service; "Science, Technology and Diplomacy in the Age of Interdependence" GPO, 1976.

This array of problems, authored by the wonderfully persistent student and scholar of public policy, Frank Huddle, was created nearly half way (1976) between the optimism expressed in the "Science the Endless Frontier" report (1945) and the present (2006). Thus we have a check-point as to whether the basic organizing precepts for the US science and technology policy are adequate for the problems we now face 60 years later. This chapter is presented in support of the argument that awareness of the necessary conditions for development of real solutions to major classical problems, even those in the midst of broad public debate, is not sufficient to bring about meaningful solutions or positive change.

Is There a Prolog?

"Science, the Endless Frontier"57 was commissioned by President Roosevelt on November 17, 1944 with this intent: "There is no reason why the lessons to be found in this experiment (the use of science during World War II) cannot be profitably employed in times of peace." The report concludes with ... "New frontiers of the mind are before us, and if they are pioneered with the same vision, boldness and drive with which we have waged this war, we can create a fuller and more fruitful employment, and a fuller and more fruitful life."

And thus the US started down a road to managing Science and Technology for the good of society, a new and brave experiment to those of us who participated in or studied science policy history. The US model was emulated and copied worldwide, as in the genesis of the Indian Institute of Technology, or the continuation of the Marshall Plan through to OECD, and the International Energy Agency or many of the functions of the UN.

We took an unfortunate but necessary diversion during the cold war, when defense-oriented budgets, and even NASA funding, were qualitatively and quantitatively formulated in a rather distorted sense from the charter empowered by the "Endless Frontier. What is meant here is that the pure social objective of the Endless Frontier was not the rational for the funding of a great portion of science during the Cold War. In many cases the commercial spin off from research justified for Defense purposes was limited in its non-military applications because of Government policies.

In the early 1970s, Dennis and Donella Meadows published The Limits to Growth58 which alerted us that the complex nature of many of these problems is such that if current trends continue, major classical problems will continue to grow in their magnitude until a catastrophe awakens us to their true nature, at which time our response most likely will be too late and too weak. In the back of this book were scenarios of hope that could be implemented with a full understanding of the situation. But few of the detractors that I encountered had read that far.

In the mid 70s Mankind at the Turning Point59 by Mesarovic and Pestel presented the hope that the reactions of society to these now classic problems would not occur as a cataclysm. The book asserted that change would occur through learning between the different regions of the world, each in different states of progress, and that this would motivate effective action. This posited that the impact of the catastrophe of Somalia, for example, or starvation among the nations in Africa, would filter back to the other regions of the world, and that effective changes would be initiated.

In the mid 80s, Meadows returned with Groping in the Dark,60 and challenged us with the idea that mankind's self-destructive capacity would limit the quality and quantity of life on Earth before we ever reached the physical limits of natural resources.

And yet, today, if we look at the problem statement from the mid point, 1976, through to today, and if we address the role of space in the solution of mankind's problems, as in this book, we have to give ourselves a grade of ...? Or perhaps we can submit a list of excused absences. Regardless, the thesis of this chapter is that we don't have much of a prolog to evaluate; there is simply not enough history of humanity dealing successfully with major classical problems for us to know with certainty how to proceed. This is particularly the case as we address the scale of resource commitment that is required for the development of space, as discussed in this book.

The major classical problems shown in the list resist solution by present policy prescriptions or approaches. A more elegant phrase, 'World Problematique,' is a concept created by the Club of Rome to describe the set of the crucial problems - political, social, economic, technological, environmental, psychological and cultural - facing humanity. http://www.clubofrome.org/.

The complexity of the world problematique lies in the high level of mutual interdependence of all these problems on the one hand, and in the long time it often takes until the impact of action and reaction in this complex system becomes evident.

A declaration of bias is necessary at this juncture, since the author of this section was a partial and imperfect replacement for Alexander King, the founder of the Club of Rome in the mid 70s, at OECD, and while I have not continuously participated in these high level discussions, except a foray into Pugwash and government advisory mechanisms, I have been a "bird on the perch" watching the machinations. It is this experience that led me to write this chapter.

What Can We Learn?

We have several examples that show how the existence of a particular technology, and a choir of advocates singing its praises, are not a sufficient condition for positive societal impact by that technology.

When you get involved in high level policy, one of the issues I believe any reasonable person struggles with is, "Why do we have to work so hard to make good things happen?" The answer, while often obvious in broad strokes of thought, requires details that open thickets of the mind which are almost impenetrable because of the multiple thorns that have to be overcome.

At a National Academy of Sciences workshop years ago on international trade, one of the Nobel laureate economists made an aside comment to me that has stuck with me over the years due to its simplicity and profoundness, and now with its relevance to this sufficiency argument.

"If every participant in society shared equally in the benefits and bore their fair share of the costs there would be few obstacles to these policies."

What I took from that part of my past is that all public policy relies on some form of Robin Hood or reverse Robin Hood formula, take from some and give to others. Our government is designed to allow special interest politics to flourish, so if some compensatory amendments are not added to basic policy prescriptions as in the statement above, any project, no matter how meritorious, is probably dead or will die over time.

So now as we address the sufficiency issues, this simple and elegant statement of an almost unassailable wisdom provides us with some key questions to help us focus on understanding sufficiency conditions.

Question 1: Who are the beneficiaries of the initiative? Who pays the costs? And how will the equity be achieved? ... both in fact and in perception.

Question 2: What scale of constituency has to be engaged to provide the resources? And how will their perceptions change over the time it takes to achieve demonstrable results? How is the balance between benefits and costs reconciled within that constituency?

Looking at the Major Classical Problems may help us find the questions we need to ask, and then we can address the tools needed to go from the questions to the character and detail of the solutions.

While none of the examples relate directly to the process of building a space hotel, or a space industry to harvest materials from the moon, it is the premise of this chapter that we may at a minimum get a grasp on some of the questions society is going to ask, and the space advocacy community is going to have to answer.

Other examples in my memory are:
* Overcoming world hunger when we already produce enough food stocks.
* Reducing dependence on imported oil (for the 30th time).
* The introduction of High Speed rail between major population centers in Europe, and the failure of that approach in the US.
* The growth of alternative energy projects, each with a legacy of advocacy but several remaining technical and economic problems, and increasing stridency for "fund me." (example: biomass fuels)
* The evolution of the peaceful use of nuclear power.
* Overcoming the math and science illiteracy in the US education system, K through 12.

A cautionary note from my experience. The notion that you can pick winners in a government policy process (politics with a small p) is almost universally unsuccessful. In my experience in Europe, the most reasoned advocates for not "picking winners" were the same Ministers who were struggling to do it. The idea appears deceptively appealing, but the track record is not there to justify actually doing so.

I have declared above that we shouldn't get a grade on our performance with Major Classical Problems because we were absent from the Problematique class. Now we're going to try to start learning and catch up before we have the final exam. We passed our mid term test, the Cold War, and our term paper on "Managing Tsunamis, Terrorism, Global Energy Markets and Hurricanes" is being graded right now. But when we sneak a peak into the professor's office he has a frown in his face and there's already a lot of red ink on our paper. Humm.

During my career I've watched crisis management morph into catastrophe management. The only remnant of change due to managing crises is to ruin your daughter's birthday, as it takes a real big catastrophe to initiate change, which even then often has little to do with the cause of the catastrophe. It's an amazing phenomenon to observe.

The learning we (the royal "we") have to accomplish must enable us to take off the blinders, particularly as we make decisions about how to participate in changes in one nation among a global society. If we craft solutions with blinders on, i.e. with too many unexamined assumptions, then all we will do is empower the Law of Unintended Consequences. Under the workings of this law, we achieve the opposite of what we intend. If history is our guide, the attempt to "focus," simplify, and stress only the key points actually sets the initiative up to fail.

Are there more questions than the two proposed above? The answer is yes.

Next, we take the example of biomass as a source of energy for society. Will it move beyond a niche solution and become a major component of the energy equation in the US and world?

Case Study - Improved Use of Biomass; aka, the Expanded Use of Flex-fuel Vehicles

To examine the issue of sufficiency, this brief case study looks at the general background of biomass and ethanol 85 as an automobile fuel, auto use as a percentage of total energy consumption, the total stock of private vehicles in the US, and the infrastructure for distribution of ethanol 85 in the Washington DC area.

Biomass/ethanol technology has a heritage that goes as far back as I can remember, as biomass production of organic fuel precursors has been on the agenda during most of my career. The process of increasing awareness has been in place since the late 60s or early 70s. Today, it is mature enough to make a dent on the production and distribution of usable fuels for autos and trucks, but how big a dent?

Over the past three decades, Brazil has worked to create a viable alternative to gasoline, and with its sugarcane-based fuel the nation may become energy independent this year. Brazil's ethanol program, which originated in the 1970s in response to the uncertainties in the oil market, has enjoyed considerable success. Many Brazilians are driving "flexible fuel" cars that run on either ethanol or gasoline, and allow you to fill up with whichever option is cheaper, which is often ethanol. Countries with large fuel bills such as India and China are following Brazil's progress closely. A similar situation is emerging in China, where the development of the infrastructure in that rapidly emerging economy struggles with fluctuations in gasoline prices. Japan and Sweden, meanwhile, are importing ethanol from Brazil to help fulfill their environmental obligations under the Kyoto Protocol, as running cars on carbohydrates instead of fossil fuels may not be a new idea, and ethanol does have drawbacks, but it offers an increasingly attractive alternative as oil prices climb.61 One of the advantages of this system is that biomass is a carbon sink, and thus theoretically could provide a net-zero green house gas energy resource.

The US is taking small steps towards the use of ethanol, but the chemical process here, relying on corn, requires more steps and is more expensive. Will our blinders deny us the nuances that make Ethanol fuels effective in Brazil and possibly in other countries?

Biochemical Conversion

Biochemical technology can be used to convert cellulose and hemicellulose polymers that occur in biomass to their molecular building blocks, such as sugars and glycerides. Using hydrolysis, sugars can then be converted to liquid fuels. Thus, agricultural crops and residues, wood residues, trees and forest residues, grasses, and municipal waste can all be converted to fuel for "flex" fuel vehicles. Claims as to how many flex fuel vehicles exist in the nations inventory since their introduction in the late 80's and during the 90's differ in number, and it is hard to get one that satisfies. After market conversions also exist. Regardless of this, even the largest number is a small share of the total.

What we need are bio-refinery processes that convert corn-based biomass feedstocks into bio-based fuels.

2010 is a DOE milestone goal to finalize a biorefinery with the potential for three bio-based chemicals. By 2012 demonstrations that achieve a 5 - 20% increase in corn's fiber yield in ethanol plants are planned.

Projected Energy Consumption for 2010		
Total energy consumption	107.87	100%
Of which motor fuel	18.70	17%

Quadrillion BTU per year

Table 35.1 Auto Use Within Total Energy Use

Given that we import 33.8 quadrillion BTU per year, a significant savings due to ethanol substitution, while good, is not going to eliminate the need to import energy.

This leads us to Question 3: Are the only effective solutions arrived at by the "thousand candles" approach of the market place? This is the notion that let a 1000 ideas be started and let the strong survive. If so, how do large infrastructure initiatives that have to begin during one generation but finish after the next generation get financed?

The total stock of personal autos on the road 2002, USA	131,072,000
Number added during 2002	8,104,000
Number alternative fuel capable, 2006	654,553

Table 35.2 Vehicle Stock

The production of flex fuel vehicles started in the late 1980s. Production in 2006 is estimated to be in the hundreds of thousands of vehicles (a negligible quantity given the total stock of autos on the road).

Question 4: When society already has a solution in place, what level of change in cost, efficiency, or function is required to motivate the change?

Question 5: When demand for any particular option is driven by a volatile market price to the consumer, how do stop-start phenomena affect the building of infrastructure? Are there politically acceptable alternatives to this aspect of market dynamics?

The number of conventional gas stations	170,000[1]
The number of alternative fueling stations	6,230
LPG sites	3,966
Compressed Natural Gas sites	1,035
Electric sites	830
E85 (Ethanol) sites	188
Biodiesel sites	142
LNG sites	62
Hydrogen sites	7

Data as of 2003

[1] www.bloomberg.com/apps/news?pid=10000039& refer=columnist_levin&sid=aH6hn6V.ZQOo - 42k

Table 35.3 The Capacity to Distribute Flex-fuels to the Consumer

Data as of 2003, which obviously do not reflect recent changes due to higher fuel prices

And now the last straw: Where can you buy ethanol for your flex fuel car? Within 100 miles of Washington DC, a Department of Energy web site (http://www.eere.energy.gov/afdc/) offers the following options: an underwhelming grand total of four public and one private filling station.

Name	City	State	Type of Access
Quarters K Citgo - Pentagon/Navy Exchange	Arlington	VA	Public - see hours
Goddard Space Flight Center	Greenbelt	MD	Private - government only
Montgomery County Agencies - Fleet Management Services	Gaithersburg	MD	Public - credit card at all times
Chevron Service Center	Laurel	MD	Public - see hours
Citgo	Annapolis	MD	Public - see hours

Table 35.4 Ethanol Gas Stations – Washington, D.C.

The goal of this chapter has not been to place this text as a roadmap, but to stimulate the reader to form your own sense some thoughts about the sufficiency conditions of any roadmap. Ask yourself, Is the proposed solution in any respect at the same scale as the problem?

My "bird on the perch" experience is presented, with all of its warts and deficiencies, in the hope that somehow we can begin to understand the dimensions of due-diligence that society must apply to the maturation and social acceptance of any major change - space or terrestrial.

It is my goal to guide the reader to recognize the need for increased awareness across a larger slice of society concerning the importance of sufficiency conditions. The challenge then becomes building tools and processes for moving from "sufficiency awareness" to the crafting of policy settings that assure sufficiency, that is, success. Will this happen? ... certainly. But will it happen by present processes? My evaluation of history says no.

Leadership in this type of process not only has to galvanize action but has to comprehend the broad reaches of the sufficient conditions. With that awareness it has to convince the public involved of the true nature of the linkages that must be considered. And finally the leadership has to be stable to endure until the appropriate elements are identified, moved into action and changes implemented. We can do this with enlightened leadership that can influence and persuade across many different constituencies. The leader in a sense has to see what others can't see, and stimulate action where otherwise there would be no action.

About the Author

Hylan B. Lyon, Jr. is President and COO of Gamma Design, Inc. In the early 1970s he was a member of the President's Science Advisor's staff for space and aviation, and since then he has held numerous business and policy assignments in high technology and aerospace in both government and industry.

Chapter 36

Theory and Action for the Future of Humans in Space

By Kenneth J. Cox, Bob Krone and Langdon Morris, Editors

"Man's venture into space is a search for the anti-theory. It is a search for the shocking. A search for the unexpected. A search for what lies beyond all of our theories—a search for the Big Surprise. Space is crammed with theories. Theories about the cosmos have been put forth by every culture known to man— from the tribes of the Hopi Indians to those of the Trobriand Islanders, from the civilizations of the Babylonians to those of the Chinese and the Gauls. Every time we launch a vehicle, we follow the tracks of theory laid by Galileo, Newton, and Einstein. Man's entry into space is the very opposite of theory. It is like Darwin's Voyage of the Beagle. It is a hero's journey, a quest for the unexpected. It is a search for challenges that defy our theories utterly. It is a hunt for the empirically shocking. A hunt for the startles and the awes from which unimagined theories are derived."
Howard Bloom in an e-mail to Bob Krone

Opportunities

What will humanity have accomplished in space by 2057, the 100 year anniversary of Sputnik? By 2069, the anniversary of Apollo? Or imagine that we are approaching 2100, and thus the end of another century of human history, at least by the Christian calendar. Will there be permanent human settlements in Earth orbit? On the moon? On Mars? Will there be mining camps on asteroids? Helium 3 farms on the moon? Solar arrays in Earth orbit beaming megawatts back to the surface? How many children will be born in zero G? How many will be born, live long and productive lives, and die without ever setting foot on the home planet? Will they appear in census statistics of any nation of Earth? Whose law will govern the treatment of their estates? What courts will have jurisdiction over their disputes?

As we were preparing the last chapters of this book Howard Bloom sent us an email in which he included the comments we have quoted above. We are sure, without question, that Howard is right, and we wonder what unimagined theories we will have derived, based on what new experiences that we cannot now foresee. What challenges will we have overcome, and what theories will those challenges have made obsolete?

We believe that space offers unparalled opportunities for humans to discover, to develop, to exploit, and to improve the human condition throughout the Solar System, on Earth, and everywhere else we venture. We believe that our adventure in space is indeed a heroic journey, not necessarily because of the bravery that space travel requires, but because when we journey to space we must confront ourselves, we must find within our selves the best and most enduring of qualities - teamwork, dedication, insight, thoughtfulness, empathy, trust, curiosity, faith, and many other attributes that we admire in others and strive for in ourselves.

We believe; we hope; and we strive for such an outcome, for we understand that the journey to space holds unmatched promise for all of humanity.

The mosaic of ideas found in this book arose from the creative minds of the forty-two contributing authors, men and women who have brought together their experience, their knowledge, and their wisdom to produce this convergence of ideas.

In these chapters we have traveled a wide path. We have examined history, mythology, attitudes, beliefs, values, leadership, governance, and policy. We have reviewed education, law, management, risk, and strategy. We have encountered biology, biotechnology, bacteriology, cognition, evolution, ecology, intelligence, colonization, and cooperation. We have examined planetary defense, lunar cities, and spacecraft crew operations. And tennis.

These themes, all of these themes and of course many others, will come into play in ways that are as countless as they are unpredictable, for these are the elements of human civilization, and when we go to space as we envision that it happen, we certainly take all of human civilization with us.

The ideas and theories presented here flow from a multitude of personal experiences in aerospace and all the industries associated with launching humans into space. We recognize that in the resulting mosaic are individual perspectives that reflect our own personal fascinations and participation in the cosmic venture that places humans in fragile machines and sends them hurtling into space. There is no one explanation that answers why we do this or why we want to continue this great endeavor, but we do.

Theory Definition

A theory, any theory, is the attempt to explain reality as it is observed, experienced, and understood today, and it is used as the basis to solve existing and future problems and to make predictions. Its broader purposes may include the desire to understand, explain, invent, improve, validate, or justify, and it also provides a framework for research. The test of theory is that it must be capable of being shared, reproduced, put into practice, and verified. Any theory, like many previous ones, may take a very long time to be fully verified, and unfolding events and discoveries could require and at times significant modifications. We evaluate theories based on how well they perform these critical jobs.

Theories fall into three categories: 1) Descriptive Theories explain "What exists;" 2) Values Theories explain and defend "What is preferred;" and 3) Normative Theories provide the foundations for "What should be."

Normative Theory is perhaps the most interesting of these, and like a ship's engine and rudder it provides direction, force, and logic to drive change. Like any good theory, it can help people and systems to avoid waste or destructive trial and error, and even to avoid failures and catastrophes by helping managers, controllers, and even citizens and voters to recognize and avoid pathological theories that lead to undesired outcomes. Any good theory helps us find order in random or chaotic events and situations, and the basic function of Normative Theory in particular is to provide a set of prescriptions for leadership.

The theory we put forth here addresses the challenges and unknowns we face when humans begin migrating to space, and recommends concepts and components of the "Mission" to begin this next great human adventure. The human experiences encountered in the actual space environment will, of course, require the "mission" to adapt and change along the way.

Theory for Space Exploration, 1957 - 2006

If we look back at the beginning of the Space Age and the very early days of NASA, the theory in evidence shows an interesting combination of two very different cultures. On the one hand, NASA in its early days was a fountainhead of innovation and creativity. Young and old scientists and engineers worked together during the Mercury, Gemini, and Apollo programs to produce a series of space craft that accomplished an unprecedented feat, the moon landings, in an astonishingly short period of time, a decade.

At the same time, the military-industrial complex was also a big part of the story, as the aerospace industry cranked out designs, components, and systems using its traditional contracting and subcontracting methods to produce billions of dollars worth of hardware, software, and know-how.

Following Apollo, however, the first elements, innovation and creativity were largely lost, replaced with a government bureaucracy that operated in the aerospace industry's preferred way of working. As that way of working became institutionalized, so too did the work. The sense of urgency was lost and the bureaucracy became stifling. Errors of communication and judgment led to the loss of two Space Shuttles and their crews, while the Space Station languished, incomplete.

Meanwhile, economic conditions and the geopolitical situation have changed radically. The Cold War ended, the war on terror began, and the US budget deficit sank deeper and deeper into red.

We can summarize the theories in effect here as follows:

1) Descriptive Theory of the space program, 1980 - 2006, What exists: Competition with the USSR compels the US to seek dominance in the literal high ground of space. The end of the Cold War leaves the space program largely without purpose.

2) Values Theory of the space program, 1980 - 2006, What is preferred: A series of NASA administrators search unsuccessfully for an overriding purpose, until Dan Goldin finally arrives at "Faster, Better, Cheaper," by which he means for NASA to succeed at less ambitious endeavors. Society is underwhelmed.

3) Normative Theory of the space program, 1980 - 2006, What should be: This dimension remains largely unaddressed, which is to say that NASA has decidedly lacked a compelling mission.

While we are critical of NASA as an organization and particularly as a bureaucracy, we do not intend to criticize any individuals in the agency. In fact, we know from exhaustive first hand experience that NASA is composed of brilliant and dedicated individuals who work extraordinarily hard and regularly produce brilliance. No, the issue is not with the people, but with the "system of NASA," with the values that the system displays, and with its underlying theories. Shaping such a system is, of course, the responsibility of leadership, beginning not only with the NASA Administrator, but also with the President, the Congress, and also with the roots of this tree, the American people themselves. We all bear some responsibility for the unsatisfying situation in which we find the space program; and as many of the authors of this book have suggested, rectifying the so-called mess will require new ways of thinking and interacting.

Clearly, then, the theoretical basis upon which this space program rests has reached the end of its useful life, and new theories will replace it.

Theory for Space Exploration and Development, 2006 and Beyond

The viewpoint shared by the many authors of this book is that now is the right time for a new theory of not only space exploration, but space development.

Let us articulate this, first of all, in terms of the three kinds of theory. Please note that we have shifted our terminology a bit here, from reference to a "space program" to reference to a "space movement." The reasons for this shift should be evident in the discussion that follows, for it addresses the transition from a government project to an enduring process of civilization..

1) Descriptive Theory of the space movement, 2006 and beyond, What exists: The perspective offered by today's mass media suggests that the world appears to be a mess, and the mess is getting worse. Cultural, ideological, and religious conflicts are global in scope, while supplies of critical resources are tightening, leading to further conflict. We appear to be entering a period of unprecedented crisis due to global climate change.

But at the same time, remarkable progress has been made in space science and space technology, and a new generation of space vehicles is emerging. Europe, Japan, China, and India are joining the US and Russia as space-faring nations, and in the US a great deal of the initiative has shifted away from government and toward private enterprise. There is serious talk about private space launches and space tourism as a new paradigm seems to be emerging: space commerce is becoming a reality.

If this descriptive theory is true, then we are at the threshold of a new era.

2) Values Theory of the space movement, 2006 and beyond, What is preferred: If we are indeed at the dawn of a new era, then what sort of era shall it be? The possibilities and perils of marketplace competition await us as entrepreneurs prepare to attempt to make new fortunes taking people away from Earth, and bringing precious resources back.

The taking away is tourism, of course, but one day it could also be emigration to permanent settlements. The bringing back could be electricity generated from solar stations in orbit, or Helium 3 from the moon, or beaming solar energy from space directly to remote areas of the Earth without costly infrastructure. Part of the reason we want to go is because we don't know all that much about what's out there!

3) Normative Theory of the space movement, 2006 and beyond, What should be: Human civilization is at risk, due primarily to the impact from human activities and actions. The purpose of going to space must therefore be to restore balance to civilization. Thus, the space movement is a means through which to address the most critical issues that humanity faces, which include, of course, violence and war - we must go to space to promote peace, and to learn how to live peacefully.

It also means we go to promote learning, and indeed there are few activities that compress the learning process as much as space travel, as it confronts every aspect of life and life support in an unyieldingly harsh environment. We have much to learn about life on Earth by attempting to live off the Earth.

It also means that we go to promote exploration, to discover whatever is there.

And what can we say about those who will undertake these journeys?

Humans who inhabit space will find life radically different from the experiences of any humans on Earth throughout our long history. But it is simply not possible from our present perspective to define or enumerate what those differences will be and how they will affect human life.

As the chapters of this book have noted so eloquently, we believe that humans living in space will be forced to deal with paradigm shifts in the physical, biological, technological, psychological, social, religious and political areas, and in so doing they will enrich Earthly civilization in countless ways that cannot be anticipated, but can be expected.

As you contemplate the implications associated with these changes, it is well to keep in mind Howard Bloom's statement, "It is a search for challenges that defy our theories utterly."

What Leonardo da Vinci observed five hundred years ago holds true today: "Learning is the only thing the mind never exhausts, never regrets, and never fears. It is one thing that will never fail us."

Leaders and citizens around the world will play critical roles in bringing forth this new era, and to them we offer the following proposal concerning the Future of Humans in Space.

Theory And Action For The Future Of Humans In Space

1. **VISION.** To explore Outer Space and to inhabit our Solar System. To benefit humanity and Earth through the capture and utilization of the resources of space. To develop the capability to diminish or eliminate threats to Earth and its humanity.

2. **PURPOSE.** To design human migration to Outer Space that will enhance humanity on earth and in space. To seek life and to direct global intelligence to achieve the goals listed in the Vision Statement. To create science and technology to support the Vision. To establish an international collaborative entity that will develop, govern and manage the leadership and resources necessary to implement the Vision. To create a new societal paradigm in the vacuum of space that will enable humans to minimize the catastrophic tragic costs that now plague human life. To do so peacefully for all humankind.

3. **REASONS.** The human urge for flight, exploration and survival, together with our curiosity about the uni-

verse, are embedded deep in human genes and human consciousness. Even if these urges were ignored, we believe the desire to improve the quality of human life, and perhaps even its ultimate survival, hinge on the successes of human exploration and habitation of space. Our generation is equipped to take advantage of the opportunity presented by the outward expansion associated with space travel. We have the capacity to design a rewarding and exciting future of collaboration that will capitalize on the lessons learned from human history on Earth. 21st Century science and technology has advanced to the point where new breakthroughs will arrive at an exponential rate. Associated with these developments there lurks the prospect of threat. We have the ability to produce tools that produce massive damage and destruction. The human migration into space will test our global society beyond anything ever experienced. In contrast to the negative effects there lies the real potential that the efforts associated with making space travel possible will produce positive benefits for humanity that reach beyond anything ever accomplished before! Because the implications of failure have such serious negative impact for human life, every effort is to be made to assure that all human potential is nurtured.

4. OBSTACLES. Insufficient understanding of the potential gains inherent in this next great adventure. Public and private leadership's unwillingness to plan and accept needed macro cost-benefits analyses. Security requirements and the resources needed to defend society. The complicated neurological, theological and psychological imprints in the genes and their affect on human behavior.

5. IMPLEMENTATION. Global education and awareness to sensitize leadership in the political, education, business, religious and media arenas to the Vision, Purposes, Reasons and Obstacles inherent in this theory. The authors of this book who share their professional expertise with thought leaders and the public.

6. IMPLICATIONS AND OUTCOMES. At the start of the 21st century, humanity has reached a critical juncture. Scenarios for two very divergent paths are clearly plausible. We can collaborate to build and excel, or we can continue to allow the expansion of waste and degradation that have the potential to destroy human life. Our future in space is one potential solution to this dilemma. It is our conviction that space migration offers the opportunity to create an "Island of Excellence" that will enable humans and our civilization to continue and thrive. The authors who created this book urge global government leadership and the public to accept the challenges involved in moving human settlement of the Solar System form potential to reality. Four questions need answer: "Why do we go?", "What will we do?" and "How will we do it?" What adjustments will be made along the way?" Our answers will determine the outcomes associated with this grand venture. History will judge our responses that define our legacy.

That this will come about is both our prediction and of course our hope. It is also our intention, and in whatever ways we can make this into reality, we pursue these with enthusiasm and dedication.

KENNETH J. COX, Ph.D.
BOB KRONE, Ph.D.
LANGDON MORRIS
1 April 2006

APPENDIX "A"

Research Agenda for the Future of Humans in Space

One of the goals throughout the development of *Beyond Earth: The Future of Humans in Space* has been to identify research needed to solve the multiple challenges linked to the permanent move of humans to space. Here we consolidate those from our authors and add other questions and hypotheses from professionals throughout the global space community.

Research questions capture subjects where there are unknowns or partial answers that require analysis. Hypotheses are different. They provide expectations for future outcomes. Hypotheses have four possible outcomes: 1) future events and research may validate them; 2) prove them to be invalid; 3) discover some mix of truth in the projection; or 4) insufficient evidence will be found to state any findings or conclusions.

This Appendix "A" summarizes the research questions and hypotheses generated throughout the development of Beyond Earth. They cover a large array of science, engineering, management , governance, policy and human factors subjects. Leadership of the Aerospace Technology Working Group (ATWG.org) is already using this list to plan future encouragement of research and publications. It is only an illustrative set of relevant research questions and hypotheses. Your additions to this list will be welcomed by ATWG leadership.

Beyond that NASA, university educators, the International Space University at Strasburg, France, space foundations and associations and national governments throughout the world will find in this list valuable ideas and concepts for planning.

Research questions and hypotheses are listed in alphabetical category order.

1. Aerospace Technology Working Group (ATWG)
2. Bacteria
3. Breakthroughs
4. Challenge
5. Children Today - Tomorrow's Space People
6. Debris in Space
7. Education
8. Energy
9. Evolution
10. Genetics
11. Global Warming
12. Governance
13. Gravity and Humans in Space
14. Intelligence
15. International Cooperation
16. Law
17. Leadership
18. Military in Space
19. Moon
20. Music and Arts
21. New Frontier
22. Popular Support for Space
23. Quality Sciences and Space Sciences
24. Risk, Safety, Reliability
25. Self Destruction
26. Spaceflight Systems
27. Space Trips for Peace
28. Win-Win Global Consciousness
29. X-Prizes

1. Aerospace Technology Working Group (ATWG)

*Hypothesis. ATWG's commitment will help transform the space programs of every space faring nation into an Earth/space movement that integrates permanent settlements with personal adventure, science, commerce, ecology, and holistic well being for everyone and everything on the Earth.

Dr. Ken Cox and the Authors of Beyond Earth: The Future of Humans in Space

2. Bacteria

* Hypothesis. Bacteria will be inside the human settlement bubble in space. These bacteria will be used for recycling, for the production of new materials from substances to be brought from the external environment, for supporting the life of other organisms, and to keep our immune system in shape and all other functions bacteria do to sustain life on Earth.

Dr. Eshel Ben-Jacob, Chapter13

3. Breakthroughs

* Hypothesis. A breakthrough is not a function of what is known, but a function of the domains in which it is known. We can know the future of space in a domain of ideas and science, we can know it in a domain of experience and action and we can know it in a domain of adventure, risk and myth. All those domains will be needed for the major human move to space.

Dr. Charles E. Smith, Chapter 8

4. Challenge
* Research Question. How to solve the greatest challenges of human life which are the mysteries of the universe that tempt us from without, but also the mysteries of ourselves that lie within the human consciousness, human knowledge, and human compassion.

Langdon Morris, Chapter 2

* Research Question. How can human civilization best integrate our global systems of government, business, academia, and faith to ensure human survival and the generation of knowledge, prosperity and spiritual well being across our solar system over the next twenty generations?

Martin Schwab, Chapters 3 and 34.

5. Children Today - Tomorrow's Space People

* Research Question. A harsh evaluation by today's youth is that they do not feel they have a part in the planning process for opening the space frontier. How should we go about giving them a voice and involving them in the planning process now?

* Research Question. How can we humans avoid fouling up places in space like the aliens did in THE WUMP WORLD?*

*THE WUMP WORLD, Bill Peet, Houghton Mifflin Company, Boston,
1970 and students in the Kids to Space survey.

* Research Question. Is a child's sense of wonder a valuable asset in our quest for answers about the Universe?

Lonnie Schorer, Chapter 18

* Research Question: Would it be feasible to utilize Satellite Space Technology to create Virtual Field Trips for kids as a way to stimulate interest in Space Technology for schoolchildren everywhere in the world? We already have the technology to do this in Orbital Space, on the Moon, and on Mars.

Dr. Elliott Maynard, President, Arcos Cielos Research Center

* Research Question: In an effort to integrate children into the vast expanse of knowledge that has been generated since the golden age of human space pioneering in the 1960s, how can more experienced generations best share the "great unknowns" of deep space pioneering with more newly arrived generations?

Martin Schwab, Chapters 3 and 34.

6. Debris in Space

* Research Question. The accumulation of debris in space is an increasing problem with potential damage to space operations. How can international research and specific plans be created to deal with this problem?

Dr. Feng Hsu, Dr. Paul Werbos and Martin Schwab, *Beyond Earth* authors.

7. Education

* Research Question. What private - public education and human capital development model can successfully integrate the international resources and interest of government, industry, and academe that will be involved with the future space enterprise?

* Research Question. What emerging telecommunications and information system technologies will transform space-related education and human capital development worldwide?

* Research Question. What economic and financial structure will be needed to support and sustain the private - public education and human capital development model developed for the future space enterprise?

Dr. Michael J. Wiskerchen, Chapter 15

8. Energy

* Hypothesis. The permanent solution to earth's energy needs has begun with the biofuels conversion for vehicles and will be finally resolved by a mix of solar energy from space with alternatives-to-oil-based energy sources on earth.

Paul Werbos, Howard Bloom, Martin Schwab, Feng Hsu, Ken Cox, and Bob Krone

9. Evolution

* Hypothesis. Some kind of movement of humanity into space is inevitable. But this great step is likely to be far more successful and meaningful if it is guided and energized by awareness of the wider evolutionary trajectories that will eventually determine the significance of humanity in the universe.

John Stewart, Chapter 22

10. Genetics

* Research Question. How to determine and accomplish needed human genetic intervention or manipulation to insure the survival of humankind off earth?

Lynn Harper, Chapter 14 and George S. Robinson, Chapter 6.

11. Global Warming

* Hypothesis. The world has seven years to take vital decisions and implement measures to curb greenhouse gas emissions or it could be too late.

Primer Minister Tony Blair, London, 7 February 2006

12. Governance

* Research Question. What can be learned from historic shifts o epochs?
* Research Question. What widespread wishful thinking hinders realistic steps towards human settlement of space, such as trust that good will, civil society and business interests can be relied upon to do most of the job.
* Research Question. What are the most critical characteristics of governance essential for human settlement of space?
* Research Question. How can those critical characteristics be realistically realized?
* Research Question. What can be done to prepare the ground for moving humanity beyond earth before a suitable governance system emerges?

Professor Yehezkel Dror, Chapter 5

13. Gravity and Humans in Space

* Research Question. How will the brain and its psychology adapt to microgravity and hypergravity?

Dr. Sherry E. Bell and Dr. Dawn L. Strongin, Chapter 11

14. Intelligence

* Research Question. Development of an information theory that is extendable to fantomark-coded messages and streaks would be crucial, as it would facilitate the invention of superior intelligent artifacts; could hold a key to communication with extraterrestrial modes of intelligence; and eventually help us understand our cosmic ancestry and the relationship between implicate and explicate orders, as envisioned by David Bohm.

Dr. Joel Isaacson, Chapter 24

15. International Cooperation

* Research Question. What aspects of the human psyche in general pose the greatest problems for creating a successful worldwide society based on mutual responsibility, enthusiasm, cooperation, and commitment to the general welfare of all participants, and how do we successfully prevent them from continuing to prevent a fully functional worldwide society capable of cooperative existence?

* Research Question. How can we stop the destruction of war that has been draining the world and all people of its resources for millennia so that we can collaborate to achieve those things we naturally desire as part of our heritage in the Cosmos?

* Research Question. How do we deal with imminent earth climatic changes which could devastate a significant portion of its land and populations of humans and wildlife so that we can fulfil our future heritage in a cooperative collaboration of international peace?

Michael Hannon, Chapter 10

16. Law

* Research Question. What is the most effective formulation of a transnational public and/or private corporation business entity to exercise independence and sovereignty to identify, recover, and commercially exploit space resources for the benefit of all humankind?

* Research Question. How to Formulate a legal operating relationship between international/transglobal military entities (administration/protection) and private entrepreneurs operating in space? This might be an appropriate variation of the English charters of the late sixteenth century (e.g. Virginia Company, Hudson Bay Company or East India Company).

* Research Question. How to formulate curricula for engineering students and graduate science students that will teach them routine and full involvement with global space law and economics, such as that being developed at the Georgia Tech engineering department?

* Research Question. How to establish a globally effective legal infrastructure to encourage and protect the process of obtaining "informed public consent" for all space activities that are designed to allow broad humankind interaction or interference with extraterrestrial life, consistent with applicable principles of metalaw?

* Research Question. How to create a new jurisprudence allowing independent personhood and legal accountability of transhumans, telepresences, and advanced artificially intelligent biorobotics functioning in near and deep space?

Dr. George S. Robinson, Chapter 6

17. Leadership

* Research Question. How do we train people to balance order, control and results focus in a context of what's good for humanity, locally and at large?

Dr. Charles E. Smith, Chapter 8

18. Militaries in Space

* Research Question. What missions in outer space (besides planetary defense) are best suited for the military systems of our world, which by achieving create political accord among the major world powers?

* Research Question. Could a negotiation framework at the presidential level be initiated by the U.S. to allow *small, annual, incremental, and reciprocal* transfers in terms of percentages from military budgets to an international civil space pioneering and defense budget or private fund?

Martin Schwab, Chapter 34

19. Moon

* Hypothesis. In 2007, the 50th anniversary of the commencement of the civil space age will take place as will the 400th anniversary of the founding of Jamestown, the first permanent "New World" English colony in America in 1607. 2007 will commence the second half-century of the civil space age, and will also mark the beginning of a program to create the first city on the Moon with the initiation of the permanent expansion of the World's human civilization beyond the Earth.

Thomas F. Rogers, Chapter 7

20. Music and Arts

* Hypothesis. Music and Arts programs throughout the world will increasingly establish programs or departments that focus the creative energy of youth toward music and the arts for humans in space.

Dr. Bob Krone, Chapter 16

21. New Frontier

* Research Question. How to create a new sense of purpose, a new set of goals, a new frontier to move once again with might and majesty, with a sense of zest that makes life worth living, through the world in which we live. One of the most challenging frontiers left to us hangs above our heads.

Howard Bloom, Chapter 9

22. Popular Support for Space

* Hypothesis. Only when regular citizens recognize the far-reaching humanitarian advantages, or can personally experience the technological advantages of the space program will a national or international space policy have broad support.

Dr. David Livingston, Chapter 12

23. Quality Sciences and Space Sciences

* Hypothesis. Quality Control and Management has been a continual emphasis for space missions. A formal merging of Quality Sciences and Space Sciences will occur for the Humans-to-Space Migration.

Dr. Bob Krone, Fellow Member, American Society for Quality (ASQ), Chapters 16, 17 & 36

24.. Risk, Safety, Reliability

* Research Question. What R&D efforts into accident theories are needed to better understand what are the complexities in accident propagations and how phenomenological events occur which often cause catastrophic system failure

* Research Question. What R&D efforts are needed to understand the human dynamics influences on the development and evolution of man-machine interfaced technological systems, and how the factors of human elements play a key role in the safety risk of all technological systems?

* Research Question. Does absolute safety exist for manned space vehicle systems? Is it possible to eliminate accident by design? What are the design philosophies and strategies that can achieve such goal?

* Research Question. How to systematically model, understand and control the interactive complexities that pose great threat to the safety of socio-technical systems?

Dr. Feng Hsu and Dr. Romney Duffey, Chapter 30

25. Self Destruction

* Research Question. The views of past and current global leaders, expressed in Chapter 3 indicate that while military relationships continue to dominate the political agenda on Earth, the lure of human space pioneering can still deliver us from our dangerous propensity for self-destruction. How can global society now build upon this consensus?

Martin Schwab, Chapter 3

26. Spaceflight Systems

* Hypothesis. NASA and global space entities should never again be confined to SINGLE and static human spaceflight architectures. If greater budgetary and capital investments were demanded, exponential increases in quality of human life would occur, based on proven records of success.

Martin Schwab, Chapters 3 and 34.

27. Space Trips for Peace

* Hypothesis. Space Trips for Peace would create crews composed of members from nations marginally friendly, hostile, or even at war with each other. Space, new to civilization and without territorial boundaries and national sovereignties, would be the ideal frontier for demonstrating that people of all cultural beliefs and religious backgrounds are able to set aside differences and work harmoniously for goals mutually considered good.

Astronaut Buzz Aldrin and Thomas F. Rogers, Chapter 10

28. Win-Win Global Consciousness

* Hypothesis. Both research and the search of human experience will be necessary to bring win-win benefits of space to the mainstream global consciousness.

Dr. Elliott Maynard, Arco Cielos Research Center

29. X-Prizes

* Research Question. How has, and could, the X-Prize, positively impact the future of exploration and development of space?

Howard Bloom, Scientist, Business Entrepreneur

APPENDIX "B"

Aerospace Technology Working Group Meetings History
(SATWG and ATWG)

Primary Themes, Meeting Dates, Location, and Host
1990 - 2005

1. Avionics Systems Architectures
November 1990; San Diego, CA
Host: General Dynamics

2. System Engineering & Integration
February 1991; Seattle, WA
Host: Boeing

3. Space Power
July 1991; Downey, CA
Host: Rockwell International

4. Standards & DoD Programs
November 1991; Nashua, NH
Host: Lockheed

5. GN&C, Tools & Processes
March 1992; Huntington Beach, CA
Host: McDonnell Douglas

6. Vehicle Health Management & Software
July 1992; Denver, CO
Host: Martin Marietta

7. Displays & Controls, Sensors
November 1992; Clearwater, FL
Host: Honeywell

8. C41, International Competitive Assessment
March 1993; Melbourne, FL
Host: Harris

9. Enterprise Integration
July 1993; Houston, TX
Host: JSC/UHCLC

10. Team USA
November 1993; Washington, D.C.
Host: Industry, AIAA, NASA

11. Space Telecommunications
May 1994; Scottsdale, AZ
Host: Motorola

12. Aerospace Education
November 1994; Boston, MA
Host: Draper Labs / MIT

13. RLV Operations Technology
May 1995; Huntsville, AL
Host: MDA / Huntsville

14. Change
October 1995; Washington, D.C
Host: Lockheed Martin

15. Cooperative Space Strategies
June 1996; AF Space Command, Colorado Springs, CO
Host: SATWG/SIIG

16. Earth Orbit Operations and Development as Enabler for Exploration
October 1996; Houston, TX
Host: University of Houston & Lockheed

17. Facilitation and Nurturing Human Exploration
April 1997; Auburn, AL
Host: Auburn University

18. Space Plane
October 1997; Seattle, WA
Host: Boeing

19. Shuttle Upgrades
April 1998; Houston, TX
Host: United Space Allies

20. Space Launch Operations for the Future
October 1998; Cocoa Beach, FL
Host: KSC

21. Space Avionics Needs for the Future
April 1999; Dryden, CA
Host: Lockheed-Martine

22. Future Space Age Technologies
October 1999; Phoenix, AZ
Host: Honeywell

23. Space Transportation: The Next Generation
May 2000; Huntington Beach, CA
Host: Boeing

24. Space Infrastructure Development
November 2000; Hampton, VA
Host: NASA-LaRC

25. Value that Space Programs Bring to Humans
May 2001; Highland Ranch, CO
Host: Lockheed Martin

26. Space Infrastructure Development: Commercial Perspectives
November 2001; Herndon, VA
Host: George Mason University

27. Space Infrastructure Development: Near Earth
May 2002; Peoria AZ
Host: Challenger Learning Center of Arizona

28. Creating the Future in Space
November 2002; Houston, Texas
Host: University of Houston Clear Lake

29. Blazing the Path for Human Space Flight
May 2003; Greenbelt, MD
Host: Goddard Visitor Center

30. Returning, with Peace and Hope
March 2004; Huntsville, AL
Host: The Boeing Company

31. Supporting the Space Exploration Vision
September 2004; Long Beach, CA
Host: California State University, Long Beach

32. Expanding the Earth/Space Vision
April 2005 Johns Hopkins University, Maryland
Host: Johns Hopkins Applied Physics Lab

33. Earth Orbit Viable Oasis Stepping Stones to Lunar, Mars and Beyond
October 2005; NASA Ames, Sunnyvale, California
Host: NASA Ames Research Center and the Silicon Valley Space Club

INDEX

C

I

J

K

W

X

Y

Z

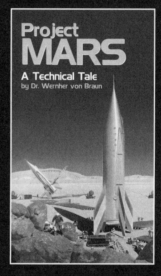

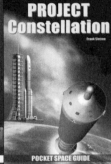

#	Title	ISBN	US$	UK£	CDN$	
1	Apollo 8	1-896522-66-1 *	$18.95	£13.95	$25.95	___
2	Apollo 9	1-896522-51-3 *	$16.95	£12.95	$22.95	___
3	Friendship 7	1-896522-60-2 *	$18.95	£13.95	$25.95	___
4	Apollo 10	1-896522-52-1 *	$14.95	£9.95	$20.95	___
5	Apollo 11 Vol 1	1-896522-53-X *	$18.95	£13.95	$25.95	___
6	Apollo 11 Vol 2	1-896522-49-1 *	$15.95	£10.95	$20.95	___
7	Apollo 12	1-896522-54-8 *	$18.95	£13.95	$25.95	___
8	Gemini 6	1-896522-61-0 *	$18.95	£13.95	$25.95	___
9	Apollo 13	1-896522-55-6 *	$18.95	£13.95	$25.95	___
10	Mars	1-896522-62-9 *	$23.95	£18.95	$31.95	___
11	Apollo 7 †	1-896522-64-5 *	$18.95	£13.95	$25.95	___
12	High Frontier	1-896522-67-X *	$21.95	£17.95	$28.95	___
13	X-15 †	1-896522-65-3 *	$23.95	£18.95	$31.95	___
14	Apollo 14	1-896522-56-4 *	$18.95	£15.95	$25.95	___
15	Freedom 7	1-896522-80-7 *	$18.95	£15.95	$25.95	___
16	Space Shuttle STS 1-5	1-896522-69-6 *	$23.95	£18.95	$31.95	___
17	Rocket Corp. Energia †	1-896522-81-5	$21.95	£16.95	$28.95	___
18	Apollo 15 - Vol 1 †	1-896522-57-2 *	$19.95	£15.95	$27.95	___
19	Arrows To The Moon	1-896522-83-1	$21.95	£17.95	$28.95	___
20	The Unbroken Chain	1-896522-84-X *	$29.95	£24.95	$39.95	___
21	Gemini 7	1-896522-82-3 *	$19.95	£15.95	$26.95	___
22	Apollo 11 Vol 3	1-896522-85-8 **	$27.95	£19.95	$37.95	___
23	Apollo 16 Vol 1	1-896522-58-0 *	$19.95	£15.95	$27.95	___
24	Creating Space	1-896522-86-6	$30.95	£24.95	$39.95	___
25	Women Astronauts	1-896522-87-4 *	$23.95	£18.95	$31.95	___
26	On To Mars †	1-896522-90-4 *	$21.95	£16.95	$29.95	___
27	Conquest of Space	1-896522-92-0	$23.95	£19.95	$32.95	___
28	Lost Spacecraft	1-896522-88-2	$30.95	£24.95	$39.95	___
29	Apollo 17 Vol 1	1-896522-59-9 *	$19.95	£15.95	$27.95	___
30	Virtual Apollo	1-896522-94-7	$24.95	£15.95	$27.95	___
31	Apollo EECOM	1-896522-96-3 *	$29.95	£23.95	$37.95	___
32	Visions of Future Space	1-896522-93-9 *	$27.95	£21.95	$35.95	___
33	Space Trivia	1-896522-98-X	$19.95	£14.95	$26.95	___
34	Interstellar Spacecraft	1-896522-99-8	$24.95	£18.95	$30.95	___
35	Dyna-Soar	1-896522-95-5 **	$32.95	£23.95	$42.95	___
36	The Rocket Team	1-894959-00-0 **	$34.95	£24.95	$44.95	___
37	Sigma 7	1-894959-01-9 *	$19.95	£15.95	$27.95	___
38	Women Of Space	1-894959-03-5 *	$22.95	£17.95	$30.95	___
39	Columbia Accident Report †	1-894959-06-X *	$25.95	£19.95	$33.95	___
40	Gemini 12	1-894959-04-3 *	$19.95	£15.95	$27.95	___
41	The Simple Universe	1-894959-11-6	$21.95	£16.95	$29.95	___
42	New Moon Rising	1-894959-12-4 **	$33.95	£23.95	$44.95	___
43	Moonrush	1-894959-10-8	$24.95	£17.95	$30.95	___
44	Mars Volume 2	1-894959-05-1 **	$28.95	£20.95	$38.95	___
45	Rocket Science	1-894959-09-4	$20.95	£15.95	$28.95	___
46	How NASA Learned	1-894959-07-8	$25.95	£18.95	$35.95	___
47	Virtual LM	1-894959-14-0	$29.95	£22.95	$42.95	___
48	Deep Space	1-894959-15-9 **	$34.95	£22.95	$44.95	___
49	Space Tourism	1-894959-08-6	$20.95	£15.95	$28.95	___
50	Apollo 12 Vol 2	1-894959-16-7 **	$24.95	£15.95	$31.95	___
51	Atlas Ultimate Weapon	1-894959-18-3	$29.95	£16.95	$37.95	___
52	Reflections	1-894959-22-1	$23.95	£16.95	$30.95	___
53	Real Space Cowboys	1-894959-21-3	$29.95	£17.95	$36.95	___
54	Saturn	1-894959-19-1 **	$27.95	£18.95	$35.95	___
55	On To Mars Vol 2	1-894959-30-2 *	$22.95	£14.95	$29.95	___
56	Getting off the Planet	1-894959-20-5	$18.95	£12.95	$23.95	___
57	Return to the Moon	1-894959-32-9	$22.95	£15.95	$28.95	___
58	Beyond Earth	1-894959-41-8	$TBA	£TBA	$TBA	___
59	Sex in Space	1-894959-44-2	$TBA	£TBA	$TBA	___
60	Lightcraft Tech. Manual	1-896522-91-2	$TBA	£TBA	$TBA	___
61	Go For Launch	1-894959-43-4	$TBA	£TBA	$TBA	___
62	International Space Station	1-894959-34-5	$TBA	£TBA	$TBA	___
63	Around the World in 84 Days	1-894959-40-X	$TBA	£TBA	$TBA	___

Go to apogeebooks.com for
further website exclusive
books and DVDs!

* Includes Bonus CDROM
** Includes Bonus DVD, NTSC Region 0
† Available only at www.apogeebooks.com